Designing Floor Slabs On Grade

Step-by-Step Procedures, Sample Solutions, and Commentary

Second Edition

Designing Floor Slabs On Grade

Step-by-Step Procedures, Sample Solutions, and Commentary

Second Edition

Boyd C. Ringo and Robert B. Anderson
Technical Editor: Mary K. Hurd

Designing Floor Slabs on Grade
Step-by-Step Procedures, Sample Solutions, and Commentary
Second Edition

published by Hanley Wood
426 South Westgate
Addison, IL 60101

Book Editor: Desiree J. Hanford
Artist: Joan E. Moran

 Library of Congress Catalog Number: 95-31988

10 9 8 7 6 5 4 3 2

ISBN 0-924659-75-0

Item No. 3020

Library of Congress Cataloging-in-Publication Data

Ringo, Boyd C.
Designing floor slabs on grade : step-by-step procedures, sample solutions, and commentary. -- 2nd ed.
p. cm.
Authors: Boyd C. Ringo and Robert B. Anderson
Includes index.
ISBN 0-924659-75-0 (pbk.)
1. Concrete slabs--Design and construction. 2. Floors, Concrete-Design and construction. I. Anderson, Robert. B.
TA683.5.S6R54 1996
624.1'8342--dc20 95-31988
CIP

ABOUT THE AUTHORS

Slab design experts and consultants, Boyd Ringo and Bob Anderson have lectured worldwide on the design of slabs on grade. They have served together as active members of the American Concrete Institute Committee 360, Design of Slabs on Grade, and Committee 302, Construction of Concrete Floors. Now they join forces to share their expertise in the how-to-do-it format of this versatile new design handbook.

Boyd C. Ringo, a leading authority on plain and conventionally reinforced concrete slabs on grade, has been involved as a professional engineer with design and construction for more than 40 years. Now in private practice as a consulting engineer in Cincinnati, he is a past chairman of ACI Committee 360, and was for many years a professor of civil engineering at the University of Cincinnati. Ringo has also worked as a structural designer and forensic investigator and has extensive experience with industrial slabs on grade.

Robert B. Anderson, a pioneer in establishing the post-tensioned slab on grade, helped in developing some of the early hardware as well as the design procedures that are accepted today. Long active in the Post-Tensioning Institute, he is now president of Robert B. Anderson Consulting Engineers, New Orleans. His practice includes commercial, industrial, and residential design as well as conventional and post-tensioned foundation design—with particular emphasis in dealing with problems of expansive and compressible soils.

CONTENTS

INTRODUCTION

How thick should the slab be?
How strong should the concrete be?
Is reinforcement needed?
Where should the joints be placed?
Can adding fibers enhance the slab's performance?
When is post-tensioning appropriate?
What can be done to control cracking?

This how-to-do-it book provides practical answers to these and other major questions that confront owners and designers when an industrial floor is needed. It is intended to simplify and improve the design of slabs on grade for commercial and residential as well as industrial uses.

"Design" includes all of the decisions, specifications, and details made and documented before construction can begin. It is based on properties of both the subgrade support and the concrete material. The process determines thickness, any necessary reinforcement, and jointing details as well as standards for construction of the slab. The authors regard design as a two-step procedure: thickness selection is done by one of the methods listed below; then other features such as joint location and treatment and construction tolerances are determined. Even though these steps are intimately related, they are commonly thought of as two separate procedures.

Drawing on their combined experience of many decades at the forefront of slab design and construction technology, Ringo and Anderson have prepared a text designed to help professionals at many different levels of slab design expertise. The book is arranged in three major parts:

Getting ready to design presents two chapters explaining the available slab design and construction methods, and outlining the input values of site and materials data needed before the design begins.

The design examples are the heart of the book, seven chapters of numerical examples worked out on a step-by-step basis for vehicle loads, rack storage post loading, column or wall loadings, and distributed uniform loads. Separate examples show how to use post-tensioning when sensitive areas such as compressible soils and plastic clays are encountered, and how to convert the post-tensioned slab to a conventionally reinforced equivalent. Problems are solved in several ways — giving the designer a choice, but always presenting the authors' recommendations as to the best way to proceed. Chapter 9, new to the second edition, presents the latest Post-Tensioning Institute method for designing hybrid slabs, ribbed and post-tensioned along a perimeter band, but having a conventional uniform thickness slab at the center.

Resource information in Chapter 10 provides needed data on joints, construction tolerances, computer alternatives, and a short course in soil mechanics. The final chapter on troubleshooting explains the steps to take when a slab is in distress and an investigation is called for. The Appendix is a recapitulation of the design aids, presenting in large, readable format all of the charts required for solutions given in the design examples. The authors expect that designers will copy these charts many times over as they draw the lines required for slab thickness determination and the selection of other variables.

Thus the user can assess his own needs and dip into the book for a whole course in slab design... or he can simply pick out the details that he needs. Undoubtedly, many readers will be able to go straight to the design examples, select a desired loading condition, and fit their own site-specific information into an already-worked example.

The authors wisely caution that there is no single or unique design procedure that can be applied to all job situations. But all of the design methods do have the common objective of crack control, and the provision of stability, flatness, and overall strength appropriate to each particular job.

Simplified methods presented

Stresses in grade slabs result from applied loads, changes in the concrete volume, and changes in subgrade support. The magnitude of these stresses depends on factors such as the degree of continuity, subgrade strength and uniformity, method of construction, quality of construction, and magnitude and position of the loadings. In the vast majority of cases, the effects of stresses can only be evaluated by making simplifying assumptions with respect to material properties and soil structure interaction. The most commonly known methods, all referred to in the examples of this book, come from the following sources:

- *The Portland Cement Association* (PCA)
- *The Wire Reinforcement Institute* (WRI)
- *The Post-Tensioning Institute* (PTI)
- *The United States Army Corps of Engineers* (COE)
- *The United States Army and Air Force* (TM/A/AF)
- *American Concrete Institute Committee 223 Standard Practice* (ACI 223)

The book provides the reader with "how-to" information on each of these methods, and also offers extensive references for those designers who want or need to know the "why" behind the design methods.

Good drawings and specifications emphasized

Experience tells authors Ringo and Anderson that the majority of floor contractors, foremen, and field workers look only at the drawing. Therefore they suggest that the key requirements should always appear on the drawing that describes the floor. But they feel that a complete set of both drawings and specifications is the best procedure by which to ensure a quality floor. The designer must specify and draw the appropriate instructions. Anything not stated or drawn cannot be followed, and cannot be expected to appear in the finished work. If the client wants a quality floor for specific objectives, a complete and detailed specification accompanied by comprehensive drawings, will help assure it. Although a good planning job takes more time than a poor one, the time spent is extremely cost-effective.

MARY K. HURD
Technical Editor
Engineered Publications

CHAPTER 1
INPUT VALUES NEEDED FOR DESIGN

1.1 — Information about the supporting soil

1.1.1 — Introduction

A slab on grade cannot be designed without numerical values that come directly from knowing what supports the slab. At the very least, a value is needed for the modulus of subgrade reaction, commonly referred to as k; however, the grade support system is more complicated than is indicated by a single value. In addition to k it is necessary to know the properties of the underlying soil and the available fill material. In other words, to design and construct a quality slab on grade, one needs to know as much as possible about the grade system that supports that slab.

The flow chart *(Figure 1)* summarizes an orderly approach to obtaining this information, and Section 10.8 provides more detail on soil characteristics and evaluation procedures.

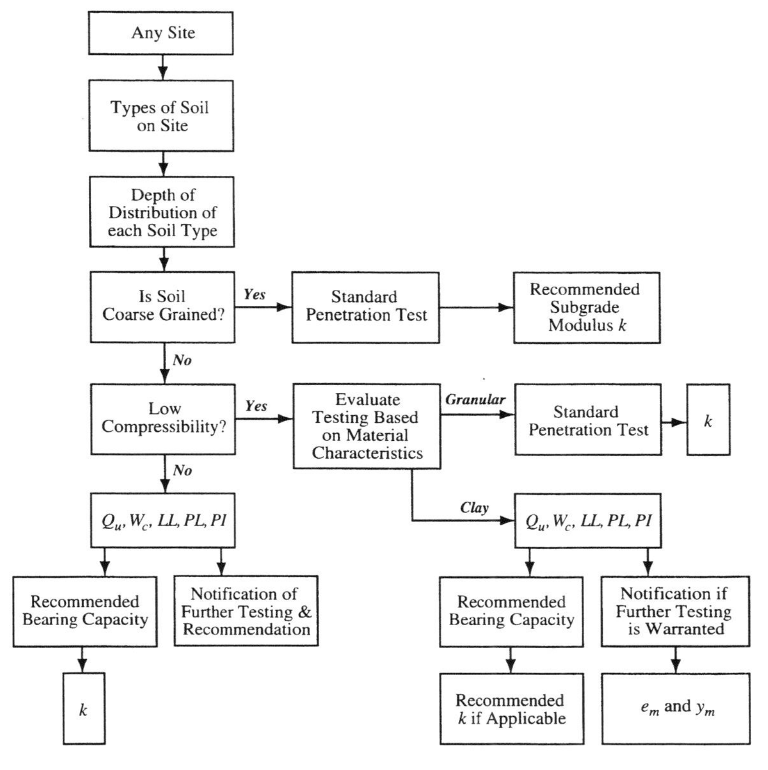

Figure 1 *Flow chart provides a guide to soils information needed for design of slabs on grade.*

1.1.2 — Working with a soils specialist

The first consideration at the beginning of any slab on grade design should be that of securing adequate geotechnical information (*Reference 1*). This should put the person responsible for the floor design into the process at the very beginning of any planning, which must include site considerations. When alternative sites are being evaluated for a project, soil conditions are often a significant economic factor.

The floor designer should be able to advise the owner as to what soils information will be needed. He should do this along with the geotechnical engineer in order to provide an optimum geotechnical report.

Commentary:
Do not omit the geotechnical specialist and do not omit the floor designer, if you want a good floor.

Too often the team effort of floor designer and geotechnical engineer is missing. This can lead either to costly overspending in obtaining soils information or to unexpected construction overruns due to omissions or errors in initial information. It must be emphasized that the slab on ground designer should be engaged either before or simultaneously with engaging the geotechnical firm.

1.1.3 — Limit risk with insufficient information

The authors have found that in much routine slab on grade design no soils information is available other than the floor designer's experience. This experience is occasionally in the job site area, but frequently is not within that geographical area. This situation often leads to relying on what previous experience dictated, such as "six inches has always worked" or "the soil is good." This may have been the situation more times than many of us care to admit. If forced into this situation in the future, the designer must protect himself by stating on the construction drawings what assumptions were made in the design process. The designer should also limit his liability by noting in writing the risks and possible consequences of inadequate soil information. Such steps not only protect the floor designer and inform the client, but often result in the client's favorable reconsideration in providing geotechnical backup.

If the floor designer is working in an area of known difficulties, unfamiliar materials or unusually heavy loadings, it might be wise to refuse to provide a design without sufficient geotechnical information.

A note such as the following could be used:

> "Due to lack of specific geotechnical information, this slab has been designed using a subgrade modulus of $k =$ ___ pci and design loading of ________________. The designer is not responsible for differential settlement, slab cracking, or other future defects resulting from unreported conditions mitigating the above assumptions."

This note may be modified as needed if other factors such as concrete strength are in question. Such precautions in both design and disclosure are simply prudent for all parties involved.

1.1.4 — Working with limited soil information

For most moderately loaded and medium sized projects, only a limited degree of geotechnical information commonly is available. This generally consists of soil classifications, for natural soil and for fill materials, with either standard penetration tests for coarse-grained soils, or unconfined compression tests and Atterberg limits on clayey materials. Of prime importance are any discontinuities in the subgrade layers which could result in slab thickness changes or in potential differential slab movements. Any discontinuities observed by the geotechnical engineer are generally pointed out. This may require further investigation to define any changes in substrata. If this is the case, a need for the joint services of the floor design engineer and the geotechnical engineer becomes obvious.

k *is used in determining the thickness of the concrete slab. That thickness is not sensitive to slight changes in the value of* k. *Obtain a slightly conservative value based on accurate soil data.*

However, assuming a relatively homogeneous site for most slab on grade designs, a value for k, the modulus of subgrade reaction, can be established using *Table 1* which is based on the Unified Soil Classification chart. Most of the examples in this book are based on use of the k value, which is a spring constant determined by soil properties. It is expressed in units of pounds per square inch per inch (psi/in.), commonly abbreviated to pci.

Major Divisions (1)	(2)	Letter (3)	Name (4)	Value as Foundation When not Subject to Frost Action (5)	Value as Base Directly under Wearing Surface (6)	Potential Frost Action (7)	Compress-ibility and Expansion (8)	Drainage Characteristics (9)	Compaction Equipment (10)	Unit Dry Weight (pcf) (11)	Field CBR (12)	Subgrade Modulus k (pci) (13)
Coarse-grained soils	Gravel and gravelly soils	GW	Gravel or sandy gravel, well graded	Excellent	Good	None to very slight	Almost none	Excellent	Crawler-type tractor, rubber-tired equipment, steel-wheeled roller	125-140	60-80	300 or more
		GP	Gravel or sandy gravel, poorly graded	Good to excellent	Poor to fair	None to very slight	Almost none	Excellent	Crawler-type tractor, rubber-tired equipment, steel-wheeled roller	120-130	35-60	300 or more
		GU	Gravel or sandy gravel, uniformly graded	Good	Poor	None to very slight	Almost none	Excellent	Crawler-type tractor, rubber-tired equipment	115-125	25-50	300 or more
		GM	Silty gravel or silty sandy gravel	Good to excellent	Fair to good	Slight to medium	Very slight	Fair to poor	Rubber-tired equipment, sheepsfoot roller, close control of moisture	130-145	40-80	300 or more
		GC	Clayey gravel or clayey sandy gravel	Good	Poor	Slight to medium	Slight	Poor to practically impervious	Rubber-tired equipment, sheepsfoot roller	120-140	20-40	200-300
	Sand and sandy soils	SW	Sand or gravelly sand, well graded	Good	Poor	None to very slight	Almost none	Excellent	Crawler-type tractor, rubber-tired equipment	110-130	20-40	200-300
		SP	Sand or gravelly sand, poorly graded	Fair to good	Poor to not suitable	None to very slight	Almost none	Excellent	Crawler-type tractor, rubber-tired equipment	105-120	15-25	200-300
		SU	Sand or gravelly sand, uniformly graded	Fair to good	Not suitable	None to very slight	Almost none	Excellent	Crawler-type tractor, rubber-tired equipment	100-115	10-20	200-300
		SM	Silty sand or silty gravelly sand	Good	Poor	Slight to high	Very slight	Fair to poor	Rubber-tired equipment, sheepsfoot roller, close control of moisture	120-135	20-40	200-300
		SC	Clayey sand or clayey gravelly sand	Fair to good	Not suitable	Slight to high	Slight to medium	Poor to practically impervious	Rubber-tired equipment, sheepsfoot roller	105-130	10-20	200-300
Fine-grained soils	Low compressibility LL<50	ML	Silts, sandy silts, gravelly silts or diatomaceous soils	Fair to poor	Not suitable	Medium to very high	Slight to medium	Fair to poor	Rubber-tired equipment, sheepsfoot roller, close control of moisture	100-125	5-15	100-200
		CL	Lean clays, sandy clays, or gravelly clays	Fair to poor	Not suitable	Medium to high	Medium	Practically impervious	Rubber-tired equipment, sheepsfoot roller	100-125	5-15	100-200
		OL	Organic silts or lean organic clays	Poor	Not suitable	Medium to high	Medium to high	Poor	Rubber-tired equipment, sheepsfoot roller	90-105	4-8	100-200
	High compressibility LL>50	MH	Micaceous clays or diatomaceous soils	Poor	Not suitable	Medium to very high	High	Fair to poor	Rubber-tired equipment, sheepsfoot roller	80-100	4-8	100-200
		CH	Fat clays	Poor to very poor	Not suitable	Medium	High	Practically impervious	Rubber-tired equipment, sheepsfoot roller	90-110	3-5	50-100
		OH	Fat organic clays	Poor to very poor	Not suitable	Medium	High	Practically impervious	Rubber-tired equipment, sheepsfoot roller	80-105	3-5	50-100
Peat and other fibrous organic soils		PT	Peat humus, and other	Not suitable	Not suitable	Slight	Very high	Fair to poor	Compaction not practical			

Table 1 *Unified Soil Classification, from* References 1 and 2.

The information in *Table 1* is intended as a guideline and should not be used as a substitute for an appropriate soils report by a soils specialist. The geotechnical firm, where engaged, can and should supply information classifying the soils according to the unified classification system *(Reference 2)*. This system is almost identical to the ASTM soil classification system *(Reference 3)*. Either is acceptable for practical design. With this information, a conservative value for k can be determined.

1.1.5 — Summary

It is not advisable to provide design services for slabs on grade when no geotechnical information is available. If the floor designer is sufficiently familiar with the site, the design can proceed; however, design assumptions along with a disclaimer statement should be placed on any drawings or recommendations.

When limited information is available from a geotechnical source, a conservative approximation of k value may be selected using the Unified Soil Classification, or its equivalent.

The floor designer should inform the geotechnical engineer, whenever one is available, of the loads anticipated and of the design procedure intended for use. This should help the geotechnical individual to provide the floor designer with the necessary information to properly execute the design.

1.2 — Types of slab loadings

1.2.1 — Introduction

Commentary:
For a quality floor, complete design is essential. This includes determination of the proper slab concrete thickness for external loadings. All decisions made or not made prior to floor slab construction are part of the design process.

The four loading types described in Sections 1.2.2 through 1.2.5 are those most commonly encountered on industrial floors. They are external loadings acting on the slab's surface. These vertical forces cause moments in the slab. They also cause shear forces in the case of substantial column or post loads on nominally-sized base plates. The slab must be designed to limit the concrete stress and provide load support with adequate reserve strength as indicated by the safety (or load) factor. The concrete slab on grade must resist these forces without showing unwanted distress.

The trend in plant use is toward heavier lift-truck capacities, higher racks with heavier post loadings, and harder vehicle wheel materials. The trend is also toward flatter floors with less evidence of cracking. It is absolutely necessary that the floor designer be supplied with accurate information concerning the applied loads in order to design and construct a quality floor.

1.2.2 — Vehicle axle loads

Most vehicle traffic on industrial floors is from lift trucks. These trucks commonly have solid or composition wheel materials. These are hard materials with relatively small contact areas. They produce higher stresses than those produced by pneumatic tires and cause more deterioration of joint edges. Although the specification sheet from the manufacturer should give specific and accurate values, approximately 95% of the total truck loading (weight plus payload) will be on the most loaded axle, usually the front axle. The slab thickness is frequently determined by the magnitude of this axle load. *Reference 4* gives total static axle loads ranging from 5600 pounds (rated payload 2000 pounds) to 43,700 pounds (rated payload 20,000 pounds).

The following specific information *(Reference 5)* should be obtained in order to determine the slab thickness for the lift truck or vehicle axle load:

Payload Capacity ____________ *(lb.)*	Model No. ______________
Vehicle Weight ____________ *(lb.)*	Wheel: ☐ Solid *or* ☐ Pneumatic
Axle: ☐ Single Wheel or ☐ Dual Wheels	Wheel's Tire Width _________ *(in.)*
Wheel Spacing *WS* ___________ *(in.)*	*or* Pressure _________ *(psi)*
Dual Wheel Spacing S_d _______ *(in.)*	Wheel Contact Area _________ *(sq.in.)*

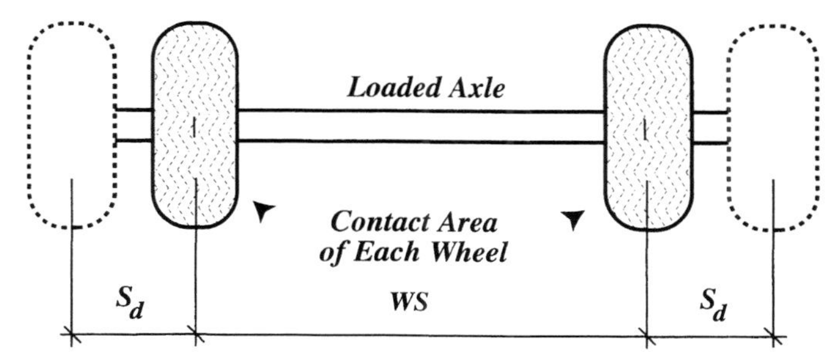

Figure 2 *Spacing of wheel loads for dual- or single-axle loading.*

Commentary:
The majority of vehicle axles are single wheel axles having only one wheel at the end of each axle, in which case $S_d = 0$.

The critical values needed for design are the axle load capacity, the spacing and number of wheels, the type of wheel material, and its contact area. The axle load capacity and the wheel material are not too difficult to obtain. The wheel contact area is more difficult to obtain and frequently must be estimated. The authors recommend a conservative estimate, which would be to assume a higher tire pressure and divide into the wheel load to get a lower contact area.

Pneumatic pressures range from 80 psi to 120 psi. The areas for solid tire materials may be estimated using equivalent pressures between 180 psi and 250 psi according to *Reference 4*. The contact area of dual wheels (two wheels, closely spaced, at each end of the axle) may be conservatively estimated by using the contact area of each of the two wheels along with the area between the two wheels. The authors further recommend that, if more accurate information is not available, 100% of the total static axle load should be used as the design value. This applies to lift truck loads centered along the axis of the vehicle and does not apply to lift trucks with swing capabilities.

The flatness or smoothness of the floor's top surfaces is important. Where vehicles move rapidly, or where use of the floor requires a smooth surface in certain areas, the floor planner should consider and specify an appropriate flatness and levelness. This is discussed further in Section 10.4.3.

1.2.3 — Uniform loads with aisles

A truly uniform load on the surface of the floor will cause no bending stresses within the general area of that uniform load. However, a truly uniform load is rare. In most cases, loads that are near-uniform are on pallets or other support. The critical bending stress in the floor is the stress on the top of the floor (due to so-called negative moment) in the middle of the aisle, located an equal distance from the uniform loads on either side of the aisle. This is particularly important since any cracking that does occur is in the middle of the aisle and is extremely visible. Wider aisles are less critical. The uniform load then can be truly uniform, or it can be due to a set of posts, as used for rack storage.

The critical values (below) needed for thickness determination due to this loading type are the magnitude of the uniform loading, in pounds per square foot, and the width of the aisle. Depending upon the design approach, it is desirable to know if the aisles and loading locations are fixed or can vary in location.

Uniform Load Weight __________ *(psf)*
Aisle Width __________ *(ft.)*
Arrangement Set in Location: ☐ Yes or ☐ No

1.2.4 — Rack and post loads

Most materials and goods in modern industrial facilities are placed on shelves as a part of rack storage. These shelves combine to produce substantial rack support loads. Rack storage has been built 90 feet and higher. This can produce individual post loads of 30,000 pounds or more. The bending stress produced is a tensile stress on the bottom of the slab. It occurs beneath the most heavily loaded post (rack support) and is increased by nearby posts, depending on the spacings.

The following specific information is required to design for this load condition:

Total load on a section of rack __________*(lb.)*
Post load __________*(lb.)*
Base plate size (area) __________*(sq. in.)*
Post spacings: X = __________*(in.)*
Y = __________*(in.)*
Z = __________*(in.)*

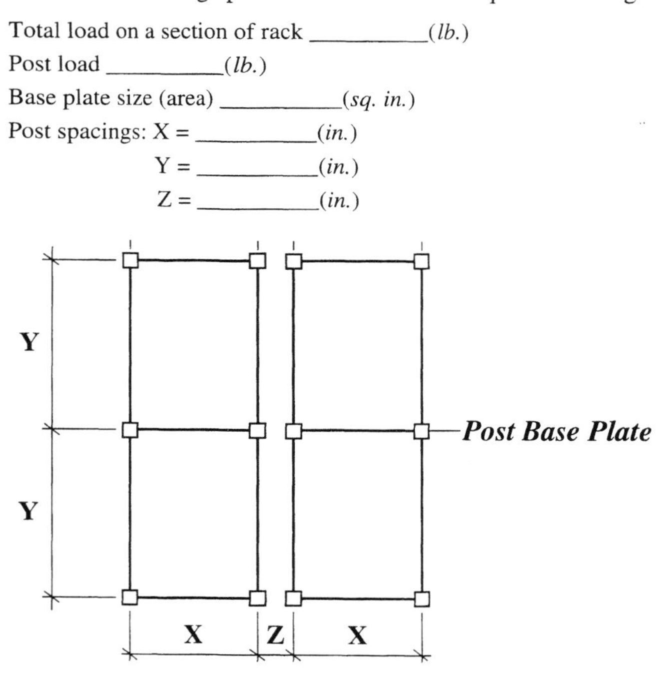

Figure 3 *Typical layout of post loads in a rack storage area.*

The important values are the forces exerted by the rack support posts on the floor, the size of the base plate used on each post, and the spacings in both directions of the support posts. Both bending stresses and shear stresses are produced. Although punching shear is not frequently considered, it can be critical when the post load is high and the base plate is relatively small.

The authors recommend that punching shear be checked. A special design case can exist for rack post loading when two posts are a few inches apart as is the case when one set of shelves is immediately adjacent to another set. This puts the loading of two posts on an approximated base plate slightly over twice the size of a single plate. The locations of any joints within this arrangement should be known since the design procedure assumes a continuous floor slab for the design. Any such joints that occur in the immediate vicinity of the storage racks must have load transfer capability.

Where high rack storage exists, it is common to find a tall turret vehicle, sometimes automated, which travels in aisles between fixed locations of storage racks. This requires construction of a smooth, flat floor in the aisle to stabilize and guide the vehicle when it is extended to upper storage levels. A suitable flatness (and levelness) must be specified for construction. This is discussed further in Section 10.4.3.

1.2.5 — Column loads

Column loading differs from rack storage post loading in that the columns normally carry more weight and the column spacings are significantly longer than the distance between posts. The authors consider the column loading as an isolated concentrated load normally unaffected by adjacent columns. If the column is supported by an isolated footing, then it is not a part of industrial floor design. However, either in rehabilitation work in an existing plant or in original design, it is possible to use the floor slab itself as the column support.

The critical values required to design for this condition are:

Total Design Load at Column Base ______ *(lb.)*
Base Plate Size (area)________ *(sq. in.)*
Slab Thickness at Column ________ *(in.)*
Distance to Nearest Joint ________ *(in.)*

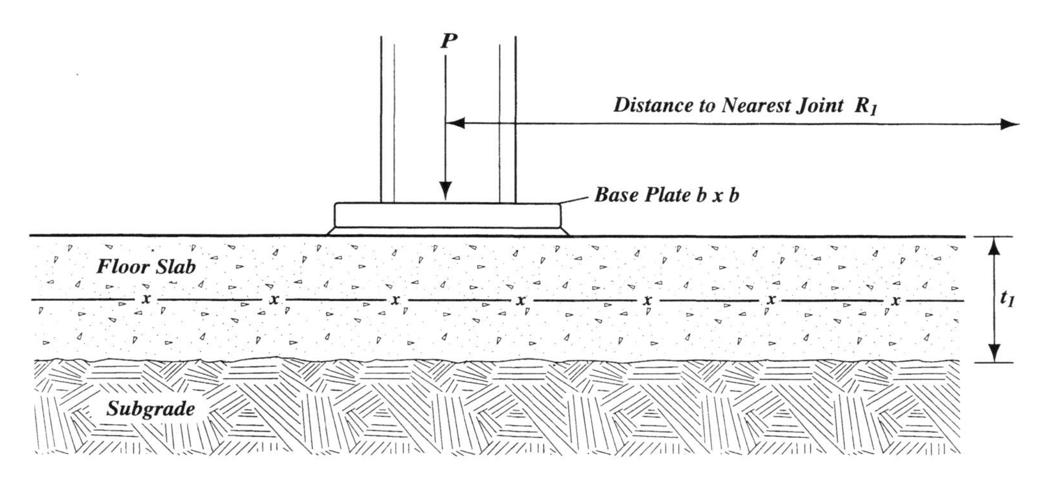

Figure 4 *When building columns impose an isolated concentrated load on the floor slab, the slab is designed as a column support.*

The designer must consider both the dead load and live load that the column imparts to the concrete slab, the size of the base plate, and the specific locations of any joints that are near the column. There may be a thickening of the slab at the column in new construction and this must be taken into account.

1.2.6 — Special or unusual loads

It is certainly not rare to find special loading situations which will control the thickness design and other features. The designer must make a special effort to determine if any such loadings exist.

Examples of such special or unusual loadings are the weight of tilt-up wall panels cast on the floor slab, machinery whose weight or configuration disrupts the symmetry of the overall floor, uplift due to use of the floor slab as a tie-down weight, and the like.

1.3 — Concrete for the slab

1.3.1 — Concrete strength properties for design

The strength of the concrete as a material affects the thickness of the floor slab and the properties of its surface. The strength also influences drying shrinkage, curling, and creep of the concrete slab. Its selection and specification are crucial to slab design and performance.

Commentary:
Select a reasonable value for f_c' that is not too low and not too high. It should be low enough not to contribute to shrinkage and curling and high enough to give reasonable strength, such as 3500 psi or 4000 psi at 28 days. Specifying the strength at a later age, such as 56 days, is reasonable if it fits the construction and use schedules. Note that a 6-inch slab at 3500 psi is stronger (more load support) and better (less shrinkage and curl) than a 5-inch slab at 4000 psi because high strength concrete tends to shrink more.

1.3.1.1 — Compression

The specifications will state the compressive strength of the concrete to be attained at specific ages. Compression tests of 6 × 12-inch concrete cylinders, tested according to ASTM C 39 and evaluated by ACI procedures, verify that the concrete as supplied and delivered meets specification requirements. The common symbol for compressive strength f_c' refers to the 28-day compressive strength.

Where post-tensioning is used in the slab on grade, values of f_c' commonly range from 3500 psi to 5500 psi. A strength of 1800 psi at 3 days is frequently required for the post-tensioning process to occur. In slabs without post-tensioning, excessive values of compressive strength are generally not desirable because higher compressive strengths are frequently achieved with higher cement factors. This, in turn, tends to increase both shrinkage and curl. For conventional floor slabs, the compressive strength could be specified at 28, 56, or 84 days, depending upon the schedule for construction and subsequent plant use. Values from 2500 psi to 4500 psi are common.

1.3.1.2 — Modulus of rupture

The tensile strength in bending, called modulus of rupture (common symbol *MOR*), is the critical property in slab design. It can be tested by third-point loading on a plain concrete beam, following ASTM C 78 for plain concrete, or ASTM C 1018 for concrete containing fibers. This value of *MOR* directly controls the thickness of the concrete floor slab and is frequently expressed in terms of a coefficient times the square root of the concrete's compressive strength, $\sqrt{f_c'}$.

Rounded gravels can give MOR test values as low as $7.5\sqrt{f_c'}$, but more commonly give values equivalent to $9\sqrt{f_c'}$. Crushed rock commonly gives MOR values equivalent to $10\sqrt{f_c'}$. A set of beam tests with third point loading is recommended for verification. The effective MOR is increased beyond these values when prestressing is used.

Coefficient values (used with the square root of compressive strength) of 9 to 11, for bank-run gravel and crushed stone aggregates, respectively, are commonly obtained by testing. PCA *(Reference 6)* recommends a coefficient of 9. The designer either assumes one of these coefficients (7.5, 9, or 11), or specifies tests to be run from the trial concrete mix to determine an appropriate value for the modulus of rupture. In the ACI building codes (*Reference 7*), ACI 318 uses a default for the *MOR* of 7.5 × $\sqrt{f_c'}$ and ACI 318.1 uses a default value of 5.0 × ϕ × $\sqrt{f_c'}$ (where ϕ is 0.65), although neither ACI 318 nor 318.1 address slabs on grade.

Assumed coefficient × $\sqrt{f_c'}$	f_c' Values						
	2000 psi	2500 psi	3000 psi	3500 psi	4000 psi	4500 psi	5000 psi
$7\sqrt{f_c'}$	313	350	383	414	443	469	495
$7.5\sqrt{f_c'}$	335	375	411	444	474	503	530
$8\sqrt{f_c'}$	358	400	438	473	506	537	566
$9\sqrt{f_c'}$	402	450	493	532	569	604	636
$10\sqrt{f_c'}$	447	500	548	592	632	671	707
$11\sqrt{f_c'}$	492	550	602	651	695	738	778
$12\sqrt{f_c'}$	537	600	657	710	759	805	849

Table 2 *Modulus of rupture, psi, for concrete strengths from 2000 to 5000 psi.*

It is not appropriate to determine the tensile strength of the slab concrete by means of briquette tests, although the splitting tensile strength test (ASTM C 496) may be appropriate for investigations.

1.3.1.3 — Shear strength

The shear strength of the concrete is rarely significant to slab design. Flexural shear need not be checked as it has not been observed to be significant. Punching shear, however, could be important for heavy values of post or column loads, especially where small base plates exist and/or in the case of thinner slabs. When the designer elects to check punching shear, the conventional ACI shear check for two-way shear is appropriate using either ACI 318, Section 11.12 or ACI 318.1, Section 7.3.7.1.

Commentary:
Check post loads for punching shear on the thinner slabs, especially when a post is placed near an isolation joint or where two posts are side by side.

1.3.1.4 — Surface durability

The upper working surface of the floor slab is critical where abrasion resistance or surface hardness is required. The strength and durability of the upper surface (the top 1/2 inch, approximately) are not necessarily equal to that of the overall slab throughout its thickness. The surface strength could be significantly reduced by improper finishing. Abrasion resistance depends on the amount of a quality aggregate or other hard material at the surface, along with the water-cement ratio within that top 1/2 inch. Abrasion resistance can also be achieved by special, often proprietary, surface treatments.

Surface durability requires careful control of water content (bleed water), finishing techniques, and curing. If significant abrasion resistance is required, surface treatments will probably be required. Surface durability is not adequately handled by merely increasing f'_c.

1.3.2 — Choosing the concrete mix

1.3.2.1 — Cements

Three ASTM C 150 types of cement are common in slab design and construction. The most common cement specified is portland cement Type I. Its selection requires no special comment. Type II may also be specified; due to its more stringent grinding specifications, Type II does not exhibit as much drying shrinkage as Type I. Shrinkage-compensating concrete may also be selected, requiring either an additive to the mix or the selection of Type K cement (ASTM C 845). Air-entrained concrete is not used for interior slabs except in the case of freezer floors or dock and staging areas subjected to a severe environment.

Consider Type II portland cement except where shrinkage-compensating concrete is used. Do not use Type III portland cement in slabs on grade except possibly for repair work.

Cement Type	ASTM Designation	Description
I	C 150	General Purpose Portland Cement
IA		Type I Air-entraining Cement
II	C 150	General Purpose Portland Cement with Moderate Heat of Hydration
IIA		Type II Air-entraining Cement
III	C 150	High Early Strength Portland Cement
IIIA		Type III Air-entraining Cement
IV	C 150	Low Heat of Hydration Portland Cement
V	C 150	High Sulfate Resistant Portland Cement
K	C 845	Expansive Cement

Table 3 *ASTM Standard Cements Used in Floor Slab Construction.*

1.3.2.2 — Aggregates

Commentary:
A uniformly graded aggregate, both fine and coarse, is recommended. Note that a gap-graded aggregate will require more cement and water, resulting in higher costs. Also, the concrete will most likely shrink more.

Designs in this book are based on the assumption that the aggregates are strong, durable and, by themselves, do not diminish the strength of the concrete mix. ASTM C 33 is the appropriate reference, although the designer may want to consider a performance specification with more stringent controls than are in C 33. It is further assumed (and recommended) that the maximum size of the graded aggregates (common symbol MCA) be as large as available and practical, and that the aggregates be well-graded.

1.3.2.3 — Mix proportions

A trial mix with locally available materials is strongly recommended, along with the appropriate tests.

Selection of the concrete proportions is beyond the scope of this book. Nonetheless, a proper design is essential to minimize the potential for drying shrinkage, as well as to produce the required strength properties.

1.4 — Safety factors

1.4.1 — Importance and meaning

Do not rely on the safety factor to prevent drying shrinkage cracks or curling effects.

A value for the safety factor must be selected before the thickness and other dimensions can be determined. This safety factor is not dictated by building codes. In the case of the slab on grade, the designer chooses the safety factor. It has a direct effect on thickness and therefore on initial costs. A factor of safety is needed because history shows that most slabs will experience distress—chipping, cracking, spalling, and wear—with an increasing number of load applications of a substantial magnitude. The safety factor is then a guard against the effects of fatigue. Another reason for including an appropriate safety factor in design is that there may be differences between the expected concrete properties (strength and/or thickness) and those which are actually obtained in the field.

The safety factor, whose common symbol is SF, is divided into the modulus of rupture *(MOR)* of the concrete (see Section 1.3.1.2) to calculate the allowable tensile bending stress appropriate for externally applied loading on the surface of the slab. This safety factor is the inverse or reciprocal of the stress ratio used in PCA literature. For example:

1. Assume a modulus of rupture *(MOR)* = 570 psi. Note that, unless tests are performed, this is commonly taken as $9\sqrt{f_c'}$.
2. Select a safety factor (SF) = 1.7.
3. The allowable bending stress is then 570/1.7 = 335 psi.
4. The equivalent stress ratio is 1/1.7 = 0.588.

1.4.2 — Common values for safety factor

For critical areas, frequent loading, and high magnitudes of load: 1.7 to 2.0 is recommended for consideration.

For important areas (but not as critical as above), loading less frequent but of substantial magnitude: 1.3 to 1.7 is recommended for consideration.

There are a number of safety factors in common use for slab on grade design. They are as follows:

SF = 2.0: This conservative value is commonly used. It is appropriate where loadings are not accurately known at the time of design, or where support conditions, or any other key items, are either not accurately known or are suspect.

SF = 1.7: This is an acceptable value consistent with the load factors used in other concrete design applications. It is used where loading is frequent and input values (design parameters) are reasonably well known.

SF = 1.4: This is an acceptable value for use under certain conditions. For example, where impact loadings do not exist, or where the loading intensities and/or the frequency of load application are less, then values between 1.3 and 1.7 may well be acceptable.

These SF values do not provide for excessive impact or shock loadings. While intended to prevent slab cracks due to applied loadings, any selected value will not automatically prevent slab cracking due to drying shrinkage of the concrete.

The choice of the safety factor is left to the designer, who must take into account the overall design situation. Safety factors are the inverse of stress ratios given in some references (*References 6, 8 and 9*). These factors are usually related to the number of load applications, which represents concrete fatigue. Safety factors are sometimes increased to account for the higher actual stresses where loads are applied to edges or joints with little or no load transfer.

Commentary:
For non-critical areas, minimum traffic applications and loadings that are low when compared to the design load: 1.3 to 1.5 is recommended for consideration.

When in doubt, use 2.0 as the safety factor.

1.5 — Slab reinforcement

1.5.1 — Reinforcing steel

1.5.1.1 — General

The term reinforcing steel refers to deformed bars, to welded wire fabric, or to steel fibers. Such steel is primarily used in slabs on grade to serve as a control of cracking due to shrinkage and other effects. However, such reinforcement is also used to provide defined amounts of structural strength, which come into existence after the slab cracks. All such reinforcement must be properly detailed on construction drawings and located properly in the slab itself. This is necessary for the reinforcement to perform its intended function.

Only when the reinforcing steel is properly sized and properly placed can its use be recommended and its cost warranted.

1.5.1.2 — Shrinkage crack control

The presence of reinforcing steel will not prevent conventional portland cement concrete from cracking due to shrinkage of the concrete as it hydrates, dries out, and hardens. Such cracking is actually caused by the restraint to shrinkage which is predominantly related to the materials of the concrete mix, the roughness of the base upon which the slab is placed, and the effectiveness of the curing process.

While the steel will not prevent such cracking, it will hold cracks tight (hairline), maintain aggregate interlock, and prevent faulting of the slabs. It is for these reasons that we refer to the purpose of the steel as that of crack control.

The best way to reduce or prevent shrinkage cracking is to pay close attention to the smoothness of the base, the quality of the concrete, the curing process, and the joint spacings.

1.5.1.3 — Subgrade drag

The so-called subgrade drag equation is commonly used to select the area of reinforcing steel to be used for shrinkage crack control *(Reference 6)*. This equation does not give areas suitable for structurally active steel. The intent is to select the steel, placed in both directions, to accept the tensile stresses that would have existed and been accepted by the concrete itself were it to have remained uncracked. The equation is:

$$A_s = \frac{FLw}{2f_s}$$

where A_s = the cross-sectional area of reinforcing steel, in square inches per lineal foot of slab width

F = coefficient of friction between base and slab

L = slab length between free ends, feet, in the direction of the intended steel

w = weight of the concrete slab, psf
(usually 12.5 pounds per inch of slab thickness)

f_s = allowable steel stress, psi

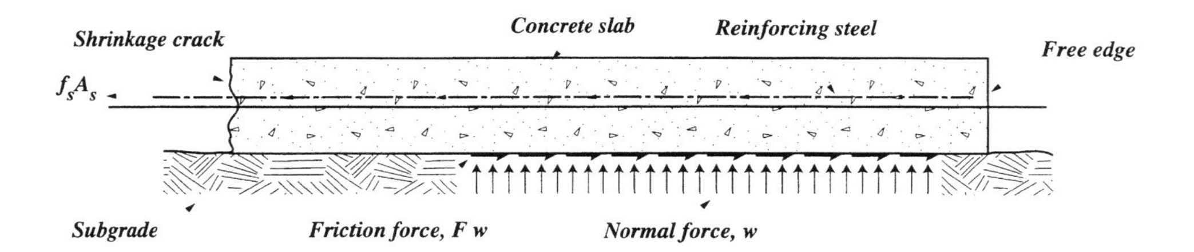

Figure 5 *Free body diagram of forces existing to produce subgrade drag action.*

Commentary:
Unanticipated rutting or unevenness in the subgrade can result in a coefficient of friction increase. This condition can result in more and/or wider shrinkage cracks. Both proof rolling and specification requirements can assist in this matter.

The steel must be located at or above the mid-depth of the slab. If no other criteria control, it is recommended that the steel be at 1/3 the slab depth from its top surface.

The coefficient of friction is commonly assumed as 1.5, but the precise value varies with the roughness of the subgrade surface. The Post-Tensioning Institute *(Reference 10)*, has a table of friction values for designers' use if a more refined approach is warranted. The steel stress should be from 2/3 to 3/4 of the yield point of the steel. A coefficient of 2/3 of the yield point is commonly used since the steel must remain elastic in order to perform its function.

The equation assumes that the slab will shrink from its free ends equally on each side of the center toward that center. A free end is an end or a joint at which there is no restraint to slab motion other than that from the base surface.

Figure 5 shows the forces considered to be present after the concrete cracks due to drying shrinkage (or decreasing temperature), when the subgrade drag forces are developed as the concrete attempts to shorten its length. Subgrade friction is assumed to be uniform and the maximum shrinkage force is mid-way between free and unrestrained ends of the slab length. The steel reinforcement, selected to resist this maximum force, spans the crack and attempts to drag the remaining concrete section. This drag is resisted by the frictional forces between the bottom of the concrete and the top of the base material. Only the weight of the concrete slab is included in the standard subgrade drag equation as causing the frictional force. Additional weight on top of the slab at too early an age will tend to further increase the frictional forces.

Specifications indicating the compaction of subgrade material can assist in the control of the density of the base. Proof rolling, a term that refers to random vehicular traffic on the base without causing rutting, thereby verifying compaction, can both enhance compaction and help assure the designer of satisfactory base tolerance.

If economic restraints pressure the designer to opt for the subgrade drag formula, it is advisable to place a note on the drawings advising the owner as to thermal gradient limitations.

1.5.1.4—Temperature method

A more desirable procedure for selecting the reinforcing steel for uniformly thick slabs on grade is the temperature method. Its similarity to the subgrade drag formula lies in the fact that the formula does not provide sufficient cross sectional area of steel to be structurally active. The formula provides an area that is capable of controlling temperature cracking. The formula is:

$$A_s = \frac{f_r(12)t}{2\,(f_s - T\alpha E_s)}$$

α, the coefficient of thermal expansion, can also be written as simply °F. The term refers to the unit change in length of one inch of concrete per degree Fahrenheit.

where A_s = the cross-sectional area of steel in square inches per lineal foot of slab width
t = slab thickness, inches
f_r = tensile strength of concrete, psi
(calculated as 0.4 × MOR)
f_s = allowable steel stress, psi
T = range of temperature the slab is expected to be subjected to, °F
α = thermal coefficient of concrete, inches per inch per degree Fahrenheit
E_s = modulus of elasticity of steel, psi

The coefficient of thermal expansion is normally taken as 6.5×10^{-6} in/in/°F. A common thermal gradient for this procedure is 50° in controlled environments and more in uncontrolled environments. The modulus of elasticity for steel is normally taken as 29×10^6 psi.

Commentary:
An even better procedure for crack control is to equate the shrinkage potential of the concrete to a needed force in the reinforcing. This results in $P = \delta A E_c$ *where* δ *is the unit shrinkage of the concrete, A is the unit concrete cross sectional area, and* E_c *is the modulus of elasticity for the concrete. Excessive panel lengths are still discouraged. This amount of steel (1.3%) is often considered too costly.*

1.5.1.5— Equivalent strength method

An alternative procedure for selecting the area of steel for slabs on grade is the equivalent strength procedure. This procedure equates the cross sectional area of concrete multiplied by the working stress to an equal amount of reinforcing steel force. The formula is:

$$A_s = \frac{36\sqrt{f_c'}\ t}{f_s}$$

where A_s = cross-sectional area of steel in square inches per lineal foot of slab width
t = slab thickness, inches
f_c' = compressive strength of the concrete, psi
f_s = allowable steel stress, psi

The above formula assumes the rupture modulus of concrete at $7.5\sqrt{f_c'}$, with a working stress of $0.4 \times MOR$.

Neither the temperature method (Section 1.5.1.4) nor the equivalent strength method for selecting steel area negates the requirement for ample contraction joints. Concrete is going to shrink and therefore it will crack. The designer's responsibility is to provide a degree of crack control that is compatible with the owner's use and budget constraints.

1.5.1.6— Slab with structurally active steel

Since the reinforcing steel absolutely must be in its detailed location, bolster (bar) supports are recommended.

The conventional design of an industrial floor intentionally selects a slab thickness with the intent that the floor will remain uncracked due to superimposed loadings on the slab's surface. A slab with structurally active steel is one with sufficient areas of steel to produce more moment capacity than that of the unreinforced, uncracked concrete slab. The steel requirements depend upon the amount of moment capacity required by the loadings *(Reference 11)*.

The behavior of the slab designed with structurally active steel reinforcement is similar to but not equal to that of the conventionally designed slab. It may be expected to remain uncracked up to the loading that produces a moment which exceeds the concrete's cracking moment capacity. At that point, the steel becomes structurally active and added moment capacity, and thus additional load support capability, exists. The added strength depends on the percentage of distributed steel, whether one or two layers are used, and the location(s).

Simply defined, a slab on grade is structurally reinforced (structurally active) when the moment capacity of the reinforced slab exceeds the cracking moment of the unreinforced slab.

1.5.1.7— Steel types and designations

The authors' choice as the most straightforward procedure for selecting steel requirements is to express the needed area in terms of either Grade 40 or Grade 60 deformed bars. From this initial selection, adjustments can be made for modified areas of welded wire fabric. Since welded wire fabric frequently achieves yield strengths as high as 75 to 80 ksi, in contrast to the 40 or 60 ksi yield strength specified for reinforcing bars, a lower cross-sectional area than that selected for reinforcing bars may be acceptable when using welded wire fabric. The designer may wish to designate a performance specification for welded wire fabric substitutions for a conventional rebar design rather than risk the exclusion of potential wire fabric suppliers. An important factor

Commentary:

In specifying welded wire fabric, the W *indicates smooth or plain wire, and* D *indicates deformed wire. Numbers following the letter* W *or* D *show cross-sectional area of the wire in hundredths of a square inch. Numbers ahead of the* W *or* D, *such as 12×12, indicate spacings of longitudinal and transverse wires, respectively. Generally, the D-wire fabric is specified with a minimum 75 ksi yield stress and the W-wire with a minimum 65 ksi yield. Smooth wire sizes range from W 1.4 to W 20; deformed wires range from D 4 to D 20.*

As Table 4 *shows, smaller steel areas are used for yield stress above 60 ksi.*

is the spacing of the significant individual elements in welded wire fabric sheets. A spacing of 12 inches should be designated when possible to provide sufficient stiffness for bolstering, while at the same time leaving room for workers to place their feet without disturbing the mesh.

As a guideline, Table 4 provides equivalent strengths for a number of the more common cross-sectional areas encountered and their welded fabric counterparts.

Deformed Bars, Grade 60			Welded Wire Fabric Equivalent	
Bar Size	**Bar Spacing**	**Steel area A_S, square inches per lineal foot of slab width**	**Plain (smooth) Wire minimum yield 75 ksi**	**Deformed Wire minimum yield 85 ksi**
#3	18" c.c.	0.08	12 × 12 — W 6.5	12 × 12 — D 6
#3	12" c.c	0.12	12 × 12 — W 9.5	12 × 12 — D 9
#4	24" c.c.	0.10	12 × 12 — W 8	12 × 12 — D 8
#4	18" c.c.	0.13	12 × 12 — W 10.5	12 × 12 — D 9
#4	16" c.c.	0.015	12 × 12 —W 12	12 × 12 — D 8
#4	12" c.c.	0.20	12 × 12 —W 16	12 × 12 — D 14

Table 4 *Common deformed bar spacings and their welded wire fabric equivalent.*

1.5.2 — Post-tensioning tendons

1.5.2.1 — Introduction

Post-tensioning tendons have been quite common as an alternative form of reinforcement in slab-on-ground design. Just as steel and welded wire fabric are used to control shrinkage, temperature, and slab integrity, post-tensioning can provide the same qualities with additional enhancements. It is vital that post-tensioned reinforcement be properly detailed on the drawings and properly installed in order to function properly. Care in detailing and installation is a small price to pay for some of the additional benefits that can be gained.

When cast-in-place slabs, whether slab on grade or structurally supported, are prestressed, post-tensioning is the method employed. With post-tensioning the tendons are tensioned after the concrete has hardened. Prestressed units made in a casting yard are generally pre-tensioned. That is, the tendons are tensioned before the concrete hardens around previously tensioned bare, unsheathed tendons.

Post-tensioning tendons are encased in a sheath, and are anchored at each end in the concrete slab *(Figure 6)*. The sheath permits the tendon, which is usually 1/2 inch in diameter, to slip along its entire length in order to be tensioned.

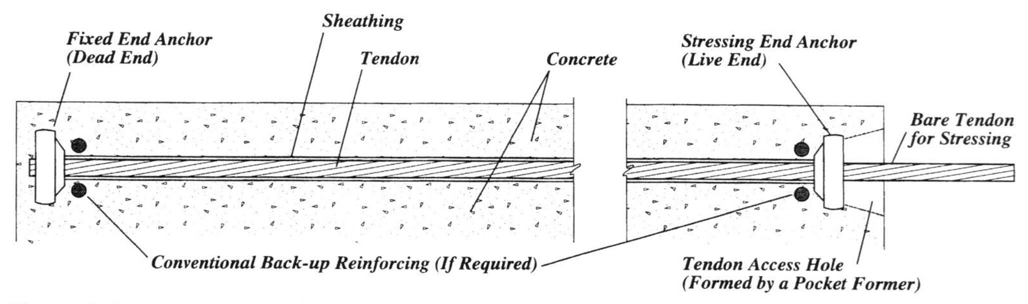

Figure 6 *Post-tensioning tendon in cast-in-place slab.*

Post-tensioning tendons may be either bonded or unbonded subsequent to tensioning. Most slab-on-ground applications use unbonded tendons. As shown in *Figure 7* the sheath generally consists of a plastic encasement with the annular space between the tendon and plastic filled with a corrosion-inhibiting grease *(Reference 10)*.

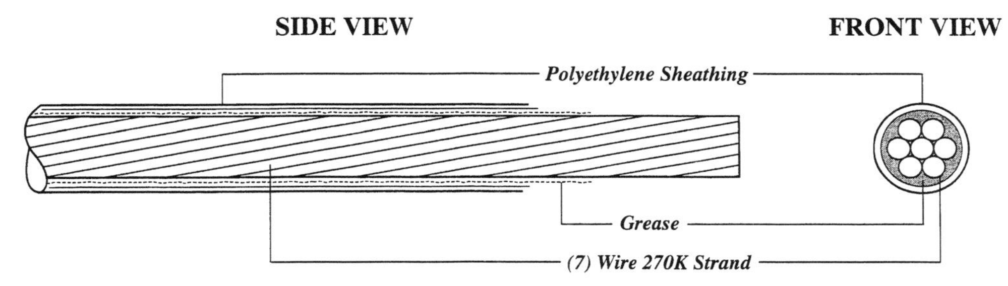

Figure 7 *Cross-section of unbonded post-tensioning tendon.*

1.5.2.2 — Crack control

Post-tensioning dynamically controls cracks by pre-compressing the concrete as a result of the tension in the prestressing tendons. Whereas reinforcing steel cannot prevent the formation of shrinkage cracks, and can only hold cracks tight, timely post-tensioning can control the amount of cracking by pre-compressing the concrete before hydration is effectively complete.

1.5.2.3— Curl control/post-tensioned slabs

One of the unique advantages of post-tensioning is its ability to limit curl in uniformly thick slabs. Curl can be offset by eccentrically post-tensioning just below the neutral axis. In addition, there are fewer joints in post-tensioned slabs; joints are generally located about every 250 feet to 350 feet. Fewer joints and better curl control result in favorable conditions where frequent forklift truck traffic occurs.

1.5.2.4— Structural effectiveness

When post-tensioning is used to reinforce an industrial slab, the pre-compression forces applied by the tendons supplement the rupture modulus of the concrete. In a conventionally reinforced slab with structurally active steel, it is necessary for the concrete to crack before the reinforcing becomes effective in resisting forces from applied loadings. In a post-tensioned slab, the average post-tensioning stress is added to the modulus of rupture, thus further increasing the slab's resistance to cracking. The post-tensioned reinforcement is considered active rather than passive.

In both slabs of uniform thickness and ribbed and stiffened slabs, active reinforcement with post-tensioning permits the use of a gross (uncracked) section modulus resulting in greater stiffness than that of a cracked section modulus. This can result in a thinner slab, thereby increasing its economic appeal.

Post-tensioning stresses in slabs on grade may range upward from a low value of 60 psi, after taking into account subgrade drag. Routinely, net post-tensioning stresses in industrial floor slabs vary from 60 psi to 150 psi after losses, with 100 psi being the most commonly used value. Tendons are routinely composed of 1/2-inch-diameter, unbonded 7-wire, 270k strand.

Commentary:
Post-tensioning tendons for industrial floor slabs are normally detailed at the neutral axis (i.e. the center of the slab). Therefore, chair supports are necessary to hold the tendons in this position.

1.5.2.5— Tendon types

Most post-tensioning tendons used in floor slabs on grade are unbonded monostrand as seen in *Figure 6*. The predominant size is 1/2-inch diameter, followed by 0.6-inch diameter strand. The 0.6-inch strand is a stiffer material and is usually found where slightly higher post-tensioning forces are needed. Table 5 lists post-tensioning tendon sizes and the values needed for design.

Nominal Diameter (in.)	Nominal Steel Area (sq. in.)	Nominal Weight (lb. per 1000 ft.)	Minimum Breaking Strength, lb.	Minimum Load at 1% Extension, lb.	Total Elongation Under Load (%)
3/8	0.085	292	23,000	19,600	3.5
7/16	0.115	395	31,000	26,350	3.5
1/2	0.153	525	41,300	35,100	3.5
0.6	0.217	740	58,600	49,800	3.5

Table 5 *Grade 270k strand for prestressed concrete.* (Reference 11)

1.6— Fiber enhancement

Both steel fibers and polymeric (often called synthetic) fibers can be used to improve certain characteristics of slabs on grade. Mixed into the concrete, they offer the advantage of an evenly distributed element within the concrete, the predominant effect being the control of plastic shrinkage cracking. They can also give industrial floors added resistance to impact, fatigue, thermal shock, and abrasion.

Steel fibers are mixed into the concrete at rates ranging from 30 pounds to 200 pounds or more per cubic yard. Values of 30 pounds to 60 pounds per cubic yard are most common. Steel fiber mixes may show an increase in modulus of rupture. It is best, however, to determine the increase by beam tests using ASTM C 1018. Slump must be carefully controlled to permit proper placement and surface finishing of the concrete.

Synthetic (polymeric) fibers are generally mixed into the concrete at rates ranging from 1 pound to 3 pounds per cubic yard, depending on the type of synthetic and the desired effect.

Table 6 provides guidelines for fiber types and dosages to secure enhanced floor slab properties.

Loading Condition or Enhancement Desired	Steel Fibers	Polymeric (Synthetic) Fibers
Shrinkage control	15-33 lbs./yd^3	1.5 lbs./yd^3
Light dynamic loading	30-50 lbs./yd^3	1.5 lbs./yd^3
Medium dynamic loading	40-65 lbs./yd^3	3-4.8 lbs./yd^3
Severe dynamic loading	65-125 lbs./yd^3	4.8-6 lbs./yd^3
High impacts	85-250 lbs./yd^3	9.6-12 lbs./yd^3

Table 6 *Fiber concentration guidelines.* (Reference 12)

CHAPTER 2

SLAB TYPES & DESIGN METHODS: THE DESIGNER'S CHOICE

2.1 — Types of slab construction

2.1.1 — Introduction

The most common approach to planning and designing a concrete slab on grade is to select the type of construction first. It is then that most designers look for the best available reference by which to determine thickness, reinforcement, joint spacings, and other details. There are six readily identifiable slab types (*Reference 8*) commonly used:

- Type A, *plain concrete slab*
- Type B, *slab with shrinkage control reinforcement*
- Type C, *shrinkage-compensating concrete slab*
- Type D, *slab post-tensioned for crack control*
- Type E, *lightly reinforced slab (with rebar or post-tensioning tendons)*
- Type F, *structurally reinforced slab*

To the foregoing six types, the authors would add a seventh:

- Type G, *fiber-enhanced concrete slab*

These seven types appear in *Table 7* and are described in the following sections.

2.1.2 — Type A, plain concrete slab

This concrete floor slab is intended to remain essentially uncracked. It contains no reinforcement whatsoever (no bars, no tendons, no wire, no fibers) and uses portland cement Type I or occasionally Type II. The slab may be strengthened at the joints by means of thickening, dowels, or keys. Relatively close spacing of the joints is the primary means of controlling shrinkage effects.

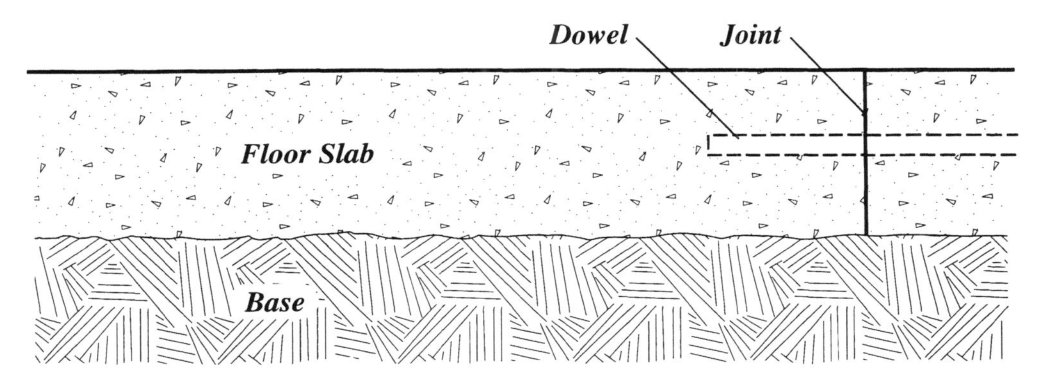

Figure 8 *Plain concrete slab contains no reinforcement but may be strengthened at the joints.*

Commentary:
The Portland Cement Association recommends joint spacings in feet equal to from 2 to 3 times the slab thickness in inches (Reference 6). *This is conservative, but relatively safe.*

2.1.3 — Type B, slab with shrinkage control reinforcement

Commentary:
Consider the use of widely spaced bars or wires, in WWF, such that spacings in each direction are 14 to 16 inches. Be sure this steel is adequately supported to guarantee its proper position.

Type B floor slabs are similar to the plain concrete slabs, but they contain a nominal amount of distributed reinforcement to act as control for the effects of shrinkage and temperature. The reinforcement must be stiff enough to be properly placed. It must be at or above the mid-depth of the slab. The slab is expected to remain essentially uncracked. The usual cement is portland cement Type I or occasionally Type II. Joint spacings are usually the same or slightly larger than for a plain concrete slab *(Reference 8)*. Steel is commonly selected by means of the subgrade drag equation.

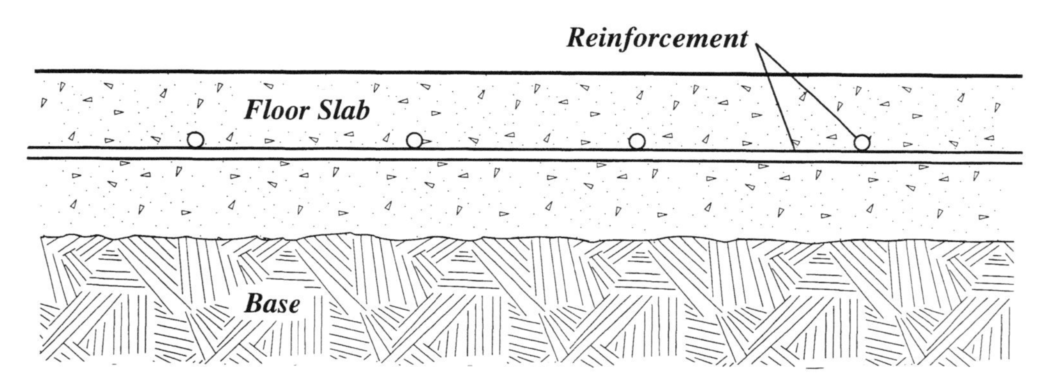

Figure 9 *Type B slab with shrinkage control reinforcement.*

2.1.4 — Type C, slab with shrinkage-compensating concrete

Control prism tests, following ASTM C 878, are strongly recommended to assure the proper amount of expansion and contraction.

These slabs require reinforcing steel, which absolutely must be properly located. The concrete, which is called shrinkage-compensating concrete, does shrink, but before shrinking it expands by an amount intended to be slightly greater than the subsequent shrinkage. This concrete may be produced with an expansive admixture or with Type K cement. Joint spacings may be significantly greater than those of the two previously described slabs. ACI Committee 223 has published the best document *(Reference 13)* to guide the designer in planning a shrinkage-compensating concrete floor.

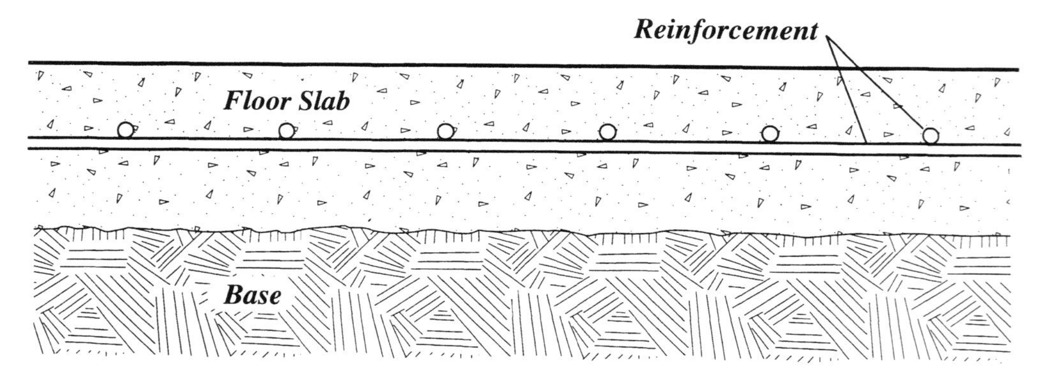

Figure 10 *Shrinkage-compensating concrete slab must contain reinforcement as recommended by ACI 223.*

2.1.5 — Type D, slab post-tensioned for crack control

These slabs contain post-tensioning tendons intended for crack control and are made with portland cement Type I or II. The prestress forces not only increase the effective modulus of rupture but they also allow a wide spacing of construction joints with no intermediate contraction (control) joints. *Reference 10* presents design aids for this type of floor construction.

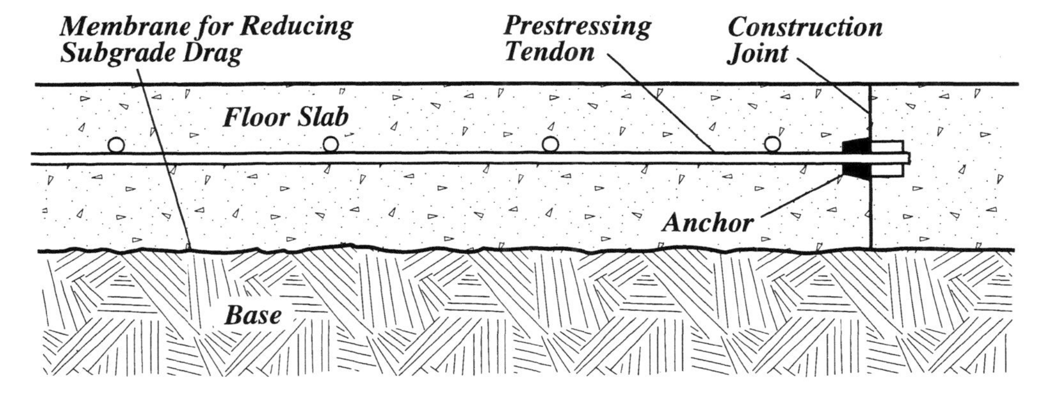

Figure 11 *Slab post-tensioned for crack control.*

Commentary:
With post-tensioned slabs, one or two layers of polyethylene sheeting, perforated or non-perforated, are often used for reducing subgrade drag.

2.1.6 — Type E, lightly reinforced structural slab

These slabs are considered to be structurally reinforced; however, the expectation is that they will remain essentially uncracked. Portland cement Types I and II are common. Type E slabs will have steel reinforcement, post-tensioning tendons, or both. They are planned to support structural loads such as columns and walls directly on the slab. They also are commonly used to resist the forces caused by swelling or shrinking of unstable soils.

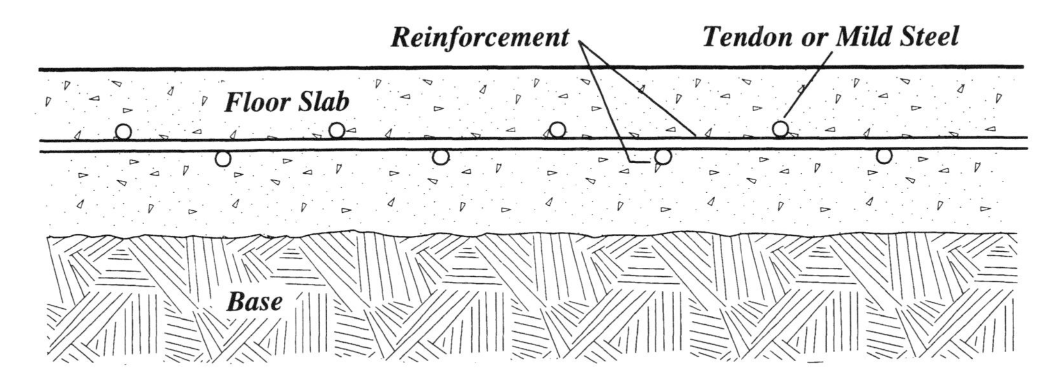

Figure 12 *The lightly reinforced structural slab (Type E) may be uniformly thick or ribbed for stiffness and may be reinforced with bars, post-tensioning tendons, or both.*

2.1.7 — Type F, structurally reinforced slab

This slab type differs from all the others in that the design intentionally allows cracking (due to superimposed loading) at some determined level of loading. These slabs are structurally reinforced with one or two layers of steel reinforcement in the form of deformed bars or welded wire fabric (sheets). Portland cement Types I and II are normally used. The location of the steel is critical to the structural capacity. Joint spacings are not critical to the design, other than for the construction process, since a certain amount of cracking is considered acceptable.

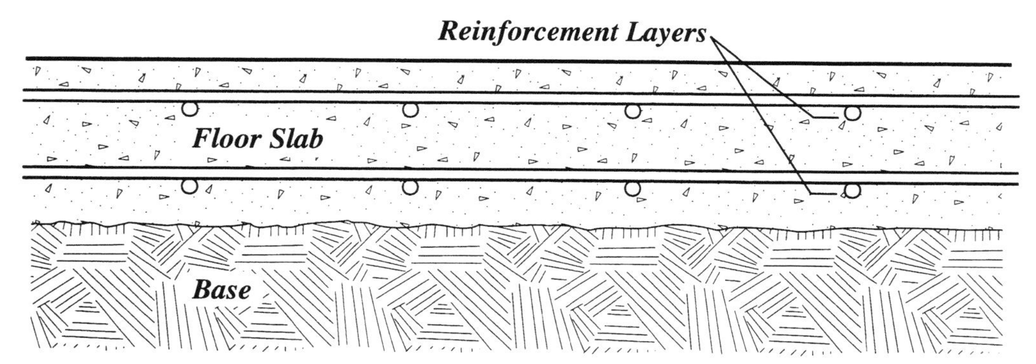

Figure 13 *Structurally reinforced slab has one or two layers of reinforcement, either bars or welded wire fabric.*

2.1.8 — Type G (Authors' designation), fiber-enhanced concrete slab

Concrete floor slabs may have their performance enhanced by the addition of specific and preplanned amounts of fibers. The selection of fiber type and material, as well as its specific dosage, are critical (*Reference 12*).

These fibers, either synthetic or steel, may be used in plain, reinforced, or prestressed concrete. Synthetic fibers, such as polypropylene, nylon, and glass, are effective in resisting the formation of early-age plastic shrinkage cracking. Steel fibers, commonly added in amounts from 30 to 60 pounds per cubic yard of concrete, are distributed throughout the concrete mix and act as crack arresters to prevent the propagation of micro-cracking.

The presence of fibers also can increase the effective modulus of rupture. This increase must be determined by a set of beam tests, in accordance with ASTM C 1018.

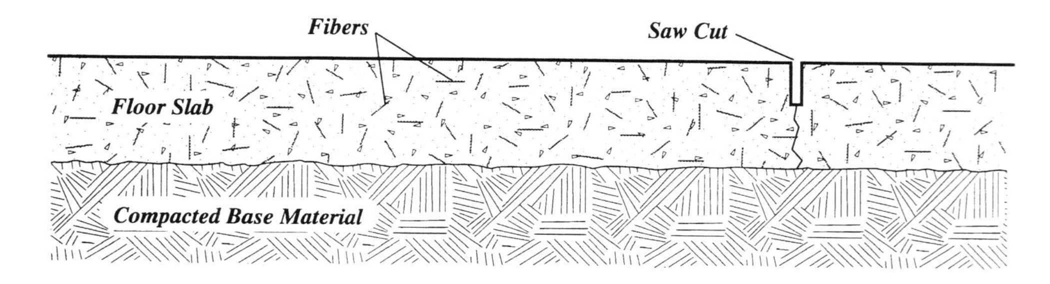

Figure 14 *Slab enhanced with fibers, either synthetic or steel.*

Slab Types		***Design Methods***						
		PCA	**WRI**	**COE**	**PTI**	**ACI 223**	**MATS**	
A	Plain concrete; no reinforcement; portland cement	*x*		*x*			*x*	*Thickness selection*
		x		*x*				*Related details*
B	Temperature & shrinkage reinforcement; portland cement	*x*	*x*	*x*			*x*	*Thickness selection*
		x	*x*					*Related details*
C	Temperature & shrinkage reinforcement; shrinkage compensating cement	*x*	*x*	*x*			*x*	*Thickness selection*
						x		*Related details*
D	Post-tensioning for crack control; portland cement	*x*	*x*	*x*	*x*		*x*	*Thickness selection*
					x			*Related details*
E	Post-tensioning and/or nonprestressed steel reinforcement; portland cement	*x*	*x*		*x*			*Thickness selection*
					x			*Related details*
F	Nonprestressed steel reinforcement; portland cement	*x*	*x*	*x*	*x*			*Thickness selection*
			x	*x*	*x*			*Related details*
G	Fiber-enhanced concrete	*x*		*x*			*x*	*Thickness selection*
								Related details

Table 7 *Correlation of Slab Construction Type With Design Method, Adapted from* Reference 8.

2.2 — Methods of slab thickness determination

There are six commonly used methods or procedures by which industrial floors are planned and the thickness determined:

- PCA, *Portland Cement Association method*
- WRI, *Wire Reinforcement Institute method*
- COE, *U. S. Army Corps of Engineers method*
- PTI, *Post-Tensioning Institute method*
- ACI 223, *ACI Committee 223 method*
- MATS *(PCA) Finite Element Analysis*

Commentary:
Although the word "design" is commonly used, "thickness determination" more accurately describes the main results of the methods and procedures listed.

All of them have proved to be effective when their recommendations and details are followed. It is therefore not possible to state that only one design method is to be used. The selection for a specific job depends on the nature of the job and on the personal preference of the floor designer. *Table 7* provides some assistance in this selection. The six methods described below are used in the design examples of this book.

2.2.1 — Portland Cement Association (PCA) method

PCA's charts and tables allow slab thickness selection for dual and single wheel axle loads, for rack support post loading, and for uniform loads with fixed or variable positions. Reinforcement is optional and is intended for shrinkage and temperature effects. Loadings are assumed to be in the interior of the slab area and, therefore, PCA recommends that joints be strengthened *(References 6* and *14).*

2.2.2 — Wire Reinforcement Institute (WRI) method

The Wire Reinforcement Institute provides a method of thickness selection for single wheel axle loads and for uniform loads with aisles. Only loadings on the interior of the slab are considered. WRI charts include the effect of relative stiffness of the slab with respect to the subgrade. Steel reinforcement is assumed in the process *(Reference 15).*

2.2.3 — United States Army Corps of Engineers (COE) method

The Corps of Engineers design charts are based on Westergaard's equations for edge stresses in slabs on grade. They are therefore appropriate for loads immediately adjacent to joints or edges. Also included is the effect of load transfer across a joint in terms of a load transfer coefficient of 0.75, which in effect reduces load support capability at the joint by 25%. Steel reinforcement is optional, although its use is implied. Loadings handled are heavier axle loadings and other vehicle loads by means of a design index. *(References 16* and *17).*

2.2.4 — Post-Tensioning Institute (PTI) method

The Post-Tensioning Institute publishes tables and charts that deal with strength requirements for loadings due to soils which expand or contract significantly. Post-tensioning tendons are the intended technique; however, since moments and shears are determined in the calculation process, steel reinforcement may also be used. These tendons may also be used for control of shrinkage and temperature effects and the effective increase in modulus of rupture may be included in the design. *(Reference 10).*

2.2.5 — ACI Committee 223 (ACI 223) method

This method does not deal directly with the selection of the slab's thickness, which must be handled by one of the other methods cited. Rather, it deals with the control of early-age expansion and the following shrinkage of the concrete slab. Reinforcing steel is required and its location is critical to performance. The predominant result of using this method is a much wider spacing of joints along with the elimination of shrinkage cracks *(Reference 13).*

Commentary:
MATS is sophisticated and powerful. It does not give the required thickness as its output, but rather analyzes for moments from which the floor designer can calculate stresses. While useful for investigating existing slabs, especially those with combination loadings, the authors feel that its arithmetic precision exceeds that required for normal construction tolerances.

2.2.6— MATS (PCA) Finite element analysis

This finite element analysis software program produced by the Portland Cement Association analyzes ground-supported slabs as well as mat foundations, and combined footings for given situations. Version 5.01 (1994) runs under Windows V3.1. It requires a mouse and 1 MB of free disk space with 500 KB of free conventional memory. It allows analysis of variable thickness slabs loaded simultaneously with uniform and concentrated static loads (*Reference 18*).

2.2.7— Other methods

In addition to the six methods cited, equations from the technical manuals of the U. S. Army and Air Force *(Reference 17)* have been used in this book. Other analytical and computer-software procedures not emphasized in this book certainly should be mentioned. Included are:

- Theoretical equations for a beam on an elastic foundation *(Reference 19)*
- An equation for negative moment in an aisle between uniform loads *(Reference 20)*
- A computer software program (PC) called AIRPORT *(Reference 9)*

There are no doubt others, but the above are most commonly used for industrial floor design at this date.

Type of Loading	*Design Methods*					
	PCA	**WRI**	**COE**	**PTI**	**ACI 223**	**MATS**
Uniform Loads & Aisles	*x*	*x*				*x*
Storage Rack Post Loads	*x*					*x*
Lift Truck Wheel: Interior Loadings	*x*	*x*				*x*
Lift Truck Wheel: Edge Loadings			*x*			
Concentrated Loads Fixed Locations	*x*	*x*				*x*
Vehicle Loads With Impact			*x*			*x*
Post-Tensioning Prestress[1]				*x*		
Shrinkage-Compensating Concrete[1]					*x*	

[1] *These are not thickness selection methods as such; however, the techniques affect directly the details of use of the other four reference methods.*

Table 8 *Slab Thickness Selection Methods Appropriate for Different Types of Loadings.*

2.2.8— Correlation of construction type with design method and loading condition

Each of the cited methods has particular loading conditions for which it is most effective, as developed in the design examples. Considering the various types of loadings, *Table 8* from *Reference 8* shows which design method is appropriate for each kind of load. *Table 7*, as already mentioned, correlates the slab construction methods with design methods appropriate for each one. *Table 7* also shows which methods aid in thickness selection and which methods have information on the related details, such as joint spacings, joint materials, and steel requirements.

CHAPTER 3
DESIGN FOR VEHICLE AXLE LOADS

3.1 — Design objectives

The vehicle axle load is commonly known as the lift truck load, although it can be any vehicle that travels on the concrete floor with its wheels in contact with the slab surface. It is a single axle with one or two wheels at each end of the axle. One wheel at each end is ordinarily found in industrial plants. This produces a pair of concentrated loads on the slab and frequently controls the slab thickness required. The vehicle axle load is a common design problem and must be considered in floor designs. *Figure 15* shows a lift truck and the geometry of the axle loading on the slab.

Commentary:
Frequently, plant management will purchase new lift trucks whose capacities are greater than the lift trucks assumed in the floor design. The designer must anticipate this as much as possible.

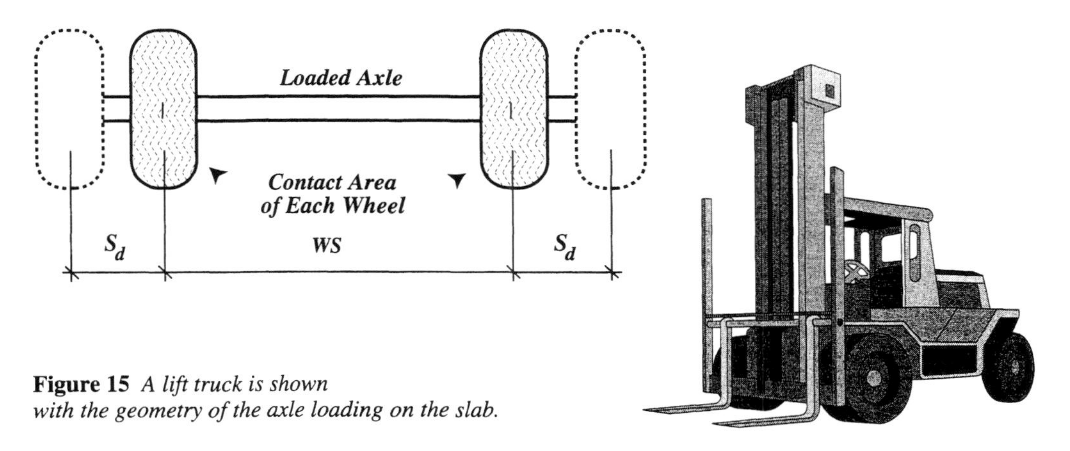

Figure 15 *A lift truck is shown with the geometry of the axle loading on the slab.*

The majority of vehicle axles are single wheel axles having only one wheel at the end of each axle, in which case $S_d = 0$.

To be able to determine the thickness of the concrete floor slab as well as the effect of any prestress or reinforcement that may be considered, a number of values are needed. Some values come from loading specifications and some come from the materials, the site, and the designer. The necessary information includes:

All decisions made or not made prior to construction are part of what most floor planners call design. It is essential that a proper determination of concrete slab thickness be included in this process, especially for heavier loadings.

For the vehicle:

- Vehicle weight in pounds
- Load capacity in pounds
- Total axle load in pounds
- Single wheels or dual wheels?
- Solid or pneumatic tires?
- Tire width in inches or tire pressure in psi
- Wheel contact area in square inches
- Wheel spacing in inches

For the site and materials:

- Concrete compressive strength in psi
- Concrete modulus of rupture in psi
- Modulus of subgrade reaction (soil) in pci
- Safety factor selected by designer

The best source of vehicle data is the specification sheet from the manufacturer. If this is not available, some values may have to be assumed in order to complete the design.

Commentary:
Do not make the mistake of neglecting the check of a flexural tensile stress on the bottom of the slab just because it is on the bottom surface. If a crack occurs, it will work its way up to the top and across the slab due to continued vehicle traffic. Note that no moments, and therefore no flexural stresses, are caused by the concrete slab itself since it is placed on the ground wet and then hardens. Bending stresses due to the concrete's dead weight are produced only if the slab curls up or the ground support settles.

The design checks the capacity of the floor slab to resist the moment in the slab beneath the loaded axle. This loading causes tension on the bottom of the slab beneath the most loaded wheel. It is sometimes called a positive moment. Since the wheel loads are normally of equal value, both on the vehicle and in the design charts, the moments are equal beneath each wheel. The example in Section 3.5 is for the special case of unequal wheel loads on the same axle.

When the slab is to remain uncracked, the objective is to limit the actual tensile stress to an acceptable value. This acceptable value, usually called the allowable stress, is the modulus of rupture divided by the selected safety factor. The thickness determination for this loading is shown in the examples in Sections 3.2, 3.3, and 3.4.

If the concrete floor does not need to remain completely crack free (that is, if hairline cracks due to loading are acceptable), then the approach can change. The objective then would be to determine the applied moment in the slab. The moment would be used to design the slab using common reinforced concrete procedures and to select appropriate areas of steel reinforcement. This can be most easily done with WRI charts as illustrated in the example of Sections 3.8 and 3.9.

When shrinkage-compensating concrete or post-tensioning is used in building the slab, then the design process is altered. The intent of these procedures is to maintain an uncracked slab by chemical prestressing or physical prestressing rather than by adjusting slab thickness and joint spacings. An additional intent is to use wider joint spacings. These design procedures are illustrated in Sections 3.6 and 3.7. Joint spacings are discussed in Section 10.2.

3.2 — Using PCA charts to design for axle loading: AUTHORS' CHOICE

The PCA design charts appear in the appendix along with larger copies of all other design graphs. Reduced size charts are used in these examples for illustrative purposes. These PCA charts were originally published in *References 6* and *14*.

3.2.1 — Single wheels, interior loading

For the first problem using PCA charts, the following values will be used:

From materials, site, and designer:

Concrete compressive strength: $f_c' = 4500$ psi
Modulus of rupture (using $9 \times \sqrt{f_c'}$): $MOR = 604$ psi
Subgrade modulus: $k = 200$ pci
Safety factor: SF = 2.0

This lift truck loading is heavier than most vehicles in industrial plants, but even higher axle loadings are occasionally found. The contact area depends on the wheel material. If this is an extremely hard material the contact area can be much smaller, sometimes as low as 10 square inches.

From lift truck specifications:

Lift truck capacity = 20 kips
Vehicle weight = 22 kips
Total axle load = 42 kips *(Assume 100% on front loaded axle)*
Axle: Single wheels at each end of axle
Wheel spacing = 42 inches
Tire pressure = 250 psi *(Rational value of 250 is used to represent a solid tire, to obtain contact area.)*
Wheel contact area = 84 square inches (one wheel) *(Calculated here from a single wheel load divided by tire pressure [21,000/250 = 84 square inches]. However, the authors recommend that you use the manufacturer's specification sheet if it is available for the vehicle.)*
Effective diameter of contact area = 10.34 inches, calculated using diameter = $2\sqrt{Area/\pi}$ *(Needed for use of WRI charts)*

The slab thickness is then selected using *Figure 16* in the following manner:

- Calculate the stress per 1000 pounds of axle load. First find the allowable stress, which is the modulus of rupture divided by the factor of safety = 604/2 = 302 psi. This is the acceptable flexural tensile stress for the concrete.

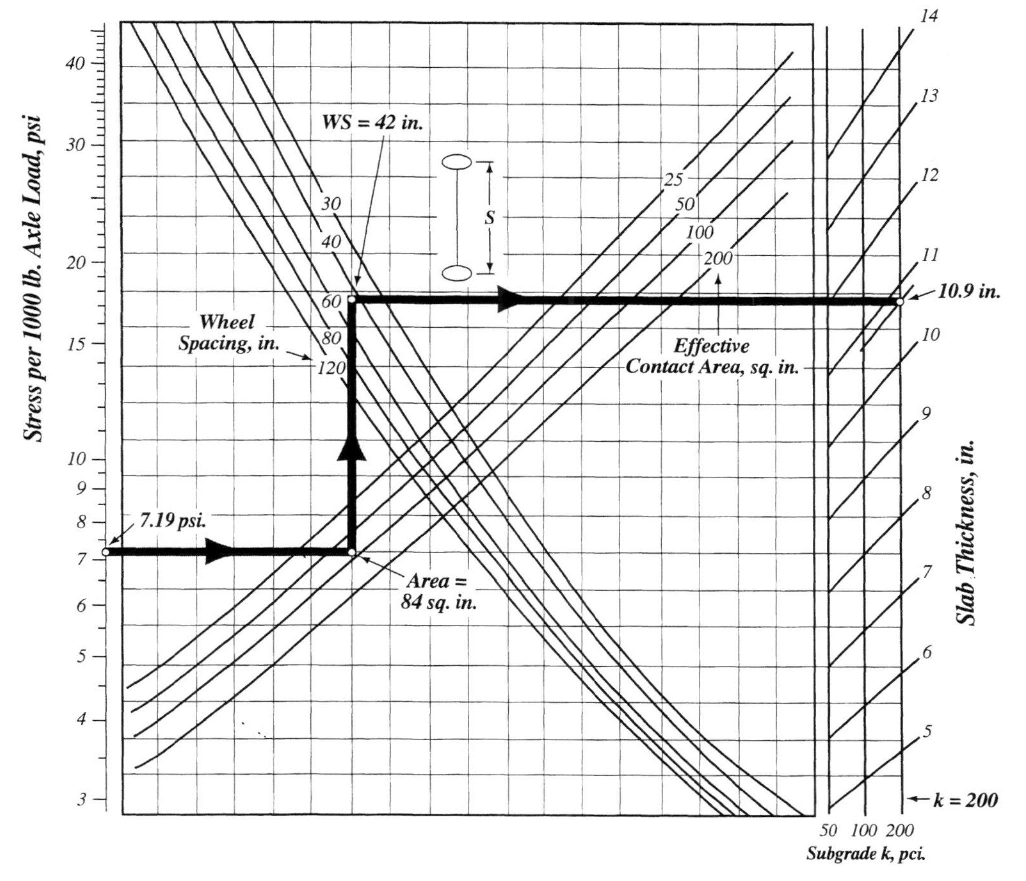

Figure 16 *Use of a PCA design chart to select slab thickness for single axle loading.*

Commentary:

All graphical design charts have limitations. Figure 16 *has no curve for wheel contact areas less than 25 square inches, and no curves for values of* k *above 200 pci nor below 50 pci. Other means are needed when variables fall outside the limitations of any design chart. Interpolation is permissible; extrapolation is discouraged.*

- Divide this by the axle load. This gives: 302/42 = 7.19 psi per 1000 pounds of axle load which is to be located on the left hand vertical axis of *Figure 16*.
- Draw a horizontal line from the 7.19 psi stress value to the given wheel contact area, which is 84 square inches, plotting as close as is graphically possible.
- Draw a vertical line to the wheel spacing of 42 inches, again plotting it as close as is possible.
- Draw a horizontal line to the right and intersect the vertical line representing the *k* of 200 pci.
- The required thickness is indicated as very close to 10.9 inches.
- A thickness of 11 inches is recommended for the first example.
- *When this example is solved by other procedures, the required thickness is 10.9 inches using WRI charts (see* Section 3.3), *11.0 inches using the AIRPORT program, and 11.6 inches using the MATS program.*

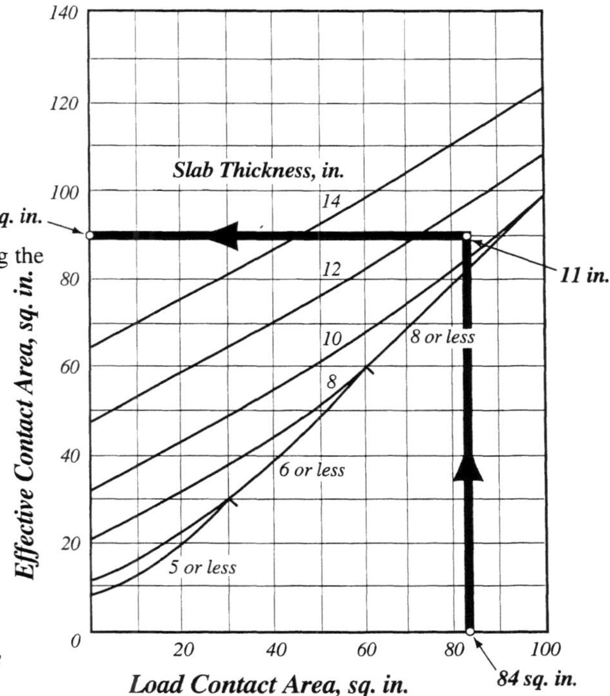

Figure 17 *Chart for determining effective wheel contact area for concrete slabs.*

Using the actual area as shown in this example (84 square inches) is always conservative. Figure 17 *(from* Reference 14*) shows how to use an effective contact area, which is slightly less than the actual area for thicker concrete slabs. Entering* Figure 17 *from the bottom with the actual wheel contact area, go upward to the estimated slab thickness, then horizontally to the left to read an effective area. In this case, the effective area would have been 90 square inches, which in turn would have reduced the theoretical slab thickness to approximately 10 3/4 inches.*

3.2.2 — Dual wheels, interior loading

For dual wheels on a single axle, a second example using the PCA charts is shown with all the input values being the same except that the lift truck will be considered to have a dual-wheel-loaded axle; that is, two wheels on each end of the axle. While most dual-wheel truck axles have a wider wheel spacing, this example will maintain the same 42-inch spacing from center to center of wheel(s).

Separation of wheels at the end, S_d = 14 inches
Clear distance between wheels, S = 28 inches
Contact area per wheel = 42 square inches

A correction (reduction) factor is needed, since this is a dual-wheel axle loading. This is found using *Figure 18* as follows:

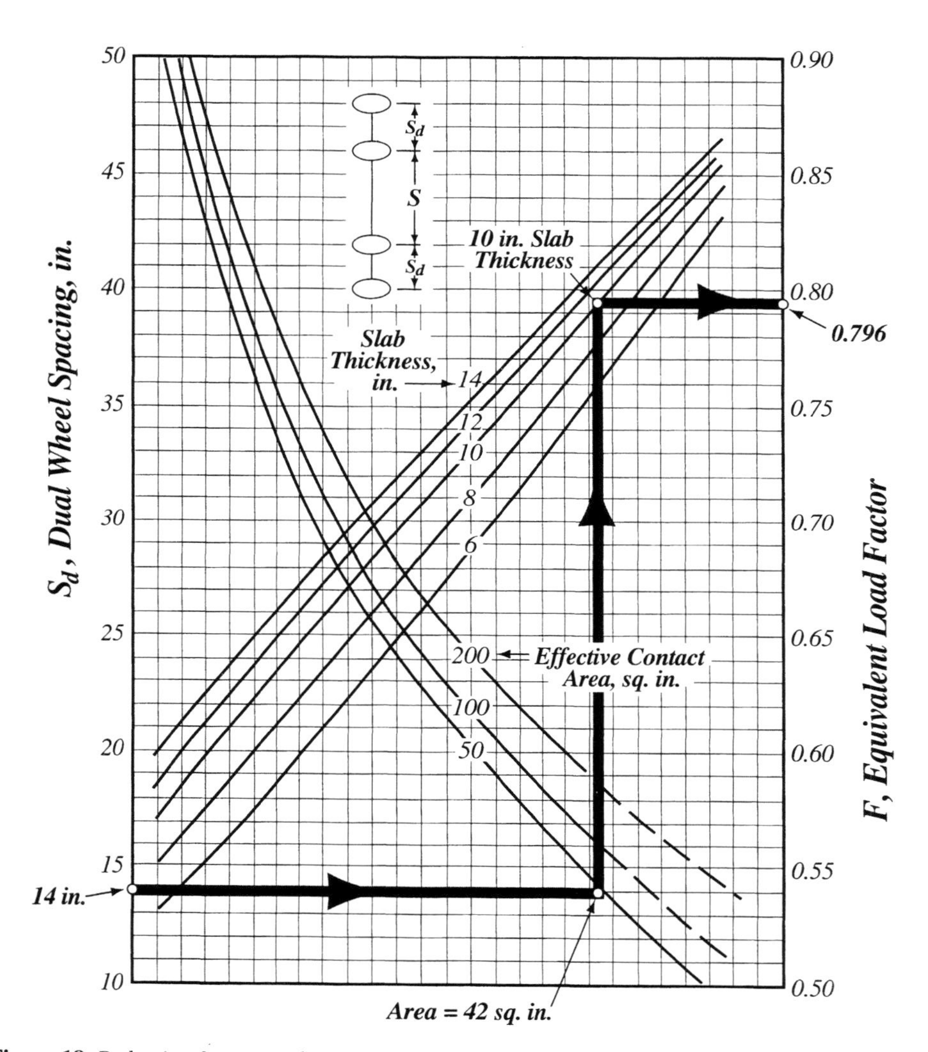

Figure 18 *Reduction factor used with PCA charts when designing for dual-wheel loads.*

Commentary:
Either of the previous two examples (11-inch slab for 84-square-inch contact area, or 10.75-inch slab for 90-square-inch effective contact area) can provide the basis for the trial slab thickness of 10 inches, since the thickness will be slightly less for dual wheels than for single wheels on an axle.

Any physical feature which increases the contact areas will reduce the required slab thickness. With all other values unchanged, dual wheels allow a reduction in thickness; however, dual-wheel lift trucks are not common in industrial facilities.

- Enter *Figure 18* from the left with the separation of wheels at the end, 14 inches. Draw a line horizontally to the contact area of 42 square inches, then vertically to the slab thickness of 10 inches (usually estimated), and then horizontally to the right-hand axis. Here the equivalent load factor F is read as F = 0.796.
- Multiply the original axle load of 42 kips by the equivalent load factor F to obtain the corrected axle load:

$$0.796 \times 42 = 33.4 \text{ kips}$$

- Then repeat the process of the previous example, using *Figure 16* to obtain a thickness of 9.4 inches.

3.2.3 — Stress increases for loads at edges or joints

PCA charts give thicknesses based on loadings at the interior of the floor slab. The same is true for WRI charts and for the AIRPORT program. An axle load from a moving vehicle will move along the floor and eventually cross a joint. When any loading moves to a joint or to the edge of the slab, the bending stress is increased. In many cases, the increase is substantial. The MATS 5.01 computer program gives the designer a way to investigate this stress increase at joints and edges.

The following statements are based on analysis of selected example problems using MATS 5.01. For loads such as moving vehicle axle loads at a free edge with no load transfer ability, the stress increase can be as much as 50% to 70%. The Section 3.2.1 example (42 kip axle load on a two-wheeled axle) shows a stress increase of 62% for the free edge loading case. Other loadings, such as rack posts, swing lift trucks, or columns (on slab as integral footing), show stress increases of even greater percentages. Where substantial load transfer exists, such as at a doweled joint, the stress increase will be less, perhaps half of the 50% to 70% range.

To account for the stress increase and maintain approximately the same safety factor, the floor designer who uses the PCA charts should consider increasing the safety factor for loadings placed at a slab edge or placed immediately adjacent to a joint.

If Example 3.2.1 is solved using a safety factor of 2.6 instead of 2.0 (assuming an increase in actual bending stress of 30%), the required thickness would be 12.7 inches instead of the 10.9-inch chart value for the same loading at the slab's interior.

Using the increased thickness over the entire floor may be too expensive. One could consider using the greater thickness for a specific distance back from the joint (both sides) and then tapering to the lesser slab thickness. The taper should be on a slope of 1 to 10 (gradual) and the distance from the joint to the beginning of the decrease in the thickness should be at least 1 times the radius of relative stiffness (See Section 10.7).

3.3 — Design for axle loading using WRI charts

The WRI charts appear in the appendix where they are reproduced at a larger and more usable scale. They were originally published in *References 15* and *21*. The procedure accounts for the relative stiffness between the subgrade and the concrete slab in its use of the *D/k* ratio determined from *Figure 19*. The procedure also requires two other charts, *Figures 20* and *21*. For this example, the following values will be used:

From materials, site, and designer:

Concrete compressive strength: $f_c' = 4500$ psi
Modulus of rupture: $MOR = 604$ psi
Subgrade modulus: $k = 200$ pci
Safety factor: SF = 2.0

From lift truck specifications:

Total axle load = 42 kips
Axle: Single wheels at each end of axle
Wheel spacing: WS = 42 inches
Wheel tire contact area = 84 square inches
Effective diameter of contact area = 10.34 inches
(These vehicle specifications are the same as for the PCA example in Section 3.2.1)

- The first step is to assume a slab thickness. This could also be the last thickness from several trial designs.

  Assume $t_s = 11$ inches

- Since a value for the concrete modulus of elasticity is not given, it must be determined from the standard relationship

  $$E_c = w_c^{1.5} \times 33\sqrt{f_c'}$$ This gives a value of 3824 ksi.

  where w_c is the weight of concrete taken as 144 pcf.

Commentary:
The equation $E_c = 57{,}000 \times \sqrt{f_c'}$ *can also also be used. Either is considered adequate since both* f_c' *and* E_c *change (increase) with time.*

- Using *Figure 19,* from the intersection of E_C and t_S, draw a horizontal line to the right to the curve representing the subgrade modulus k. Then draw a line down to the horizontal axis and read the value for *D/k*.
- Select *D/k* = 22.

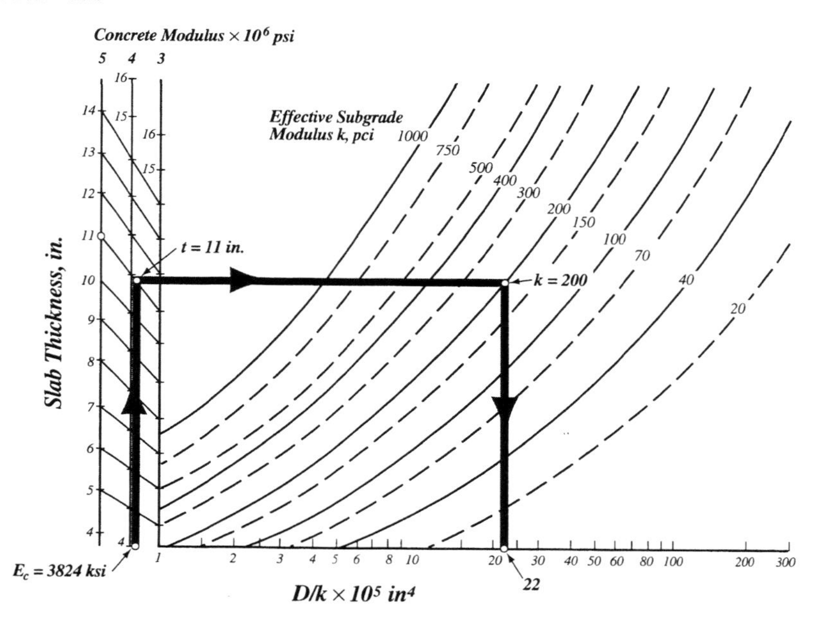

Figure 19 *WRI chart for determination of the D/k value.*

- Now use *Figure 20* to determine the applied moment due to the vehicle's axle load. This involves two steps. Enter this chart from the bottom at the equivalent loaded diameter, which is 10.34 inches, draw a line vertically to the *D/k* curve, which is 22, and then draw a line left to the moment axis. This is the applied moment for the first wheel. Its units are inch-pounds per inch per 1000 pounds of wheel load.
- Unit moment = 255 inch-pounds per inch per 1000 pounds of wheel load.

Commentary:
Do not miss the difference between the PCA charts and the WRI charts. The PCA charts are in terms of axle loading and the WRI charts are in terms of wheel loading. Also there is a far greater range of k *values available for use in the WRI charts than in the PCA charts.*

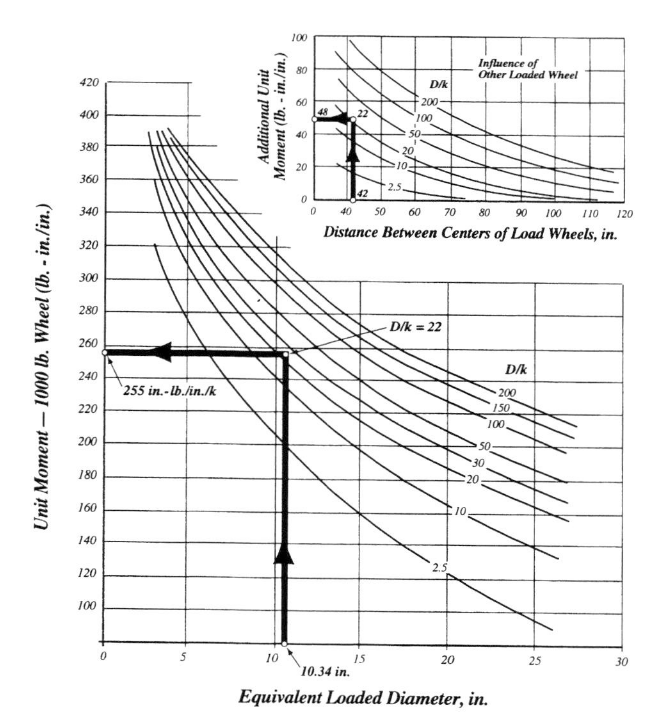

Figure 20 *WRI determination of applied moment due to vehicle axle load requires two steps.*

• Use the upper and smaller portion of *Figure 20* to find the added moment due to the second wheel at some distance away.

• Use a wheel spacing of 42 inches. Draw a vertical line from WS = 42 up to the *D/k* value of 22. Draw a horizontal line to the left to select a moment of 48, which is still in units of inch-pounds per inch per 1000 pounds of wheel load. Add the two values.

• Unit moment = 255 + 48 = 303 inch-pounds per inch per kip

• Calculate the total applied moment due to this lift truck.
Moment = (303)(21) = 6360 inch-pounds per inch (or foot-pounds per foot)

Commentary:
This moment can be used directly to determine the required slab thickness, as is done in this example. It can also be used to design a reinforced concrete slab of the same or of a thinner thickness. This is illustrated in Section 3.8.

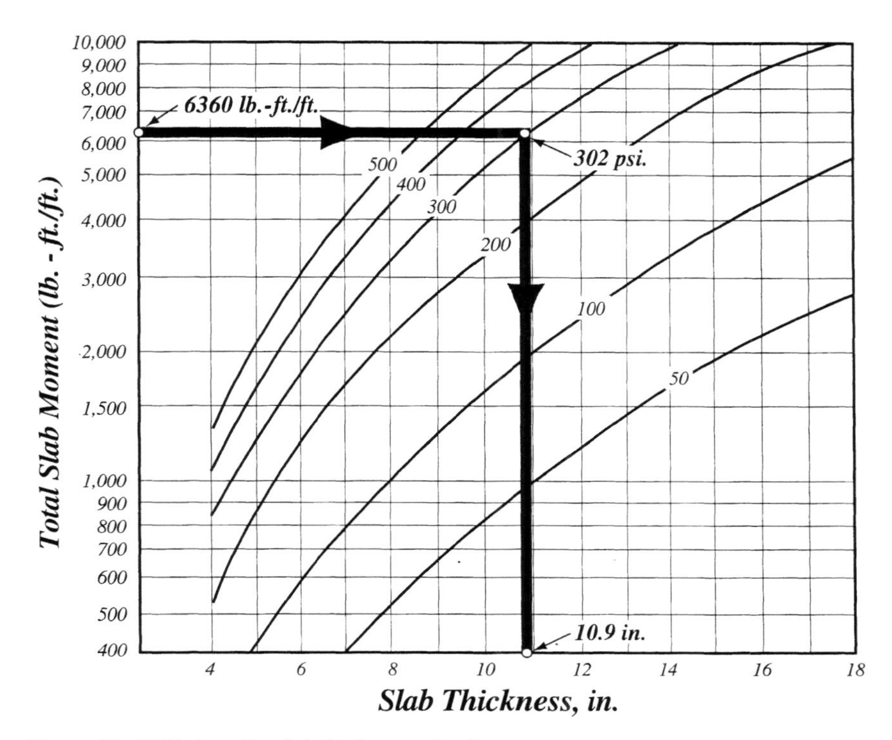

Figure 21 *WRI chart for slab thickness selection.*

Note:
From page 26, allowable stress 302 psi = 604/2 (modulus of rupture divided by selected safety factor).

• Use *Figure 21* by starting on the left-hand vertical axis with the applied moment, 6360 foot-pounds per foot, and draw a horizontal line to the curve representing the allowable stress, which is the modulus of rupture divided by the safety factor: 604/2 = 302 psi. Then, draw a vertical line down to the required thickness, which is very close to 10.9 inches.

• An 11-inch slab is recommended.

3.4 — Design for axle loading using COE charts

The charts referred to as the COE charts appear in larger size in the appendix. While identified as COE material (due to the fact that American Concrete Institute documents have referred to them in this way), they are taken from a United States Army and Air Force technical manual *(Reference 17)*. Only their designs for vehicle loadings are included here.

The COE approach differs from the other design methods cited. It bases the designs on categories of loadings with a design index determined by usage rather than on individual and specific vehicle specifications. *Reference 17* divides vehicles into three general classifications as follows: lift trucks, pneumatic or solid-wheel vehicles, and tracked vehicles (usually military vehicles). The designs are based on loading the interior slabs on the assumption that interior joints are either tied or doweled and therefore possess a certain degree of load support from slab panel to slab panel.

Three charts are included. They indicate the three design checks expected when using the COE materials as follows:

Commentary:
The safety factors, using Figures 22, 23, and 24, vary in magnitude and can be as low as 1.3

Figure 22 is for forklift truck axles of 25 kips or less. However, the chart uses a design index to represent a class of vehicles combined with the frequency of the vehicles' use. The chart is intended for trucks, cars, buses, and small lift trucks.

Figure 23 is for heavier forklifts whose axle loads range from 25 to 120 kips. This chart uses the specific vehicle, but has assumed the basic configuration of a single axle with a pair of dual-wheels, a 58.5-inch wheel spacing, a 13.5-inch dual-wheel spread, and an unidentified standard contact area. These variables are built into the chart solution. To obtain a required thickness, the chart uses the modulus of rupture, the axle load, the subgrade modulus, and the number of vehicle passes for the expected life of the slab.

Figure 24 is for associated pavements, open storage areas, roads, and streets. It uses a general category of vehicles with a design index, the modulus of rupture, and the subgrade modulus to obtain the required thickness. As in the other COE charts, the safety factor is built into the process.

In a five day work week with two eight-hour shifts, 50 operations per day is equivalent to 1 operation every 20 minutes, and 250 per day is equivalent to 1 operation every 4 minutes.

To use *Figure 22* or *24*, it is necessary to select a value for the Design Index (DI). A guide to this selection is in *Table 9*. The index is based on a combination of vehicle axle load and the number of times per day the load passes over a specific location on the floor. Daily passes over a number of years represent the effect of fatigue. The Design Index can be determined directly from *Table 9* as long as the axle load and daily passes fit the table.

The Departments of the Army and the Air Force also use a description of one or more vehicles to select a category of vehicle, which in turn helps with a selection of the Design Index. *Reference 17* gives assistance on this procedure.

While the charts are relatively easy to use, there are two values that are not so simple to determine accurately. They are ***design index*** *and the* ***category of vehicle****. For industrial floor design, the* ***design index*** *is difficult to select. It ranges from 1 to 10, where values of 5 to 7 seem to represent normal plant activity fairly well. The* ***category of vehicle*** *is not specific to individual vehicles and is not easily selected.*

Design Index Values		
Axle Load Forklift Truck (kips)	**Assumed Daily passes (uses) throughout 25 years**	**Design Index (DI)**
10	50	4
10	250	5
15	10	5
10	250	7
15	100	7
15	250	8
25	5	8

Table 9 *Design Index Values, based on varying axle loads and number of load applications, in order of increasing severity (*Table 5.1, Chapter 5, Reference 17*).*

For each of the three examples using COE charts, the following values will be used:

Concrete modulus of rupture: $MOR = 604$ psi
Subgrade modulus: $k = 200$ pci

3.4.1 — COE, light lift truck

The first example is for a relatively light lift truck with an axle load of 15 kips and an assumed 100 uses (passes) per day for a 25-year life. The cited reference classes this as a design index of 7.

- Enter *Figure 22* at the flexural strength *(MOR)* of 604 psi, draw a line horizontally to the $k = 200$ pci curve, then down to the design index of 7 and then horizontally to the right-hand axis where the required thickness is 6.7 inches.
- A slab thickness of 7 inches is recommended.
- *When this example is solved by other procedures, the required thickness is 6.5 inches using WRI charts, 6.5 inches using PCA charts, 6.4 inches using the AIRPORT program, and 7.6 inches using the MATS program.*

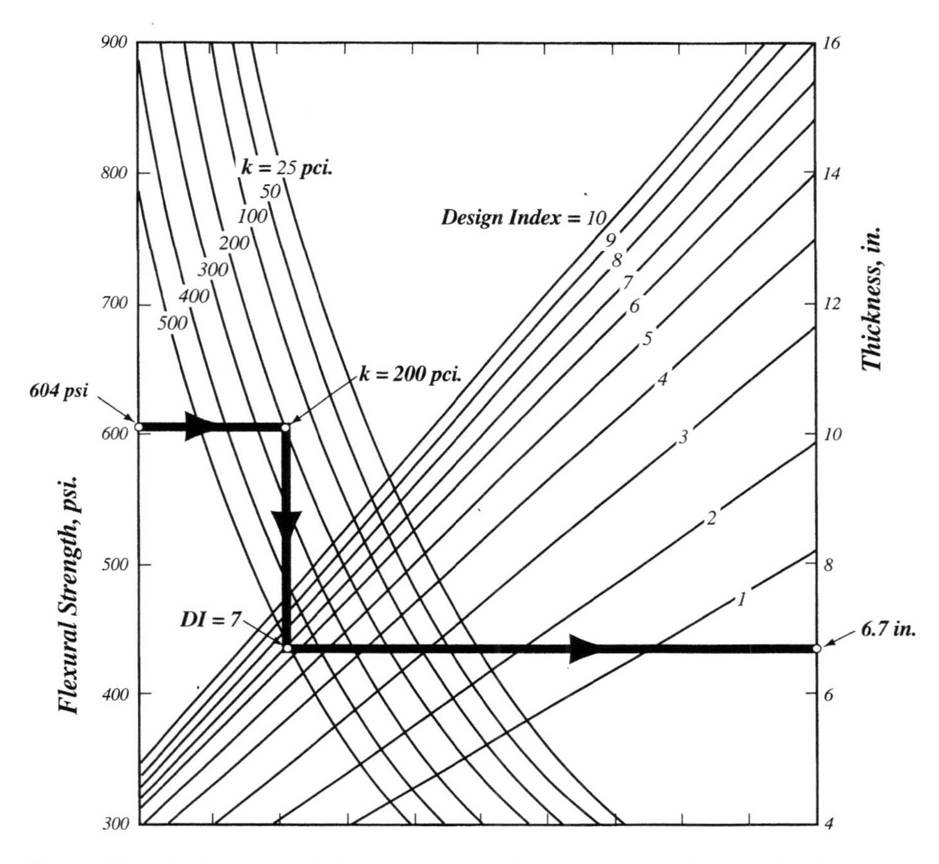

Figure 22 *COE chart for slab thickness selection for relatively light lift truck loading.*

3.4.2 — COE, heavy lift truck

The second example is for a heavy lift truck with an axle load of 42 kips. The number of vehicle passes will be calculated. It will be based on 1 pass in each 5 minutes, for two 8-hour shifts, 5 days each week, for 50 weeks per year and for a 20-year expected life. This results in 960,000 vehicle passes. The 1,000,000 line will be used.

Commentary:
Note that the selected number of vehicle passes will affect the thickness required. For 5,000,000 passes, 8.3 inches of slab thickness is required, while 10,000 passes call for only 6.3 inches. The number selected for vehicle passes could be considered as representing fatigue in the concrete.

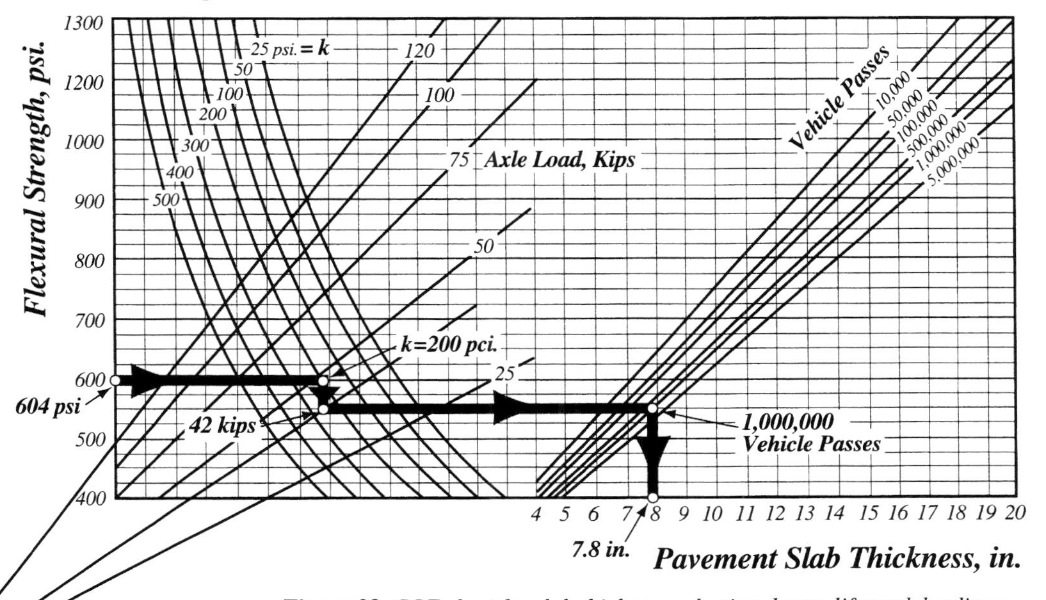

Figure 23 *COE chart for slab thickness selection, heavy lift truck loadings.*

- Enter *Figure 23* with the flexural strength of 604 psi, draw a horizontal line to the curve for k = 200 pci, then vertically to a line for the 42 kip axle load (this line must be drawn in). Then draw a line horizontally to the line for vehicle passes and then vertically to the slab thickness, which is very close to 7.8 inches.
- An 8-inch slab should be adequate, although an 8.5-inch slab might be a better choice.

3.4.3 — COE, outdoor paving areas

The third example using *Figure 24* is for general vehicles using associated outdoor paving areas (storage, parking, etc.). This will be a design index of 5 for Category IV vehicles—that is light traffic with no more than 25% of it being trucks and no more than 10% possessing three axles.

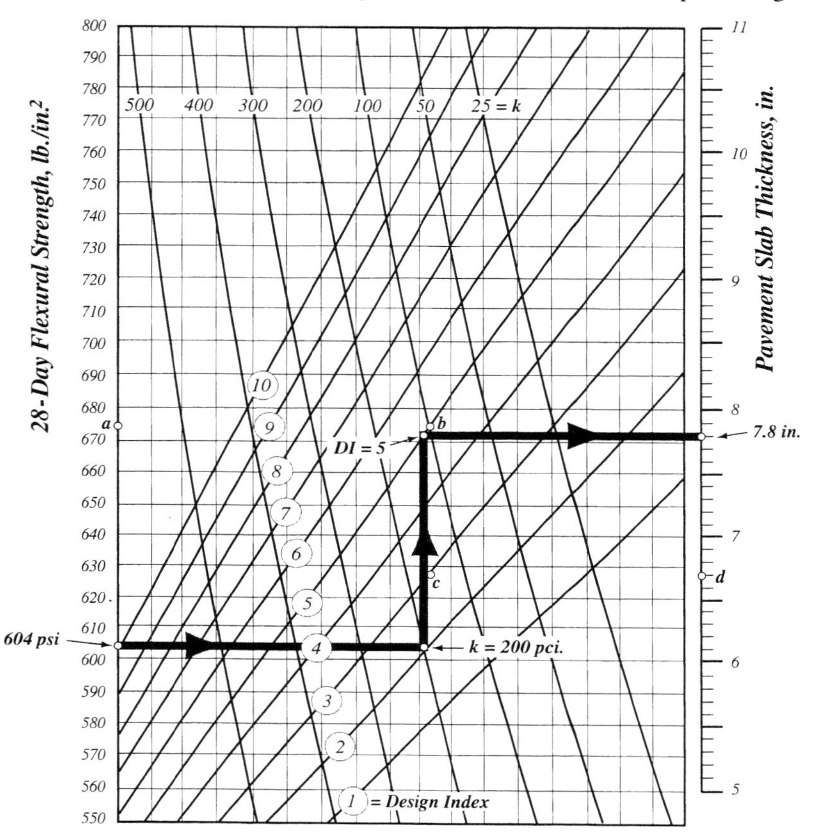

Figure 24 *COE thickness selection for outdoor paved areas serving general vehicles and parking.*

Commentary:
Although the chart indicates as little as 5 inches for slab thickness, the COE reference recommends no less than 6 inches as the selected thickness.

- Enter *Figure 24* with the *MOR* of 604 psi, draw a line horizontally to the curve representing k = 200 pci, then a line vertically to the design index of 5, and then horizontally to the required thickness of 7.8 inches.
- An 8-inch slab is recommended.
- *When this example is solved by other procedures, the required thickness is 7.8 inches using WRI charts, 6.9 inches using PCA charts, and 6.0 inches using the AIRPORT program.*

3.5 — Adapting PCA charts for use with swing vehicles

This is a special case. It is not uncommon to find a lift truck with lateral swing capability such as shown in *Figure 25*. This lift truck usually has a lower-rated capacity; however, when the forks are set in a "full-right" (or "full-left") position (a 90° swing), the majority of the load is on the right-hand front wheel (cross-hatched in *Figure 25*). This subjects the slab to a load from a single wheel instead of an equal-wheel axle loading.

There is no available chart to handle this design problem in a straightforward manner. Unless a sophisticated computer solution is used, the charts must be adapted to solve the problem. That is the approach of this example.

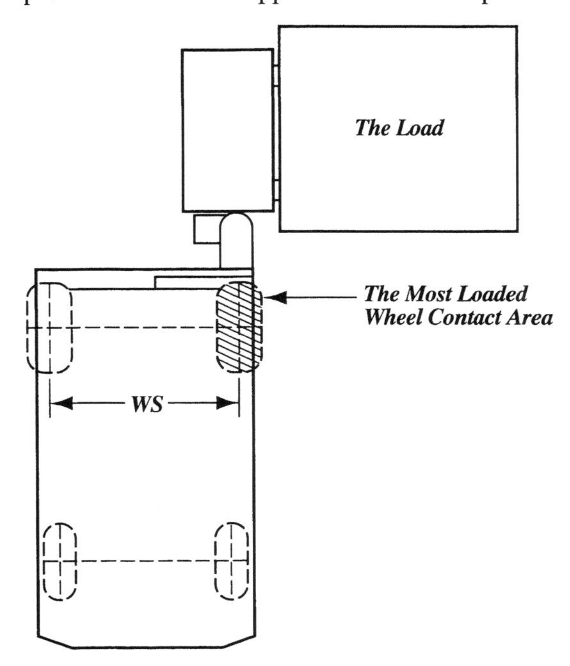

Figure 25 *Lift truck with lateral swing capability.*

Commentary:
There are computer software packages that will either accurately solve the problem (Reference 18) *or give approximate solutions* (Reference 9). *Any finite element program intended for slabs on grade should produce very good results. The authors feel that the most practical solution, as well as the easiest and fastest, is the one shown here.*

The PCA axle loading chart, *Figure 26*, from *References 6* and *14*, will be used by adapting it to handle a single wheel loading. This is done by using a large enough wheel spacing so as to essentially eliminate the effect of the second wheel, regardless of the load imparted by the second wheel. This is necessary since the chart assumes two equally-loaded wheels.

These values are taken from an actual swing vehicle. The wheel tread is a very hard material. Resultant contact pressure is 930 psi (high, although it is as stated by the manufacturer). The wheel's contact area is sometimes referred to as the footprint. Values should come from the manufacturer's vehicle specification sheet wherever possible.

Values for this example will be:

From materials, site, and designer:

Compressive strength: $f_c' = 4500$ psi
Modulus of rupture: $MOR = 604$ psi
Subgrade modulus: $k = 150$ pci
Safety factor: SF = 1.7

From the lift truck:

Payload capacity = 6000 pounds
Lift truck weight = 18,950 pounds
Total weight = 24,950 pounds
Wheel spacing: WS = 41 inches

In the swing position (full 90° swing to the right):

Load on critical wheel = 19,170 pounds
Contact area (one wheel) = 20.6 square inches

- To adapt the design chart, it is necessary to double the wheel load so as to have a chart-usable axle load.
  $2 \times 19{,}170 = 38{,}340$ pounds ("axle load")
- To then eliminate the effect of the second wheel, use a fictitious wheel spacing of 3 times the radius of relative stiffness ℓ. (See Section 10.7 for explanation of the radius of relative stiffness).
- For a 10-inch slab on a subgrade k of 150 pci, the radius of relative stiffness ℓ is 38.8 inches.
- Three times ℓ is $3 \times 38.8 = 116$ inches. Note that the chart's last curve is for WS = 120 inches. Contact area = 20.6 square inches. Note that the chart's last curve for area is 25 square inches.

• Using *Figure 26* as explained in Section 3.2, with the "axle" load of 38.34 kips, the contact area of 20.6 square inches, the allowable stress of 355 psi (stress per kip of axle load is then 355/38.34 = 9.26), and the subgrade modulus of 150 pci, the resultant slab thickness is 9.5 inches.

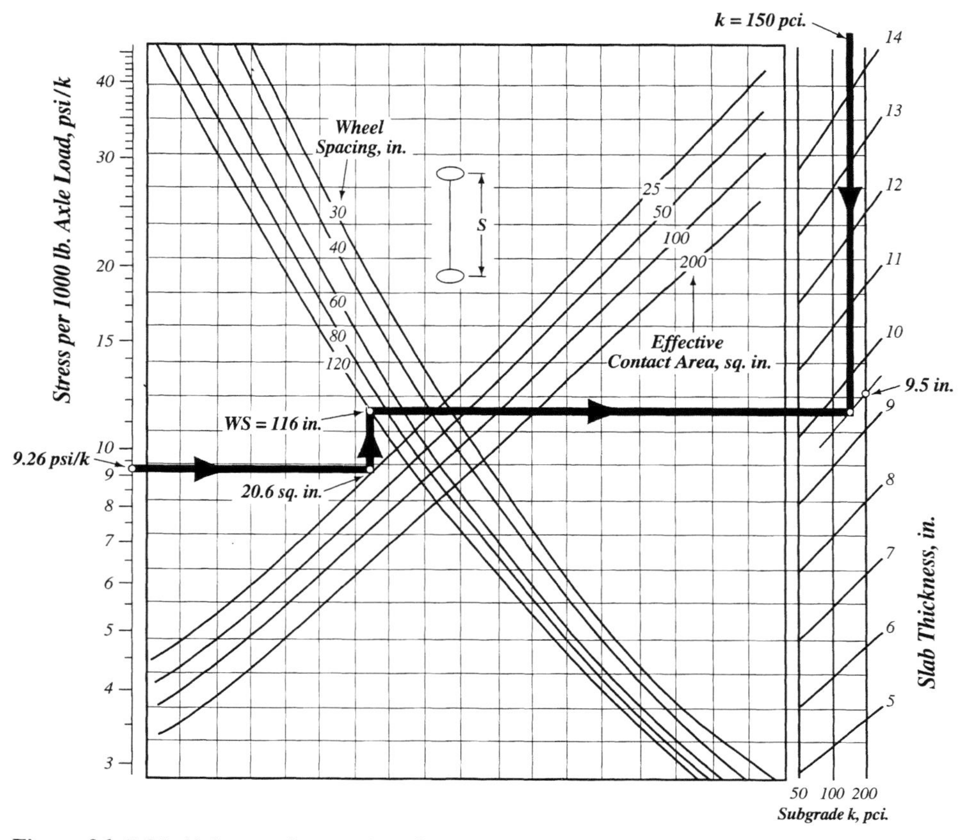

Figure 26 *PCA thickness selection chart for single-axle loading.*

Commentary:
When a second wheel load is located at a distance equal to or more than three times the radius of relative stiffness from the first wheel, it need not be used in the analysis. It contributes no additional moment (Reference 9).

• A 10-inch slab is then recommended.

• *When this example is solved by other procedures, the required thickness is 11.0 inches using WRI charts, 10.0 inches using the AIRPORT program, and 9.1 inches using the MATS program.*

3.6 — Using shrinkage-compensating concrete for slabs subject to axle loads

Shrinkage-compensating concrete is used to control cracking *(Reference 13)*. Thickness selections are made in the same manner as in previous examples; that is, using the PCA charts, the PCA tables, the WRI charts, or the COE charts. The modulus of rupture *(MOR)* of the concrete remains unchanged and the thickness selected is also unchanged (assuming the same loading and site values). The joint spacings, however, are greater than those specified for normal portland cement concrete slabs. See Chapter 10 (Section 10.2.4) for a discussion of joint spacing.

Use of this concrete allows a much wider spacing of joints than is appropriate for portland cement concrete. The total length of joints is dramatically reduced. A reduction of 80% is common.

The concrete may be produced using an additive or using Type K cement. This concrete behaves differently from conventional portland cement (PC) concrete. Shrinkage-compensating concrete expands first and then shrinks. Its total shrinkage is approximately the same as for PC concrete. It must have reinforcing steel properly placed and expansion joints are required. The best reference is the report from ACI Committee 223 *(Reference 13)*. ACI 223 shows how to determine the required expansion, the correct percentage of steel, and the compressive requirements of the expansion joints.

Slab thickness selection is essentially the same as previously described, and no separate example is given here. If advantage is taken of the presence of reinforcing steel, it is possible to reduce the thickness of a shrinkage-compensating concrete slab using the concept of structurally active reinforcement. This concept is discussed in Section 1.5 and illustrated in Section 3.8.

Commentary:
The selected percentage of reinforcing steel is important to the performance of shrinkage- compensating concrete floors. If used for shrinkage compensation, the percentage of steel should not be greater than 0.6. When higher percentages are used for structural purposes, then higher expansion levels, measured according to ASTM C 878 procedures, would be required (Reference 8).

3.7 — Using post-tensioning tendons for a slab with axle loads

Post-tensioning tendons provide crack control by means of the compression applied to the concrete. Post-tensioning can also be used to reduce the slab thickness by means of an increase in the effective modulus of rupture *(MOR')*. The prestress level is variable. After losses due to anchorage effects, tendon creep, and concrete creep, the resultant prestress can be expected to be from 75 psi to 100 psi (common) to as much as 150 psi. The spacing of the tendons is commonly between 30 and 40 inches; this allows the workers to easily walk on the subgrade between the tendons.

The following example using post-tensioning will follow the example in Section 3.5, which is the solution for a swing lift truck using an adapted PCA chart. The original example used a *MOR* of 604 psi, a *k* of 150 pci, and a SF of 1.7 as material, site, and design values. The swing lift truck had a design wheel load of 19.17 kips on an area of 20.6 square inches. The chart-usable axle load was 38.34 kips.

The only change from the procedure described in Section 3.5 is the change in the modulus of rupture. *Figure 26* can be used as the design chart.

Experience has shown that the MOR *is actually increased beyond the value calculated by using* $9 \times \sqrt{f_c'}$. *The prestress prevents propagation of the microcracking, which is part of the phenomenon that sets the tensile strength of the concrete* (MOR).

- Assume a prestress of 125 psi.
- Effective modulus of rupture *(MOR')* = 604 + 125 = 729 psi
- Allowable stress = 729/1.7 = 429 psi
- The stress per kip of axle load = 429/38.34 = 11.2 psi
- Use a wheel spacing of $3 \times \ell = 116$ inches, as is explained in the original example in Section 3.5, with the wheel contact area of 20.6 square inches.
- The required thickness is 8.5 inches.

Because of the influence of prestress in increasing the effective modulus of rupture, a thickness of 8 or 8.5 inches may well be appropriate instead of the 10 inches selected in Section 3.5. The decision is a matter of judgment and depends in part on the designer's confidence in the site conditions and the loadings.

3.8 — Designing structural reinforcement for a slab with vehicle loading

Steel reinforcement is placed in a slab on grade for one or more of four basic reasons. Regardless of the reason, the steel must be accurately positioned in the slab. The authors feel that it must be positively supported so as to be in its intended location. The four basic reasons are as follows:

Commentary:
Proper positioning of steel reinforcement within the concrete slab is essential. This is critical for reinforcing bars and welded wire fabric sheets. Where the steel belongs depends on why the steel has been specified; however, single layers of steel must be at or above the slab's middepth. The authors recommend that the steel be supported during construction by slab bolsters or other devices.

1. To act as crack control. Here, the steel is commonly selected by use of the subgrade drag equation. It is a nominal amount of steel. The selection of steel areas by subgrade drag is dicussed in Section 1.5.1.3.

2. To act as the required steel with shrinkage-compensating concrete slabs. When used here, the steel is essential for slab performance since joints are usually at a much wider spacing than is normally used. The steel is selected following the requirements of ACI Committee 223 *(Reference 13)*.

3. To provide reserve load capacity. For this purpose, the slab remains at the thickness based on the uncracked and unreinforced slab section. The steel area is increased such that the moment capacity of the reinforced and cracked section is greater than the allowable moment for the slab *(Reference 11)*.

4. To allow use of a thinner slab. Here the safety factor provided for load capacity is due to a substantially larger area of steel (properly located) than that indicated by the subgrade drag equation. This slab is expected to develop hairline cracks (generally less than 0.01 inch wide) due to loading. The process is illustrated in the example which follows.

One example is shown here using reinforcement in two layers to reduce the thickness of the floor slab and to provide the load support and reserve safety factor (load factor) (Item 4 above). Each layer is at approximately the quarter point of the slab's depth. The swing lift truck problem in Section 3.5 will be solved again, but using active reinforcement this time. The variables used were:

From materials, site, and designer:

Modulus of rupture: $MOR = 604$ psi
Subgrade modulus $k = 150$ pci
Safety factor (original design): SF = 1.7

Swing lift truck (has 90-degree swing capability):

Total vehicle weight = 24,950 pounds
Critical wheel load (full swing) = 19,170 pounds
Contact area (one wheel) = 20.6 square inches

The solution of Section 3.5 (adapting PCA chart for swing vehicles) resulted in a slab thickness of 9.5 inches. Now, for this example we will assume that the thickness of 9.5 inches is too much for a new floor slab in an existing plant facility where other elements (machinery, etc.) make the thicker floor unacceptable. We will reduce the floor thickness as far as is practical (frequently a field decision) and provide reinforcing steel for the loading. In brief, the following procedure is used:

Only the WRI charts give the moment directly from the chart. Other charts are used in reverse to determine the moment. Since the slab is uncracked due to loadings, the applied moment is calculated from the allowable bending stress and the slab section modulus.

1. Calculate the applied moment due to actual load.
2. Determine the thickness required for a safety factor of close to 1.10.
3. Using this new thickness, check the slab's actual safety factor with respect to load-induced cracks.
4. Select the load factor to give the design moment (greater than the applied moment).
5. From the design moment, which is an ultimate moment, select the required steel areas and spacings.

In detail, here is the design process. For a plain concrete slab, presumed to be uncracked, the section modulus SM is $bt^2/6$ and the relationship between section modulus, bending stress, and moment is

$$M = SM \times f_b$$

The usual way to do these calculations for slabs is to use a 1-foot-wide slab strip, which results in $SM = 12t^2/6$.

• Applied moment M_{app} due to actual load: from the solution in Section 3.5, with the original 9.5-inch slab, the stress produced by the loading was 355 psi. The basis for this was an uncracked slab.

Section modulus for 9.5-inch slab = $12t^2/6$ = 180.5 in^3

Applied moment M_{app} = (335)(180.5) = 64,078 inch-pounds

= 5.34 foot-kips

This is now the applied moment regardless of slab thickness.

• Thickness required for a safety factor close to 1.1:

Allowable stress = 604/1.1 = 549 psi

Section modulus required = 64,078/549 = 116.7 in^3

Since the section modulus = $12t^2/6$, the value of t^2 = 6(116.7)/12 = 58.36,

and the required thickness is t = 7.64 inches

• Select the desired practical thickness and calculate its safety factor with respect to load-induced cracks. This SF value should be greater than 1.0. Logical selections here are either 8 inches, 7.75 inches, or 7.5 inches, depending on requirements of the site and its facility.

Select t = 7.5 inches

M_{app} = 5.34 foot-kips per foot of width of slab

SM = 12(7.5)^2/6 = 112.5 in^3

Actual stress = f_b = 64,078/112.5 = 569.6 psi

Safety factor (SF) = 604/569.6 = 1.06 (Say OK)

• Select load factor and ultimate design moment. The authors feel that the load factor should be equal to or greater than 1.7 for this process. A value of 2.0 is generally recommended and is selected here.

Design moment $M_u = 2.0 \times 5.34$ = 10.68 foot-kips,

which is per foot of width of slab in both directions.

Commentary:
This safety factor should be greater than unity. If the entire process of construction is carried out with good quality control, it is highly unlikely that any cracks will develop at the design loading in the early age of the slab. This performance is determined by fatigue. The slab could be even thinner. If this option is selected, hairline cracks will almost certainly develop. Of greater importance, however, is that for the thinner slabs, it may be necessary to use only one layer of steel reinforcement. This reduces the effectiveness of the process.

• Select the required steel area and steel spacing. Moment capacities are given in the appendix for certain situations *(page 257)* as taken from *Reference 11*. This example uses slab thickness for which no table is available, therefore, the steel will be selected by calculations as follows:

The ultimate moment capacity is given by:

$$M_u = \phi A_s f_y (j_u d)$$

This moment capacity can be provided by conventional reinforcing bars or by sheets of welded wire fabric. In either case, the bars or the wire must be stiff enough to allow workers to stand on the supported steel or must have a wide enough spacing to allow stepping between the bars or wires. For the bars, the yield point f_y is 60,000 psi and for the wire, the yield point may be taken as 72,000 psi (or more).

As used in conventional ACI design, the ϕ factor is a capacity reduction factor (0.90 for bending) representing the normal variations in the construction process. The term $j_u d$ *is the moment arm used for calculating the moment capacity of the slab. It can be calculated* (Reference 7), *but is frequently assumed to be 0.9 as is done here. In the authors' opinion, this is a reasonable value.*

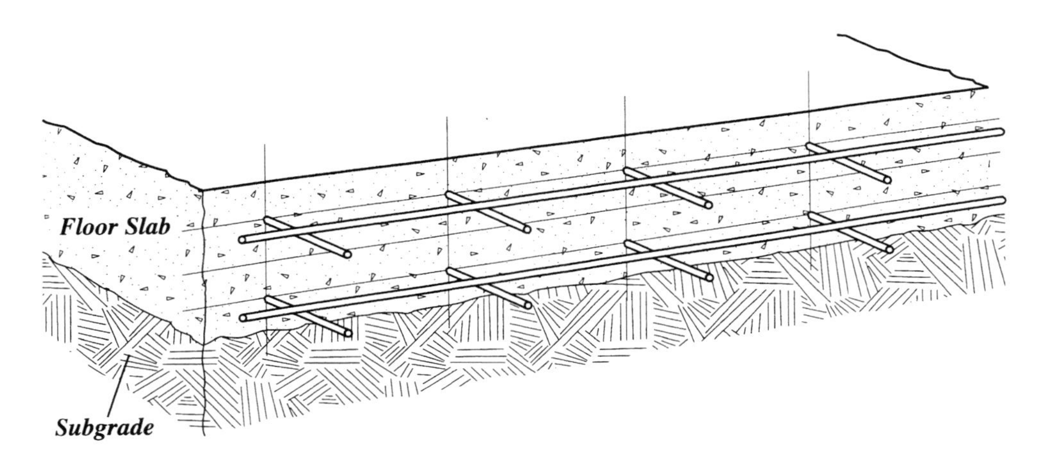

Figure 27 *Two-way reinforcement in two layers in slab on grade.*

Commentary:
The 3-inch cover required by ACI 318 is not necessary here since slabs on grade are excluded from coverage by 318. If the designer feels that 318 must apply, then a 3-inch cover may be appropriate. The material beneath a grade floor is a prepared subgrade, compacted and with site drainage. A cover of 1 inch is common. Some prefer a cover of 1.25 inches since the slab is allowed to be 1/4 inch thinner than its design value according to the ACI 117 (1990) specification.

• Determine the design depth d (this is not the slab thickness), which depends on the amount of cover and the diameter of the selected bar or wire. Steel will be designed using two layers of bars, both top and bottom of the slab. See *Figure 27.*

Assume a 0.5-inch-diameter bar or wire (each way).

Select a 1-inch clear cover.

Design depth $d = 7.5 - 1.0 - 0.5 = 6.0$ inches

• Selection of steel using steel reinforcing bars:

Yield point $f_y = 60$ ksi
Substituting in the equation for ultimate moment:

$$0.9 \times A_s \times 60 \times 0.9 \times 6.0 = 10.68 \times 12$$

$A_s = 0.440$ square inches/foot of slab width

#4 rebars require a spacing of 5.45 inches cc, based on $A_4 = 0.20$ square inches:

$$12 \times 0.20/0.440 = 5.45 \text{ inches}$$

#5 rebars require 8.45 inches cc
#6 rebars require 12.0 inches cc

• Steel recommended: #4 rebars at 5.5 inches cc (both ways)

3.9— Welded wire fabric as a structural element

In Section 1.5.1.7, the advantages of welded wire fabric were discussed. The following example shows conversion of the requirement of 0.44 square inches per foot of slab width to a welded wire fabric equivalent.

Higher yield strengths up to 85 ksi are available. In designating welded wire reinforcing, the designer should indicate what minimum yield strength is desired.

• Selection of welded wire fabric for the slab in Section 3.8 where A_s for Grade 60 steel was determined to be 0.44 square inches per foot of slab width in a slab whose design depth $d = 6$ inches.

Assume a yield point of 75 ksi for welded wire fabric and substitute in the equation for ultimate moment:

$$M_u = \phi\ A_s f_y (j_u\ d)$$

$$10.68 \times 12 = 0.9 \times A_s \times 75 \times 0.9 \times 6$$

$$364.5\ A_s = 128.16$$

$$A_s = 0.35 \text{ square inches per foot of slab width}$$

W18 wire has a cross sectional area of 0.18 square inches per wire, so two wires will provide the needed area per foot of slab width. Therefore specify W18 at 6 inch spacing, designated as W18 × 6.

Table 10 provides a wire size comparison and A_s per linear foot of slab width for 4 inch, 6 inch, and 12 inch spacing. *W* refers to smooth wire and *D* refers to deformed wire. Moving the decimal of the size number two places to the left gives the area of the wire in square inches. The *W* and *D* designations are preferred in the industry; however designation by gauges remains popular. Both are given for reference.

W&D Size No.		Wire Gauge No.	Nominal Diameter (In.)	Area (Sq. in.)	A_S - Square Inches per Lineal Foot				
					Center to Center Spacing of Wires				
Smooth	**Deformed**				**4"**	**6"**	**12"**	**14"**	**16"**
W20	D20		0.505	0.200	0.600	0.400	0.200	0.171	0.150
		7/0	0.490	0.189	0.567	0.378	0.189	0.162	0.142
W18	D18		0.479	0.180	0.540	0.360	0.180	0.154	0.135
		6/0	0.462	0.168	0.504	0.336	0.168	0.144	0.126
W16	D16		0.451	0.160	0.480	0.320	0.160	0.137	0.120
		5/0	0.431	0.146	0.438	0.292	0.146	0.125	0.110
W14	D14		0.422	0.140	0.420	0.280	0.140	0.120	0.105
		4/0	0.394	0.122	0.366	0.244	0.122	0.105	0.092
W12	D12		0.391	0.120	0.360	0.240	0.120	0.103	0.090
W11	D11		0.374	0.110	0.330	0.220	0.110	0.094	0.083
W10.5			0.366	0.105	0.315	0.210	0.105	0.090	0.079
		3/0	0.363	0.103	0.309	0.206	0.103	0.088	0.077
W10	D10		0.357	0.100	0.300	0.200	0.100	0.086	0.075
W9.5			0.348	0.095	0.285	0.190	0.095	0.081	0.071
W9	D9		0.338	0.090	0.270	0.180	0.090	0.077	0.068
		2/0	0.331	0.086	0.258	0.172	0.086	0.074	0.065
W8.5			0.329	0.085	0.255	0.170	0.085	0.073	0.064
W8	D8		0.319	0.080	0.240	0.160	0.080	0.069	0.060
W7.5			0.309	0.075	0.225	0.150	0.075	0.064	0.056
		1/0	0.307	0.074	0.222	0.148	0.074	0.063	0.056
W7	D7		0.299	0.070	0.210	0.140	0.070	0.060	0.053
W6.5			0.288	0.065	0.195	0.130	0.065	0.056	0.049
		1	0.283	0.063	0.189	0.126	0.063	0.054	0.047
W6	D6		0.276	0.060	0.180	0.120	0.060	0.051	0.045
W5.5			0.265	0.055	0.165	0.110	0.055	0.047	0.041
		2	0.263	0.054	0.162	0.108	0.054	0.046	0.041
W5	D5		0.252	0.050	0.150	0.100	0.050	0.043	0.038
		3	0.244	0.047	0.141	0.094	0.047	0.040	0.035
W4.5			0.239	0.045	0.135	0.090	0.045	0.039	0.034
W4	D4	4	0.226	0.040	0.120	0.080	0.040	0.034	0.030
W3.5			0.211	0.035	0.105	0.070	0.035	0.030	0.026
		5	0.207	0.034	0.102	0.068	0.034	0.029	0.025
W3			0.195	0.030	0.090	0.060	0.030	0.026	0.023
W2.9		6	0.192	0.029	0.087	0.058	0.029	0.025	0.022
W2.5		7	0.178	0.025	0.075	0.050	0.025	0.021	
W2.1		8	0.162	0.021	0.063	0.042	0.021		
W2			0.160	0.020	0.060	0.040	0.020		
		9	0.148	0.017	0.051	0.034			
W1.5			0.138	0.015	0.045	0.030			
W1.4		10	0.134	0.014	0.042	0.028			

Table 10 *Wire size comparison and sectional areas of welded wire fabric. (American standard customary units)*

CHAPTER 4
DESIGN OF SLABS FOR RACK STORAGE POST LOADS

4.1 — Design objectives

Storage racks create positive moments in the floor slab by means of the loads transmitted through the supporting posts. The rack locations are commonly fixed within the facility and have a given number of shelves (or supporting arms). The weight of objects such as pallets, boxes, goods, and metal items on the shelves creates the loading to the supporting posts. Therefore, there are concentrated loads to be supported by the floor slab. These concentrated post loads form a pattern according to the shelf rack arrangement. This is illustrated in *Figure 28* which shows the normal rack layout pattern.

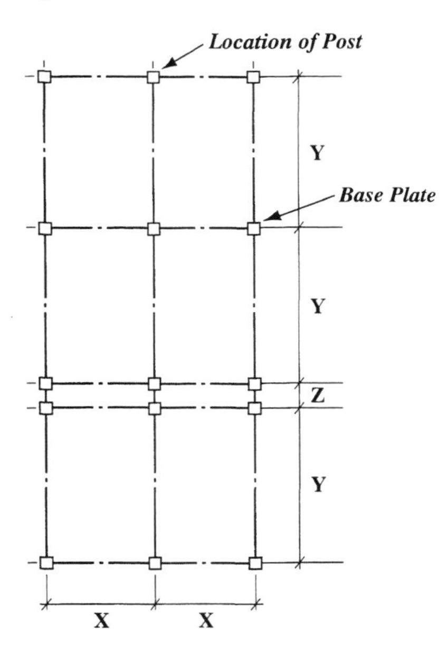

Figure 28 *Normal layout of rack storage posts on a slab on grade.*

The design for this type of loading checks the thickness (or base plate size) as limited by the tension in the bottom of the floor slab beneath the most heavily loaded single post. The procedure includes the added effects of nearby post loadings of the same magnitude. This tension is due to a positive moment which by definition puts tension on the bottom of the slab. The procedure is also used to similarly check a closely spaced pair of posts as defined by the dimension Z, where that situation exists. This does not check negative moments (tension on top) in the aisles between rack systems, which must be done as shown in *Chapter 6*.

4.2 — Information needed to solve the post load problem

To solve for the required design details, it is necessary to have accurate information about the pattern of the racks and their anticipated loadings. The usual design problem is to find the required slab thickness. Occasionally, the slab thickness is set and either the permissible loading or the required size of the post's base plate is needed. For any of these, the following values are needed:

Commentary:
This post load may be based on either the anticipated weight (psf) for each shelf, or the maximum possible load based on the post's or shelves' structural capacity.

From the loading specifications:

Post spacing X (inches), which is the short distance between posts
Post spacing Y (inches), which is the long distance between posts
Post spacing Z (inches), which is the end spacing between rack units
All of these spacings are center to center of post.
Base plate size (effective contact area), in square inches
Post load, P, pounds or kips

From the material, the site, and the designer:

Compressive strength of concrete, f_c', in psi
Modulus of rupture of concrete, *MOR*, in psi
Subgrade modulus, k, in pci
Safety factor, SF (unitless)

4.3 — Using PCA charts to design for rack storage post loading: AUTHORS' CHOICE

It is straightforward and reasonable to interpolate between chart values or to solve the problem with a k *value above and below that which exists, and then use a weighted average for the solution. It is very difficult to accurately extrapolate (work outside the boundary values) from the charts and this extrapolation is not recommended.*

It is common to store materials directly on the floor beneath the shelves. The material on the floor does not add to the post loading and usually does not increase the positive moment beneath the post. On the other hand, it does add to the uniform loading when the negative moment in the aisle is being checked, as in the example in Chapter 6.

The PCA charts used for the solution of this problem come from *References 6* and *14*. Larger copies of these charts for use by the designer appear in the appendix section. At this book's publication, the only charts available were for subgrade modulus k values of 50, 100, and 200 pci.

For the first example, a common arrangement of racks and posts will be used as follows:

From materials, site, and designer:

Concrete compressive strength: $f_c' = 3000$ psi
Modulus of Rupture: $MOR = 493$ psi
Subgrade Modulus: $k = 100$ pci
Safety Factor: SF = 1.7

From loading specifications:

Short post spacing: $X = 48$ inches (4 feet)
Long post spacing: $Y = 96$ inches (8 feet)
Spacing between rack units: $Z = 16$ inches
Base plate area: 20 square inches (4 x 5 inches)
Five shelves at 100 psf per shelf
Post load: $P = 100 \times 4 \times 8 \times 5$
$= 16{,}000$ pounds or 16 kips

The required slab thickness is selected from *Figure 29* in the following manner:

- First calculate the allowable stress per 1000 pounds (1 kip) of post load. All post loads are equal. Allowable stress = modulus of rupture divided by the factor of safety, or 493/1.7= 290 psi. Since each post load is 16 kips, the allowable stress per 1000 pounds of post load is 290/16 = 18.1 psi per kip.

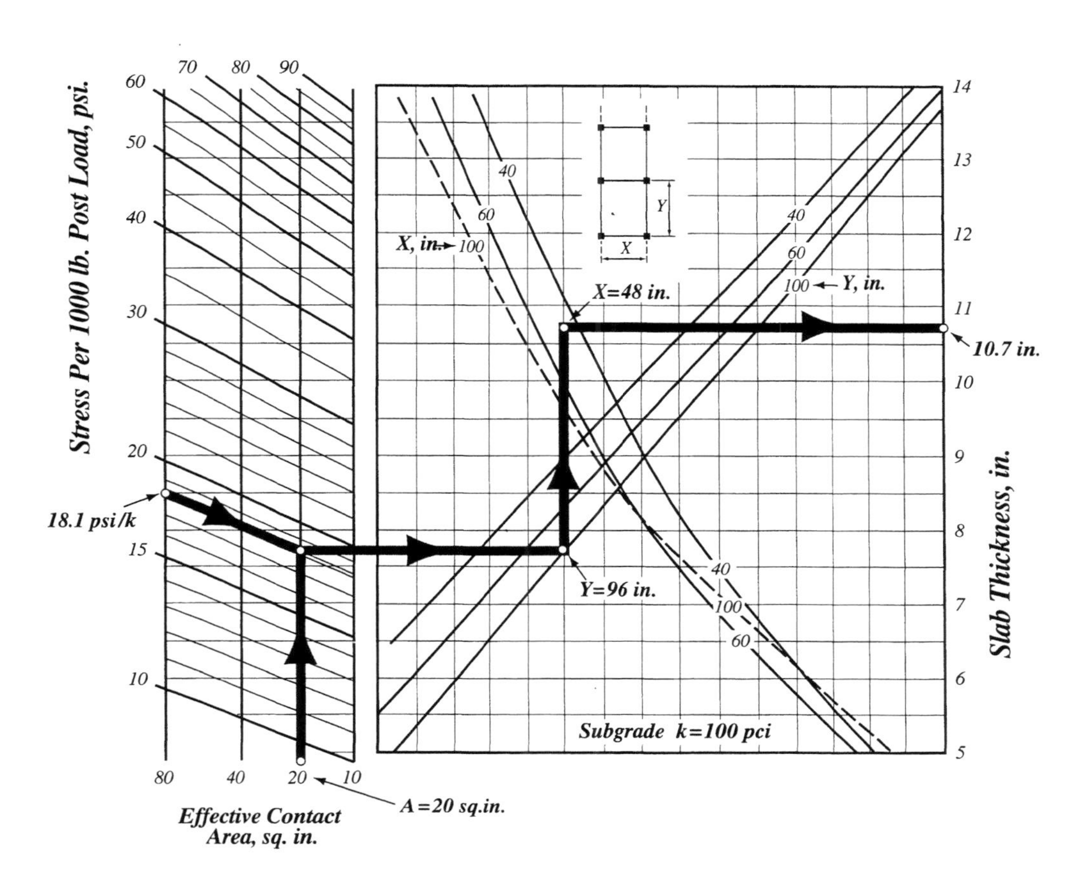

Figure 29 *PCA chart for slab thickness selection when using post loading with subgrade k = 100 pci.*

Commentary:
For the same problem, if the base plate were increased to 64 square inches (8x8 inches), the slab thickness indicated by the chart would be very close to 9.4 inches. On the other hand, if the same problem (base plate 4x5 inches) were checked using a shelf load of 50 psf (instead of 100 psf), the required slab thickness would be very close to 6.9 inches.

- Enter the *Figure 29* chart at the base plate area of 20 square inches and move vertically up to the slanted line at 18.1 psi per kip.
- Draw a straight line to the right to the line representing $Y = 96$ inches. **The chart must be used by going to the *Y*-line first and then to the *X*-line.** The reverse is incorrect.
- Draw a line vertically to the curved line representing $X = 48$ inches, as close as is graphically possible.
- Finally, draw a horizontal line to the right where it intersects the vertical axis representing thickness. For this example, the intersection is at a thickness of 10.7 inches.
- A slab thickness of 11 inches is recommended.
- *When this example is solved by other procedures, the required thickness is 10.4 inches using the AIRPORT program and 10.6 inches using the MATS program.*

4.4 — Using PCA charts when actual k does not equal chart k: AUTHORS' CHOICE

Since charts exist for only three specific k values *(Reference 14)*, here is a rational procedure which can be used when a field value of k exists for which no rack load chart is available. This could be the case for a k of 300 pci or a k of 150 pci, as examples. These are handled in two different ways depending on whether the field k is between chart k values or outside of them (higher or lower).

The PCA documents are the only known source of design charts for patterned post loadings.

When *k* is between values for which the charts were developed:

Interpolation or averaging can be used. The authors suggest determining the required thickness for the chart k value immediately below the actual field value and then repeating that for the chart k value immediately above the field value. Then, average the two thicknesses and consider rounding it up to the next acceptable thicker number. For the original example in Section 4.3, and using the PCA charts, the following results:

- Assume the field k to be 150 pci.
- For $k = 100$ pci, the thickness required is 10.7 inches.
- For $k = 200$ pci, the thickness required is 9.4 inches.
- The average thickness is then 10.05 inches.
- A thickness of 10 inches is recommended.

In all slab design problems, when a variable such as k, WS, X, etc. falls between plotted curves, interpolation or averaging can be used with confidence. Linear (straight-line) interpolation is almost always acceptable.

When k is less than lowest k value or higher than the highest k in the charts:

Extrapolation can be risky and is not recommended. Only two options remain. One option is to use a computer program such as MATS, or possibly AIRPORT. *(References 9* and *18).* These computer programs are briefly discussed in Section 10.8. The other option is to use the vehicle axle charts from WRI references. PCA charts cannot be used since they have the same limitations on k as the post-loading charts. This process then adapts a dual-post loading to be analyzed as a two-wheel axle loading. Both are done as concentrated loadings with contact areas and both are static loadings. Therefore, the latter can closely approximate the solution needed for the former.

The example to demonstrate this uses original values in Section 4.3, but this time for a field k of 300 pci. *Figures 30* and *31* show the work.

- Assume actual or field $k = 300$ pci
- Post load P = 16 kips
- Contact area (for one post) = 20 square inches
- Spacing of loads = 48 inches. The closer of the two post spacings is the correct choice to serve as load spacing.

Commentary:
The WRI chart results are sensitive to numerical assumptions. They must be used cautiously. However, if the post loading appears to control a thickness selection where the k *value is high or low and where a computer solution is not available, then this procedure should give acceptable results.*

Since WRI charts are used (because of the wider range of k values), the slab thickness must be assumed in order to start the solution. If it is assumed too high, the design moment will be too conservative (too high). If it is assumed too low, the design moment will be unconservative (too low). Therefore, it is logical to repeat the process of assuming a thickness and then calculating the required thickness until reasonable agreement is achieved. Although there is no established recommended difference between the first and last thickness of a specific cycle, the authors suggest that the thicknesses differ by no more than 15 percent.

- Assume a thickness of 10 inches.
- Using a compressive strength of 3000 psi and the ACI equation of

$$E_c = w_c^{1.5} \times 33\sqrt{f_c'} \qquad \text{gives } E_c = 3320 \text{ ksi.}$$

- From *Figure 30, D/k* = 9.7
- Determine the equivalent diameter of the 20-square-inch loaded area as 5.05 inches. For rectangular base plates, the WRI charts require use of the diameter of a circle of the same area.
- From both parts of *Figure 31,* the applied moment under one post is 300 + 25 = 325 foot-pounds per kip of post load.
- The applied moment is then

$$M_{app} = 325 \times 16$$
$$= 5200 \text{ foot-pounds per foot of slab width}$$

- The required slab thickness can then be determined from a WRI chart, or, as is illustrated here, by calculation:

  Allowable stress = MOR/SF = 493/1.7 = 290 psi
  Section Modulus required = (5200)(12)/290 = 215 in.^{3}
  From the actual section modulus of $12(t^2)/6$, the required thickness is 10.4 inches.
- A thickness of 10 1/2 inches would be recommended.
- *When this example is solved by other procedures, the required thickness is 9.8 inches using the AIRPORT program and 10.0 inches using the MATS program.*

Designing for a patterned post loading by using a two-wheel axle loading as the replacement for the post pattern leaves out the effects of the other posts. Only the one post and its nearest equally-loaded post are included. If other posts are more distant, this is reasonable; however, the process must still be considered as approximate. The authors believe the results are reasonable, although not exact.

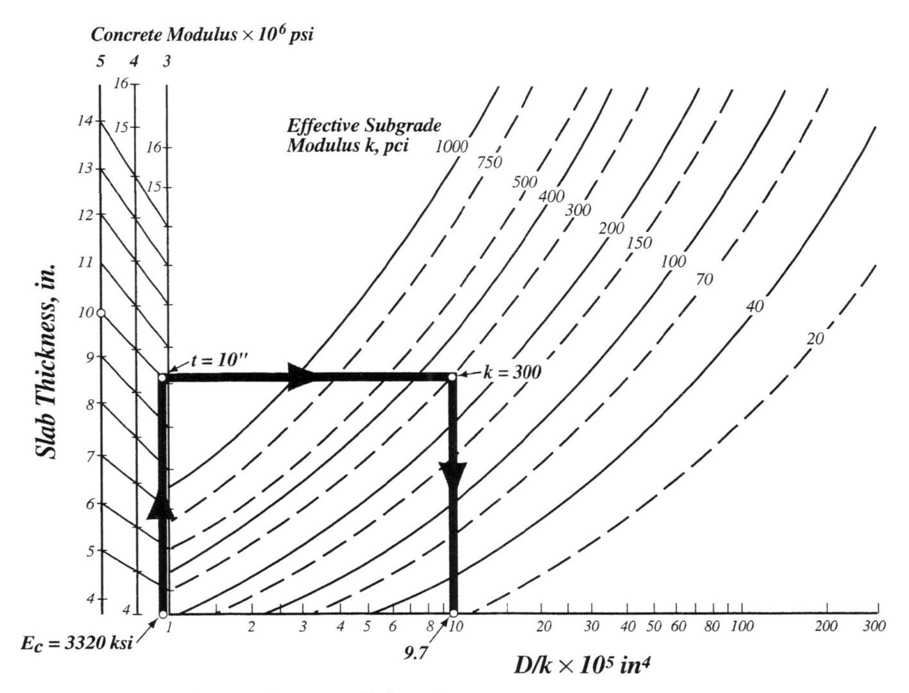

Figure 30 *Use of WRI chart to determine* D/k *ratio.*

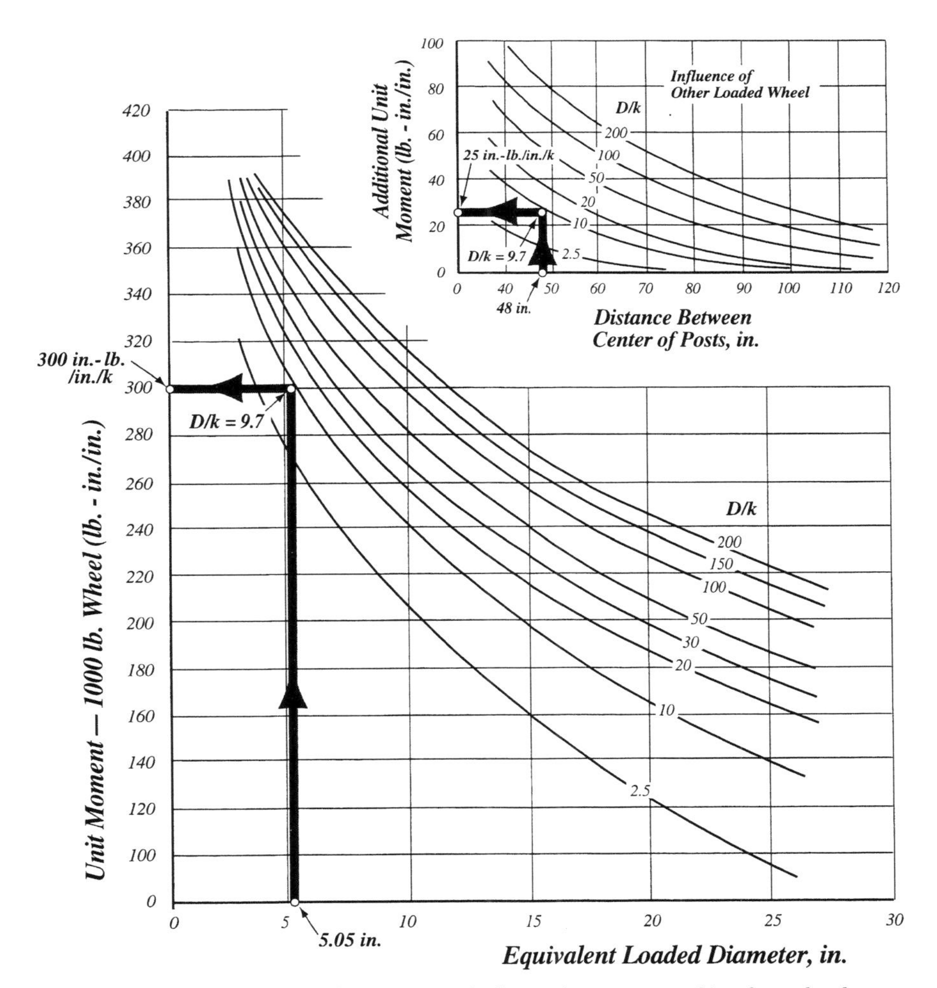

Figure 31 *Use of WRI chart to determine equivalent unit moment per kip of post load.*

4.5 — Using shrinkage-compensating concrete, post-tensioning tendons, or active steel reinforcement

The examples in Sections 3.6, 3.7, and 3.8 explain the purposes and uses of these three techniques in slab design. They also illustrate the calculation process. The use of post-tensioning tendons or structural steel reinforcement is shown in detail. The use of shrinkage-compensating concrete is explained, but the design for joint spacings and joint materials is discussed in Section 10.2.

CHAPTER 5
SLAB DESIGN FOR COLUMN OR WALL LOADINGS

5.1 — Scope of the problem

This design problem deals with a concentrated load or a line load supported directly on the industrial floor slab rather than by an isolated footing or by a grade beam. Concentrated loads from building elements supported directly by the slab are frequently found in industrial facilities. While a slab on grade is normally thought of as an isolated slab or a "floating slab," it is common to find these concentrated column and wall loadings. Either of these loads can be substantial and can control the thickness or other details of the slab on grade design.

The column load: It is common to find a column load that is not part of a patterned post and rack system. Thus, it is considered an isolated and concentrated load transmitted to the slab through a base plate. The column can be required due to renovations within an existing plant or can be part of the original design. If allowed by the appropriate building codes, it can be a column which provides support to some part of the structure. In the case considered here, the slab forms an integral footing as shown in *Figure 32 (Reference 22)*, and is considered as an interior loading in this book. The slab on grade itself serves both as the floor slab and as the footing for the column; thus, it is known as an integral footing. Further, the column is not adjacent to an edge or to an isolation joint.

Commentary:
In Chapter 13 of ACI 318-89, "Building Code Requirements for Reinforced Concrete," slabs on grade are specifically excluded; however, the wording is "which do not transmit vertical loads from other parts of the structure to the soil" (Section R13.1 of Reference 7*). In Chapter 1 of ACI 318.1-89 (Revised 1992), "Building Code Requirements for Structural Plain Concrete," slabs on grade are not considered within the definition of structural plain concrete. Again, the statement includes "unless they transmit vertical loads from other parts of the structure" (Section 1.2.2R of Reference 7). Therefore, the designer must use appropriate judgment to determine whether or not the particular column-to-slab loading condition is governed by ACI 318 or ACI 318.1.*

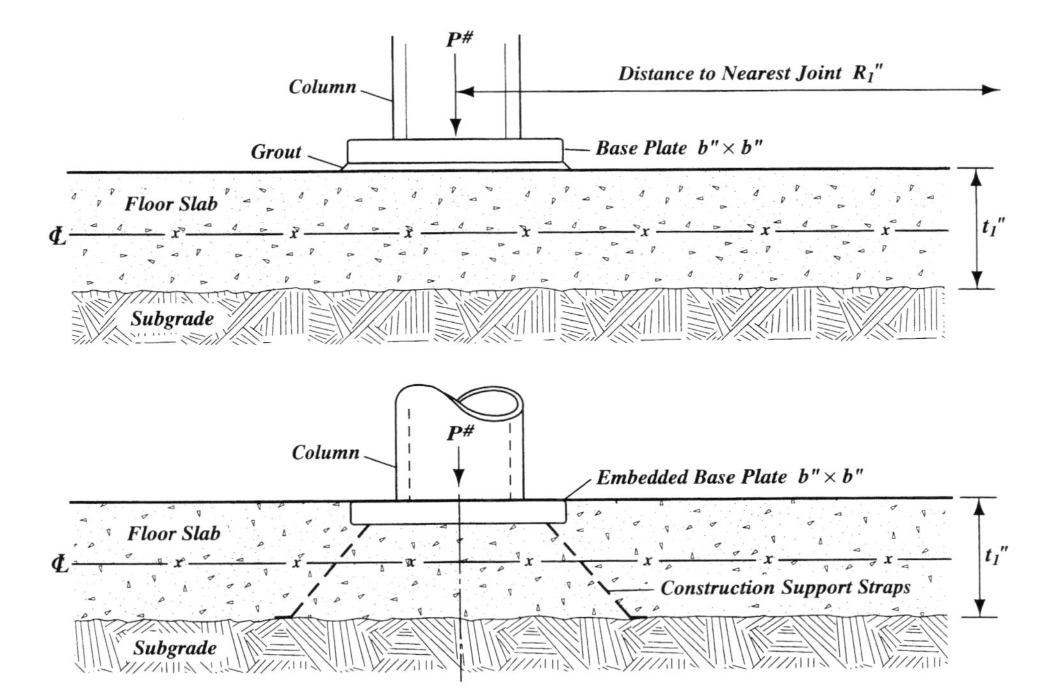

Figure 32 *The slab on grade may form an integral footing for column loads.*

The wall load: Walls are frequently placed directly on and supported by the floor slab. These walls place a line loading on the floor at or near the center of a slab unit or at or near an edge or joint. The weight applied ranges from a few hundred to a few thousand pounds per lineal foot. The basic problem is illustrated in *Figure 33 (Reference 17)*. The most common position for a wall is at a joint within the floor area or at the floor slab edge.

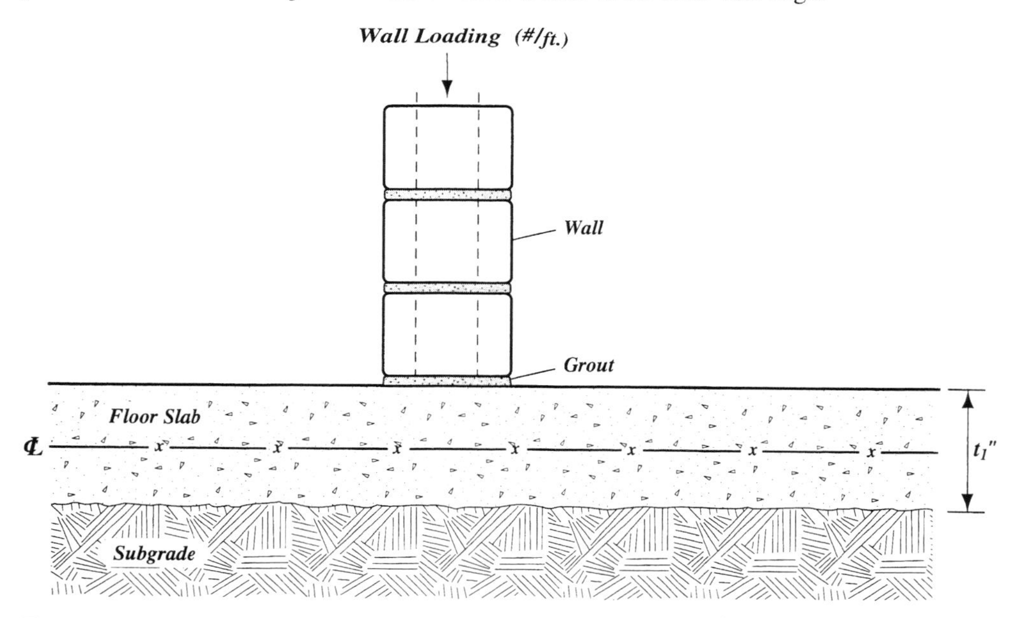

Figure 33 *Walls place a line load on the slab on grade.*

5.2 — Design objective

Commentary:
Do not neglect to check these loadings merely because the tension and presumably the crack are on the bottom of the slab. Such a crack may start on the bottom (out of sight), but it will certainly work its way to the top and become visible.

The designer must check the tension on the bottom of the slab due to the positive moment produced by the load. The column load, being applied to the interior of a floor slab panel, is supported by two-way plate action of the continuous concrete slab. The objective is to keep the floor uncracked, and the allowable stress is selected with that in mind. Punching shear is not commonly a problem for column loads; however, when the slab is thin and the load is high, base plate size or slab thickness may be dictated by punching shear. The wall loading also is concentrated, but only with respect to its width. It is long enough to be considered a continuous line load. This load produces tension on the bottom of the slab. The tensile stress on the bottom will control the concrete's thickness for a constant thickness slab.

5.3 — Input values needed for design

To be able to determine either the floor thickness required, the base plate size required, or the loading to be allowed, the following information is required:

Column:

Column load (P in pounds)
Base plate size (b inches × b inches)
Thickness of slab exclusive of column load area (t_l in inches)
Only needed when a change in floor thickness beneath the column is under consideration.

Wall:

Wall load p, in pounds per lineal foot

Thickness of slab exclusive of wall load area (t_l in inches)

Only needed when a change in floor thickness beneath the wall is under consideration.

Material, site, and designer:

Concrete's design compressive strength f_c', in psi

Modulus of rupture of the concrete MOR, in psi

Modulus of subgrade reaction k, in pci

Safety factor or load factor (selected by designer)

Note: Some constants are assumed within the equations. These are identified in the examples.

5.4 — Using equations to design for column loads: AUTHORS' CHOICE

When a column is supported directly on the industrial floor, the floor itself becomes a footing integral with the concrete floor slab. The equations used to solve this problem come from *Reference 22*. They are altered here in format for easier use.

Two features are critical. First, the slab must be thick enough to adequately support the column on its base plate. The equations check bending in the slab. Punching shear should also be checked using ACI Building Code procedures *(Reference 7)*. Whether or not punching shear might control concrete thickness or base plate size depends primarily on the magnitude of column loading and the intended base plate size. The example problem shows the punching shear check, but with minimum explanation since it is in ACI 318 *(Reference 7)*. The second feature is the location of nearby joints in the slab, whether construction, contraction, or isolation in their intent, as well as the size of a thickened portion of the slab, if used. These locations are based on a multiple of the radius of relative stiffness ℓ, which is discussed and tabulated in Section 10.7. No joint can be too close to the column. The recommendation for these distances is also shown in this example.

Figures 34 and *35* show the detail of the footing and slab unit with the necessary dimensions identified. *Figure 34* is for the slab with a constant thickness. *Figure 35* is for a slab with a thickened portion to serve as the footing, but still integral with the floor slab.

Commentary:

The columns are assumed to be "long" and subject to axial loading only. No moment is assumed to be transmitted to the slab. While the design is intended to maintain an uncracked slab, unforeseen cracking of the slab will present no danger to column support or to wall support.

Column loads, when sufficiently high, can increase the friction beneath the slab. This is especially true if the integral footing has a greater thickness than the surrounding floor slab. This can increase the tendency to crack due to drying shrinkage.

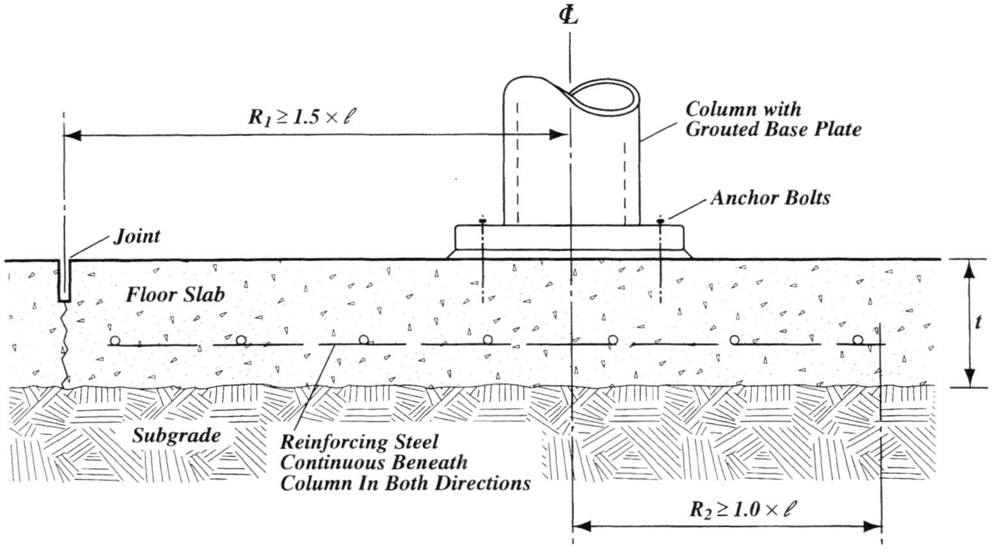

Figure 34 *Some slabs of constant thickness provide integral column footings.*

Commentary:
Where ACI 318.1 (plain concrete building code) may apply, Section 6.3.5 requires that the intended construction thickness be 2 inches more than the design check thickness (t), to allow for unevenness of excavation and for some contamination of concrete adjacent to the soil.

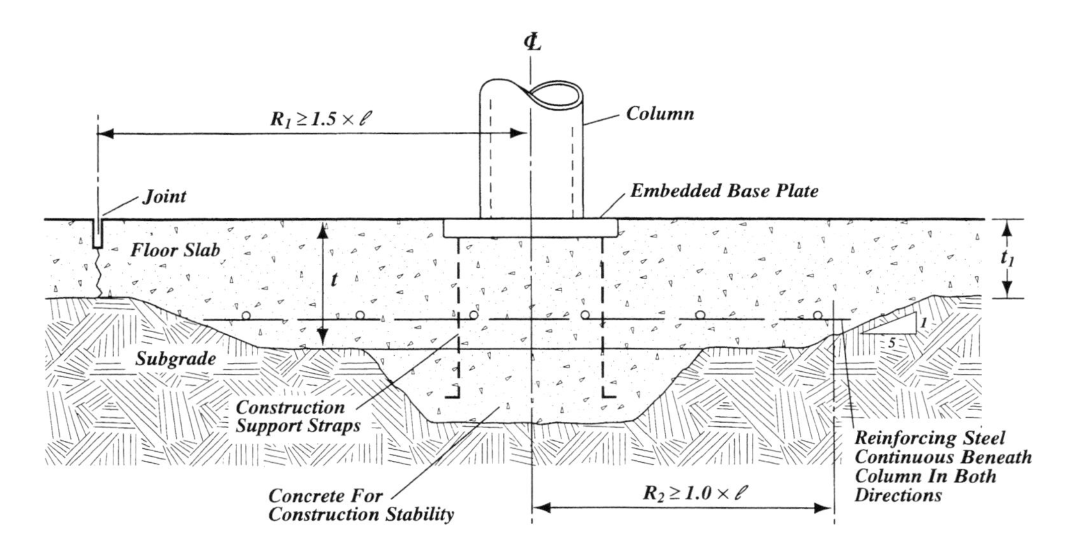

Figure 35 *Slab on grade must be thickened to provide integral column footing for certain loading and base plate combinations.*

The equations below are needed to solve this design problem. The thickness of the slab, designated here by t, is that thickness beneath the column and in the immediate vicinity of that column on all sides. The slab may or may not be thickened with respect to the average slab thickness. Note that the value of t solved for also appears within the log term. It can be solved by assuming the t within the log term, solving for t and then repeating the cycle until agreement between the t values is within 1/4 inch. The easier way is to assume a t value in both places within the equation and solve for the permissible column load. As for other solutions in this book, the authors feel that these equations are solved most easily by using a spreadsheet on a personal computer. In this way variables can be altered easily and results are rapidly displayed for the designer's use.

$$t^2 = P_u \times A \times \log_{10}\left[\frac{Bt^3}{C}\right]$$

where

P_u = the factored column load in pounds

$A = 0.03 \div \sqrt{f_c'}$

$B = 915{,}000 \times \sqrt{f_c'}$

$C = k \times b^4$

t = slab thickness beneath column, inches

b = base plate dimension, inches

For simplicity in this problem, E is assumed to be 4,000,000 psi, concrete weight is assumed to be 145 pcf and Poisson's ratio is assumed to be 0.2.

$$P_{all} = \frac{P_u}{LF}$$

where LF is the appropriate load factor. It is essentially correct to assume that the load factor is equal to the safety factor, SF.

As shown in *Figures 34* and *35*, the distance to the nearest joint must be no less than 1.5 times the radius of relative stiffness ℓ in each direction. Further, if the slab is to be thickened at the integral footing, this thickened portion should extend no less than 1.0 times that radius ℓ in each direction.

As an example, the following values are assumed, including the minimum size of the square column base plate. The column is a steel WF section. Note that the slab designer selects an appropriate load factor or safety factor.

- Slab thickness $t = 8$ inches
- Concrete strength $f_c' = 4000$ psi
- Subgrade modulus $k = 200$ pci
- Load factor or safety factor = 1.7
- Base plate $b = 10$ inches
- Radius of relative stiffness $\ell = 30$ inches

This example solves for the allowable column load (live load plus dead load total) that the slab can support. (It does not check the column itself.) The example then checks punching shear and locates the joint dimension requirement.

Solving the equations on the opposite page gives the following:

- Allowable column load $P_{all} = 19{,}030$ pounds (Ultimate column load = 32,350 pounds).
- *When this example is solved by other procedures, the allowable column load is 18,100 pounds using PCA charts, 16,700 pounds using the AIRPORT program, and 21,000 pounds using the MATS program.*

Using the relationship Shown in *Figure 35*, the distance to the nearest joint location is R_1, which is $1.5\ell = 1.5 \times 30 = 45$ inches. The integral footing is 60 inches square, based on $R_2 = 1.0\ell = 1.0 \times 30 = 30$ inches.

Check for punching shear capacity V_c using the procedures of the ACI Building Code *(Reference 7)* and its ultimate load procedures with its Equation 11-38. Assume the placement of 1/2-inch-diameter steel with at least 1 inch clear from the bottom of the footing. This gives a design d value of 6.5 inches for the 8-inch slab.

- $V_c = 4 \times \sqrt{f_c'} \times b_o \times d$.
  This is the punching shear equation (11-38) from the ACI Building Code *(Reference 7)*.
- In this equation, d = slab thickness minus cover minus one bar diameter, or $8 - 1 - 1/2 = 6.5$ inches
- b_o is the perimeter of what ACI assumes is the critical section, at a distance $d/2$ from the base plate.

$$b_o = 4 \times (d + b)$$
$$= 4 \times (6.5 + 10) = 66 \text{ inches}$$

- Therefore, $V_c = 108{,}500$ pounds, much greater than P_u

This indicates that punching shear is not critical to the slab thickness and that the base plate size of 10 x 10 inches is adequate for the column load. This check has been based on slab action only.

Commentary:
Even though the slab and the integral footing are designed to remain uncracked, it is highly recommended that reinforcing steel be placed in the lower portion of the integral footing. The steel can be reinforcing bars in both directions or sheets of welded wire fabric. It should be located at the bottom of the footing with a clear cover of at least 1 inch, depending on the applicable building code. If the applicable building code requires more cover, the square area defined by $R_2 \times R_2$ can be deepened to provide that additional cover. The steel should extend to the limit of the integral footing as defined by R_2 in Figures 34 *and* 35.

Where appropriate, punching shear may be checked using Section 6.3.7.1 of ACI 318.1-89 (Revised 1992), with proper selection for the value of width (b).

5.5 — Adapting PCA charts to design for column loads on slabs

The solution for a single concentrated load on a slab on grade is not readily available for designers. Most charts, tables and equations are set up for multiple concentrated loads such as vehicle axles and rack support posts.

The problem can be solved, however, by making use of the axle load charts published by PCA, WRI, or COE. These charts appear in examples in this book. When two loads (wheels or columns) are spaced as much as three times the radius of relative stiffness ℓ apart (see Section 10.7), the effect (added stress) of one on the other is nil. The designer then can use the chart that has been selected with a load spacing of at least $3 \times \ell$ and obtain the solution for a single load. Other examples in this book show this process.

There are appropriate solutions in several references. Most are based on Westergaard's equations and appear in pavement or highway references, as well as in selected publications of ACI, CRSI, and PCA.

5.6 — Design for wall loading on slabs using the ARMY-TM equations: AUTHORS' CHOICE

Commentary:
The authors recommend solving the equations via a spreadsheet on a personal computer for easiest use. Tabulated values in this book were created in this manner from the equations given.

Equations for allowable wall loads on a given floor slab are found in a 1987 technical manual of the Departments of the Army and the Air Force *(Reference 17)*. The equations also are given in this section. The procedure does not directly solve for the required slab thickness. Instead, the equations give the allowable wall load (pounds per lineal foot) for a given concrete thickness, along with the other values needed.

Two equations are given in *Tables 11* and *12*, along with calculated solutions for common situations. These equations are presented in their original form so that the designer can use them, selecting any desired safety factor (SF). In the publication by the Departments of the Army and the Air Force, the allowable bending stress is set at $1.6\sqrt{f_c'}$. When the modulus of rupture is taken as $9\sqrt{f_c'}$, the resultant safety factor is therefore preset to 5.625.

Table 11 shows allowable loading for interior walls, and *Table 12* shows allowable loading for walls at the edge of a slab, based on a safety factor of 2.5, which the authors recommend. Values in parentheses in each table were calculated using a factor of safety of 5.625 as in the Army/Air Force Technical Manual *(Reference 17)*.

The wall is considered narrow as compared to its length. The actual value of the wall width is not used in the equations although it is certainly needed to calculate the weight of the wall. As long as the wall is at least three times the radius of relative stiffness ℓ (as discussed in Section 10.7) from a joint or edge, it is reasonable to consider it as an interior line loading.

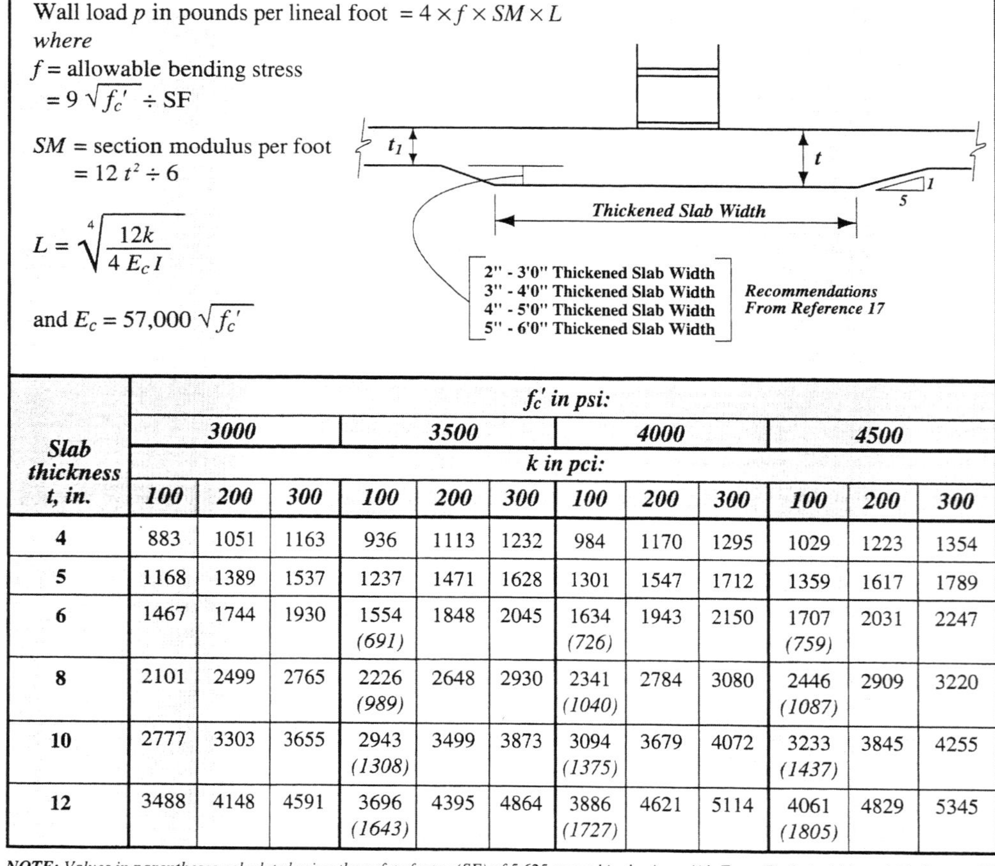

Slab thickness t, in.	f_c' in psi: 3000			3500			4000			4500		
	k in pci: 100	200	300	100	200	300	100	200	300	100	200	300
4	883	1051	1163	936	1113	1232	984	1170	1295	1029	1223	1354
5	1168	1389	1537	1237	1471	1628	1301	1547	1712	1359	1617	1789
6	1467	1744	1930	1554 *(691)*	1848	2045	1634 *(726)*	1943	2150	1707 *(759)*	2031	2247
8	2101	2499	2765	2226 *(989)*	2648	2930	2341 *(1040)*	2784	3080	2446 *(1087)*	2909	3220
10	2777	3303	3655	2943 *(1308)*	3499	3873	3094 *(1375)*	3679	4072	3233 *(1437)*	3845	4255
12	3488	4148	4591	3696 *(1643)*	4395	4864	3886 *(1727)*	4621	5114	4061 *(1805)*	4829	5345

NOTE: Values in parentheses calculated using the safety factor (SF) of 5.625 as used in the Army/Air Force Technical Manual (Reference 17) *instead of the safety factor of 2.5 as indicated for the rest of the table.*

Table 11 *Allowable interior wall loadings for selected slab conditions, based on the displayed equation from* Reference 17.

Table 11 shows a set of allowable wall loads for certain values. The values that control the allowable loading are:

- Concrete compressive strength, f_c' in psi
- Safety factor, SF
- Subgrade modulus, k in pci
- Slab thickness, t in inches

  The preliminary value for the slab thickness may be determined from other factors, or the equations can be solved several times until an acceptable thickness is determined.
- *Table 11: For concrete strength of 4000 psi, k of 100 pci, and slab thickness of 6 inches, 1634 pounds per lineal foot is allowed. Solving this example by the MATS program gives an allowable load of 1940 pounds per lineal foot.*

The drawing in *Table 12* shows the wall located at the edge of the floor panel, or at an isolation joint, along with the dimensions needed to solve the problem. *Table 12* also shows the equations to be used. These equations are for a constant thickness slab beneath and near the position of the wall. As previously noted, the equations have been presented in their original form so that the designer can set all reasonable variables, such as the safety factor.

Table 12 also presents a set of allowable wall loads for certain selected values. The values that control the allowable loading are:

- Concrete compressive strength, f_c' in psi
- Safety factor, SF
- Subgrade modulus, k in pci
- Slab thickness, t in inches

  The preliminary value of the slab thickness may be determined from other factors, or from the equations. For a constant thickness slab, this initial value is not necessary.

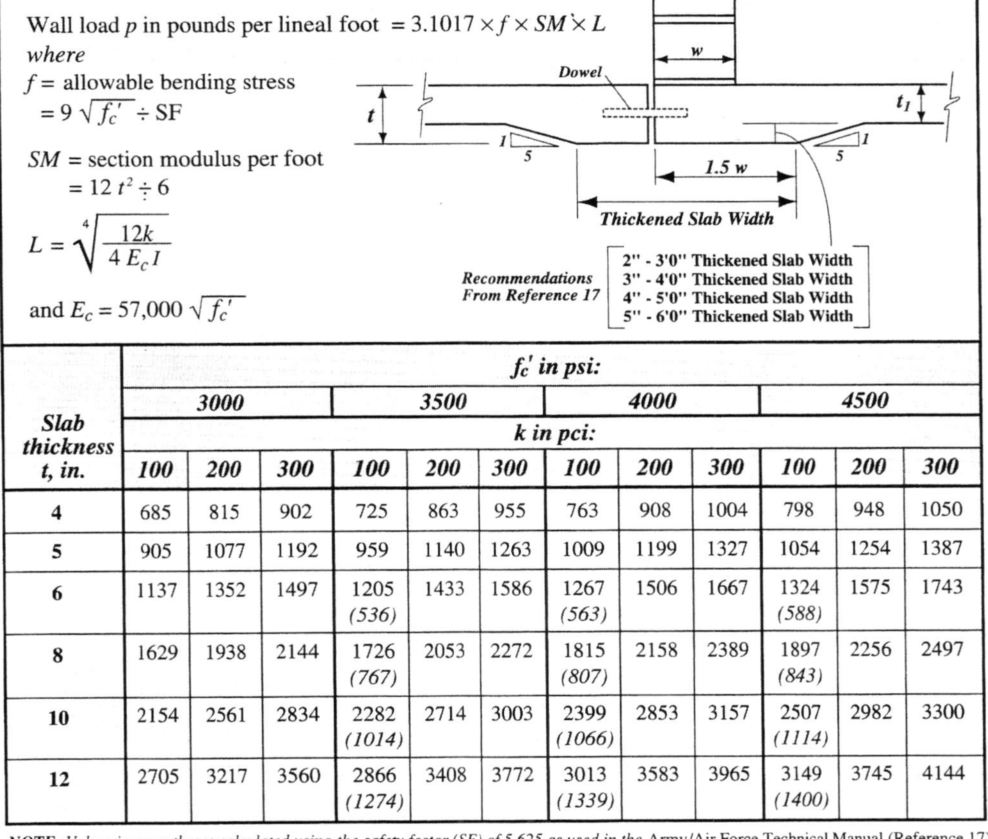

Wall load p in pounds per lineal foot $= 3.1017 \times f \times SM \times L$

where

f = allowable bending stress $= 9\sqrt{f_c'} \div \text{SF}$

SM = section modulus per foot $= 12\,t^2 \div 6$

$$L = \sqrt[4]{\frac{12k}{4\,E_c I}}$$

and $E_c = 57{,}000\sqrt{f_c'}$

Slab thickness t, in.	f_c' in psi: 3000			3500			4000			4500		
	k in pci: 100	200	300	100	200	300	100	200	300	100	200	300
4	685	815	902	725	863	955	763	908	1004	798	948	1050
5	905	1077	1192	959	1140	1263	1009	1199	1327	1054	1254	1387
6	1137	1352	1497	1205 (*536*)	1433	1586	1267 (*563*)	1506	1667	1324 (*588*)	1575	1743
8	1629	1938	2144	1726 (*767*)	2053	2272	1815 (*807*)	2158	2389	1897 (*843*)	2256	2497
10	2154	2561	2834	2282 (*1014*)	2714	3003	2399 (*1066*)	2853	3157	2507 (*1114*)	2982	3300
12	2705	3217	3560	2866 (*1274*)	3408	3772	3013 (*1339*)	3583	3965	3149 (*1400*)	3745	4144

NOTE: Values in parentheses calculated using the safety factor (SF) of 5.625 as used in the Army/Air Force Technical Manual (Reference 17) *instead of the safety factor of 2.5 as indicated for the rest of the table.*

Table 12 *Allowable edge wall loadings for given slab conditions, based on the displayed equation from* Reference 17.

Commentary:

The wall loads, when sufficiently high, can increase the friction beneath the slab and therefore have an influence on either joint spacing or required crack control steel following the subgrade drag equation.

Table 12 *shows a 23 percent lower load capacity than* Table 11, *due to lack of continuity in the floor slab on both sides of the wall. The use of dowels, or any effective load transfer system, will restore a substantial portion of this continuity.*

5.7—ACI 318.1 Building Code Requirements for Structural Plain Concrete and Commentary

This building code (*Reference 7*) is for plain concrete used for structural purposes. Plain concrete is defined as either unreinforced concrete or concrete containing less reinforcement than the minimum specified by ACI 318 for reinforced concrete. As a building code, ACI 318.1 is not likely to apply to floor slabs on grade. In the commentary (Section R1.2.2., pages 4 and 5), it is stated that slabs on grade are not considered within the context of structural plain concrete unless they transmit vertical loads from other parts of the structure to the soil. However when 318.1 does apply, there are certain sections that are critical to slab on grade design. Of particular importance are sections on allowable flexural tensile stress (Section 6.2.1) and reduction in thickness for design checks (Section 6.3.5).

CHAPTER 6
SLAB DESIGN FOR DISTRIBUTED UNIFORM LOADING ON BOTH SIDES OF AN AISLE

6.1 — The design objective

A common design problem encountered in industrial facilities is the floor that must support a loading on its top surface on either side of an aisle. The loading is considered uniform although it may in fact be on pallets or on rack shelves. The aisle width is assumed to be known and the aisle and loading are assumed to be at least twice as long as the width of the aisle. *Figure 36* shows the general layout of the loading.

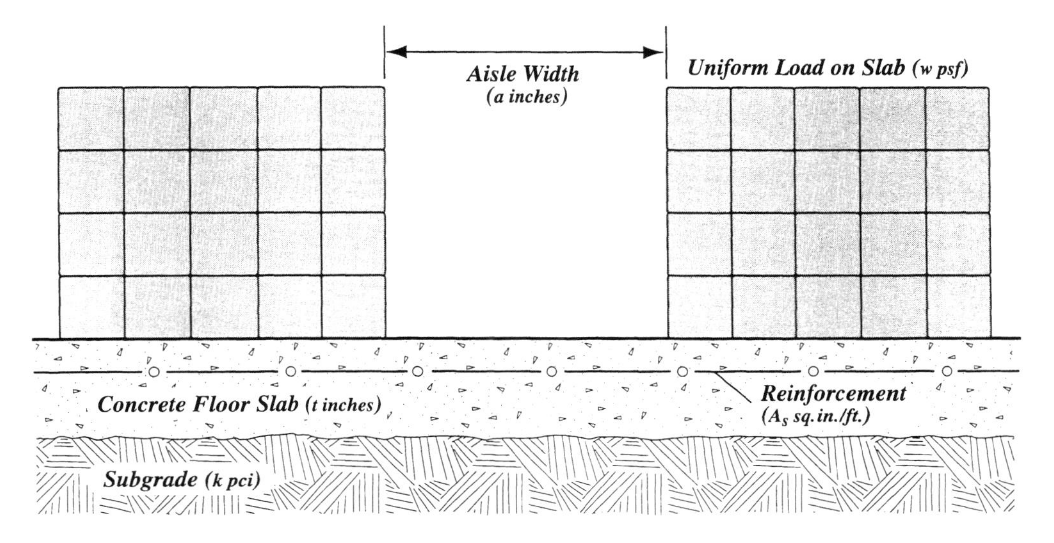

Figure 36 *Slab on grade supporting uniform loading on both sides of an aisle.*

This design checks the ability of the slab to resist the moment in the slab at the center line of the aisle. This moment is caused by the uniform loading which exists on both sides of the aisle at the same magnitude and at the same time. The moment creates tension at the top surface of the slab.

If the slab is to remain uncracked on the top of the aisle surface, then the design must limit the actual tensile stress to an allowable value determined by dividing the modulus of rupture by the safety factor. The required slab thickness can be determined by using the WRI charts *(Reference 15)* or the PCA tables *(Reference 6)*. These techniques are illustrated in Sections 6.3 and 6.4.

If the concrete need not remain completely uncracked, that is, if tight hairline cracking is acceptable, the approach changes. The objective then is to determine the moment in the slab and design the slab for it, using conventional reinforced concrete procedures to select appropriate steel areas. This is done using WRI charts and techniques illustrated in Section 6.7.

If construction is with shrinkage-compensating concrete or post-tensioning, then the design procedure is altered somewhat. The intent with both of these techniques is to maintain an uncracked slab, by means of chemical or physical prestressing rather than by slab thickness. (See Sections 6.5 and 6.6).

Regardless of which design approach is selected, flatness and levelness of the aisle may be as critical to floor performance as is crack control. (See Section 10.4.3).

6.2 — Values needed to solve the problem

To determine the thickness of the concrete slab and the reinforcement or prestress required, the following information is needed:

From the loading specifications:

Magnitude of uniform loading, w in psf
Width of aisle, a in inches
Approximate slab thickness, t in inches (needed only when WRI charts are used)

From materials, site and designer:

Concrete's design compressive strength, f_c' in psi
Modulus of rupture of the concrete, *MOR* in psi
Modulus of subgrade reaction, k in pci
Safety factor (selected by designer)

Commentary:
These WRI charts are sensitive to the position of the values and lines within the graphs. They are difficult to use when progressing from an assumed thickness all the way through the process to the required thickness. Equality between these two numbers is required for a correct solution. The authors recommend using the graphs as described here, working from an assumed thickness at both ends of the process. This makes it relatively easy to converge to an allowable uniform load.

While this example assumes a trial slab thickness, it is generally best to calculate the slab thickness based on other loadings, such as vehicle axle loads or rack-supporting post loads. Most industrial slabs are from 6 to 12 inches thick.

6.3 — Using WRI charts to design for distributed uniform loading on both sides of an aisle: AUTHORS' CHOICE

The necessary charts are reproduced large size in the appendix. They come from a report of the Wire Reinforcement Institute *(Reference 15)* and from the work of Panak and Rauhut *(Reference 21)*.

A problem will be solved using the following input data:

From the loading specifications:

Magnitude of uniform loading: 2500 psf
Width of aisle: 90 inches (7 feet 6 inches)
Approximate slab thickness: 10 inches assumed to start

From the materials, site, and designer

Concrete design compressive strength, f_c' = 4000 psi
Concrete modulus of rupture *MOR* = 570 psi
Modulus of subgrade reaction, k = 300 pci
Factor of safety: designer's choice is 1.7

The design starts with *Figure 37* where the value of D/k is determined. This is the relative stiffness of the slab compared to the subgrade

• Calculate the concrete modulus of elasticity E_c from standard ACI equation:

$$E_c = 57{,}000 \times \sqrt{f_c'}$$

With $f_c' = 4000$ psi this gives $E_c = 3.605 \times 10^6$ psi, or 3,605 ksi.

Commentary:
E_c may alternatively be calculated using
$E_c = 33\, w^{1.5}\sqrt{f_c'}$
as done on Page 29.

Using the assumed slab thickness of 10 inches, enter *Figure 37* on the left, proceeding diagonally to an intersection with the modulus of elasticity. Then move horizontally to the intersection with the curve for $k = 300$ pci. Proceeding down to the D/k axis, read $D/k = 11 \times 10^5$ in.^{4}.

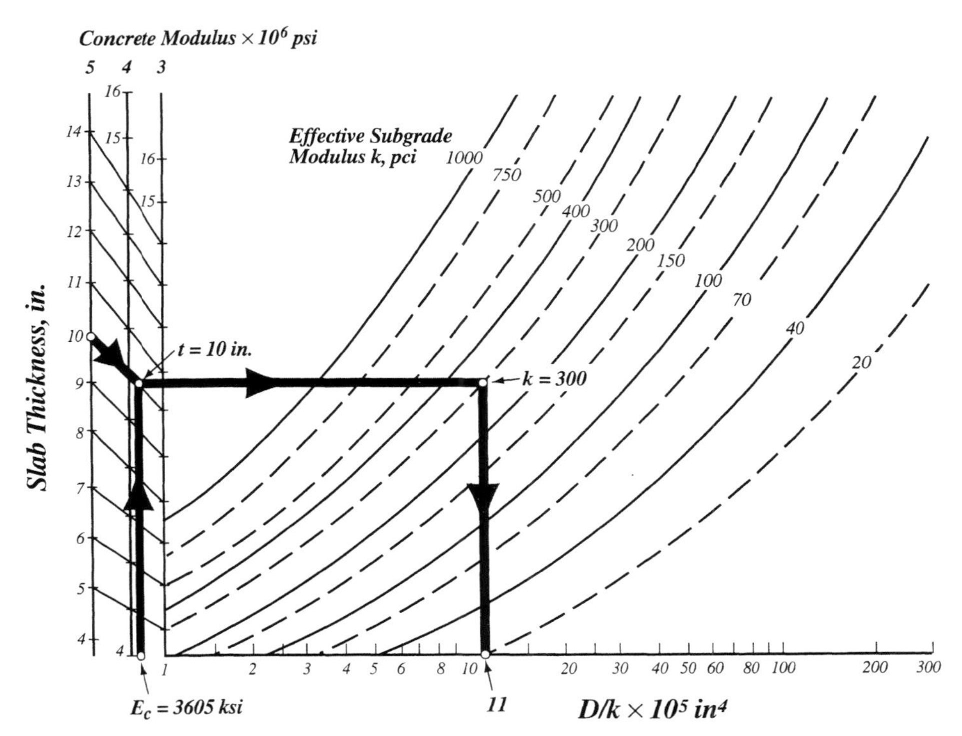

Figure 37 *WRI chart for determination of* D/k, *which represents relative stiffness between slab and subgrade.*

- The next step requires *Figure 38*, starting on the left side. Enter at the bottom with the aisle width of 90 inches. Move vertically up to as close to the D/k value of 11 as is possible. Then draw a line to the right until it intersects the right boundary line (labeled A) of the left-hand part of *Figure 38*.
- Now move to the right-hand part of *Figure 38*, entering from the bottom with the slab thickness of 10 inches. Draw a vertical line up to the allowable stress curve, again as close as is possible. The allowable stress is the modulus of rupture divided by the safety factor, in this case:

$$570/1.7 = 335 \text{ psi}$$

- Draw a line horizontally to the left to the left-hand boundary (labeled B) of the right-hand part of *Figure 38*. This left-hand boundary shows slab moments (foot-pounds per foot of slab width) based on the uncracked section modulus of the concrete slab and the allowable bending stress selected for the slab.
- A straight line (usually inclined) is now drawn between the two boundary lines A and B. This straight line crosses a vertical line at the permitted uniform loading for the slab, in kips per square foot. In this example, the permitted value of w is very close to 2600 psf. Therefore, the assumed slab thickness of 10 inches is acceptable for the uniform load of 2500 psf.
- The 10-inch-thick slab is confirmed and is then recommended. *Solving this example by the MATS program gives a required thickness of 10.6 inches.*

The softer the subgrade and/or the thinner the slab, the more it bends under loading, and the greater must be its moment capacity. This method is quite effective since it is the only procedure which includes the relative stiffness of subgrade and slab. It is also quite effective for a specific aisle width and location when used as described here.

Commentary:
These WRI procedures generally assume the use of steel reinforcement. The amount is commonly selected using tables or charts based on the subgrade drag equation, which is appropriate where slab and steel are intended to be non-structural. The purpose of the steel is to hold the concrete portions together. The steel must be located at or above the middepth of the slab. A location near the third-point (down from the top) consistent with joint depth is commonly used.

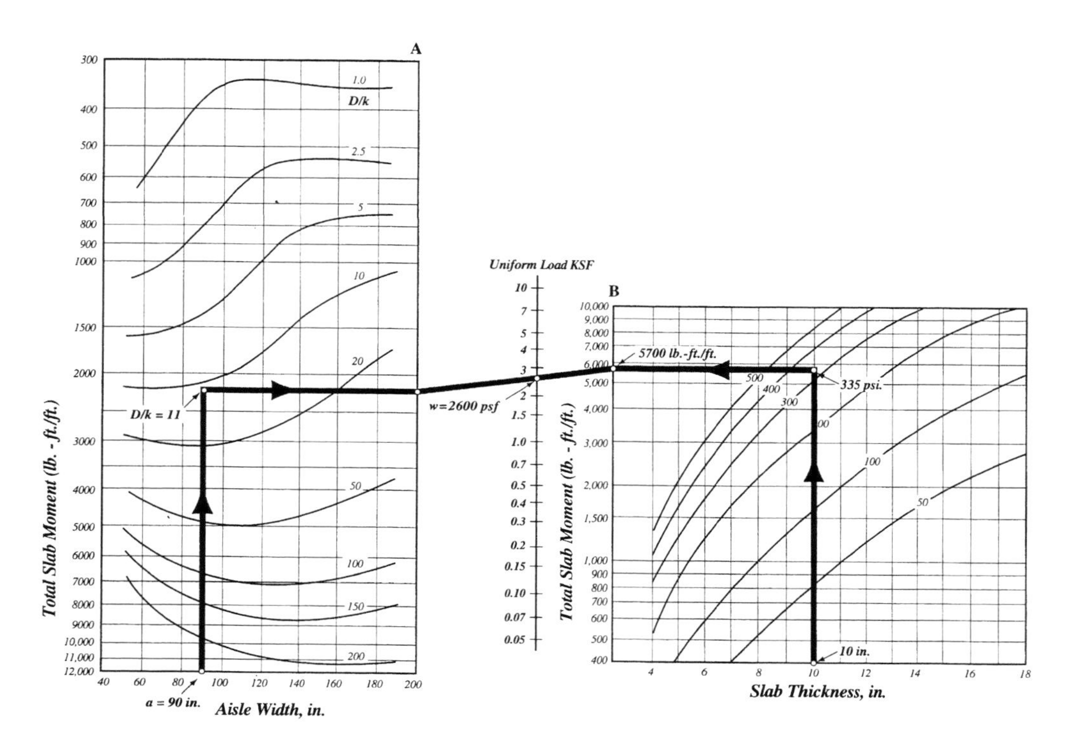

Figure 38 *WRI chart for determination of permitted uniform loading on the slab.*

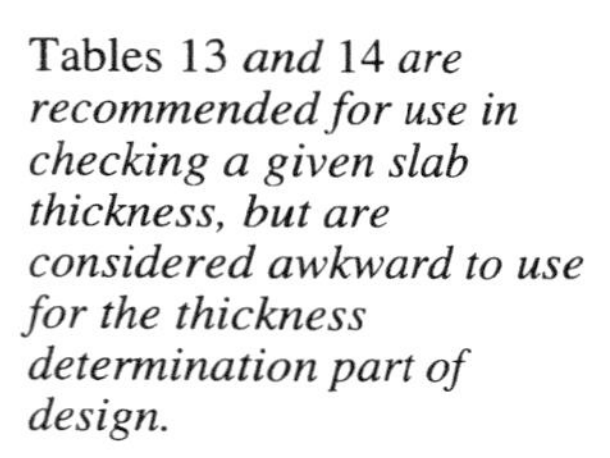
Tables 13 *and* 14 *are recommended for use in checking a given slab thickness, but are considered awkward to use for the thickness determination part of design.*

6.4 — Using PCA tables to design for uniform loading on both sides of an aisle

These PCA tables taken from *References 6* and *14*, are reproduced larger size in the appendix. *Table 13* is for a fixed layout; that is, a specific location for aisle and a uniform loading of relatively constant magnitude. *Table 14* is for a variable layout for aisle location as well as a distributed uniform loading which may vary in magnitude.

To show the use of *Table 13*, assume:

1000 psf uniform loading
12-foot aisle width
335 psi allowable concrete stress (= *MOR*/FS)

Note that the table has a load limit of 1630 psf, and is valid for a modulus of subgrade reaction *k* of 50 pci, only. The assumed loading of 1000 psf is commonly found in moderately to lightly loaded warehouse situations. Using *Table 13*:

- The solution is found by starting with an aisle width of 12 feet. Move down the column. There are four rows with loading close to, but slightly higher than 1000 psf.
- The closest appears to be 1025 psf, indicating an 8-inch slab with a working stress of 350 psi. It is reasonable to use straight-line ratios between table values to refine the design. The thickness is also affected by the subgrade modulus, which in this case is only 50 pci.
- An 8-inch slab is recommended.

As the example shows, *Table 13* can be simple to use if the problem input fits the table closely. The table is for a distributed loading up to a maximum of 1630 psf in a fixed position; it uses working stress as its basis, and includes the effect of the aisle width. The

Slab thickness, in.	Working stress psi.	Critical aisle width ft.**	Allowable load, psf+					
			At critical aisle width	At other aisle widths				
				6 ft. aisle	8 ft. aisle	10 ft. aisle	12 ft. aisle	14 ft. aisle
Subgrade k = 50 pci*								
5	300	5.6	610	615	670	815	1,050	1,215
	350	5.6	710	715	785	950	1,225	1,420
	400	5.6	815	820	895	1,085	1,400	1,620
6	300	6.4	670	675	695	780	945	1,175
	350	6.4	785	785	810	910	1,100	1,370
	400	6.4	895	895	925	1,040	1,260	1,570
8	300	8.0	770	800	770	800	880	1,010
	350	8.0	900	935	900	935	1,025	1,180
	400	8.0	1,025	1,070	1,025	1,065	1,175	1,350
10	300	9.4	845	930	855	850	885	960
	350	9.4	985	1,085	1,000	990	1,035	1,120
	400	9.4	1,130	1,240	1,145	1,135	1,185	1,285
12	300	10.8	915	1,065	955	915	925	965
	350	10.8	1,065	1,240	1,115	1,070	1,080	1,125
	400	10.8	1,220	1,420	1,270	1,220	1,230	1,290
14	300	12.1	980	1,225	1,070	1,000	980	995
	350	12.1	1,145	1,430	1,245	1,170	1,145	1,160
	400	12.1	1,310	1,630	1,425	1,335	1,310	1,330

* *k* of subgrade, disregard increase in *k* due to subbase.

** Critical aisle width equals 2.209 times radius of relative stiffness. Critical aisle width has maximum negative bending moment (tension in top slab at aisle centerline due to loads on each side of aisle). For other aisle widths, bending moments are not maximum.
Assumed load width = 300 inches; allowable load varies only slightly for other load widths. Allowable stress = one half flexural strength.

+ There is an explanation in *Slab Thickness Design for Industrial Concrete Floors on Grade* for what appear to be anomalous allowable loads.

Table 13 *Allowable distributed loads on slabs with unjointed aisle and fixed layout, based on* References 6 and 14.

critical aisle width is that width at which the bending stress in the center of the aisle at the top of the slab is a maximum. Theoretically this critical width is 2.209 times the radius of relative stiffness *(Reference 14)*.

Table 14 differs from *Table 13* in that the subgrade modulus values are 50, 100, and 200 pci and the results are controlled by the concrete's flexural strength (modulus of rupture) with the safety factor built into the table. *Table 14* has a 2285-psf limit on uniform load.

Assume a flexural strength of 550 psi for this example. Assume also that the subgrade modulus is 100 pci and the load to be used is 1000 psf.

- Move down the 550 psi column and locate the first load capacity at or above 1000 psf for a *k* of 100. This occurs at a 10-inch slab thickness, which is higher than other design procedures due to the assumption of a variable layout.
- A 10-inch slab is recommended.

Table 14 gives the allowable uniform loading, but for a variable layout of that loading. It uses concrete flexural strength (modulus of rupture) and subgrade modulus as input values. A safety factor of 2 is assumed within the table. As with *Table 13*, *Table 14* can be simple and easy to use as long as the input values fit the table, which has a limit of 2285 psf. *Table 14* consistently gives somewhat thicker slabs than *Table 13*.

Commentary:
The procedures in Sections 6.3 and 6.4 check thickness and strength of the slab solely at the aisle's center. Joint spacings and joint types, along with their load transfer ability, are not involved. It is assumed that no joint exists within the aisle nor under the loading within about 3 to 4 feet of the edge of the aisle. The stress is maintained at an acceptable level by the thickness of the concrete slab, along with the reinforcement selected. If the loading, presumed uniform in this check, is in fact applied to the slab by posts supporting rack shelves, the tension on the bottom of the slab beneath the post is not checked by these procedures. For this latter check, refer to Chapter 4.

Slab thickness in.	*Subgrade k* pci*	*Allowable load, psf***			
		Concrete flexural strength, psi			
		550	***600***	***650***	***700***
5	50	535	585	635	685
	100	760	830	900	965
	200	1,075	1,175	1,270	1,370
6	50	585	640	695	750
	100	830	905	980	1,055
	200	1,175	1,280	1,390	1,495
8	50	680	740	800	865
	100	960	1,045	1,135	1,220
	200	1,355	1,480	1,603	1,725
10	50	760	830	895	965
	100	1,070	1,170	1,265	1,365
	200	1,515	1,655	1,790	1,930
12	50	830	905	980	1,055
	100	1,175	1,280	1,390	1,495
	200	1,660	1,810	1,965	2,115
14	50	895	980	1,060	1,140
	100	1,270	1,385	1,500	1,615
	200	1,795	1,960	2,120	2,285

* *k* of subgrade; disregard increase in *k* due to subbase.

** For allowable stress equal to 1/2 flexural strength.
Based on aisle and load widths giving maximum stress.

Table 14 *Allowable distributed load on slabs with unjointed aisles and variable layout, from* References 6 and 14.

Commentary:
The use of this concrete allows a much wider spacing of contraction joints than is appropriate for standard portland cement concrete. The designer is cautioned against spacing joints too far apart. Joints (construction joints in this case) spaced at approximately 50 to 60 feet each way are considered good, although greater spacings have been successfully used. The joint spacing selected must depend on whether or not hairline cracks are to be permitted. Critical to the prevention of cracking are any possible restraints to slab motion, such as posts, pits, etc.

6.5 — Using shrinkage-compensating concrete for slabs with distributed uniform loading on both sides of an aisle

The use of a shrinkage-compensating concrete, produced with either Type K cement or an additive, is a change in material rather than loading or design assumptions. The concrete behaves differently, but the thickness selections are done using the same methods as with normal portland cement concrete.

No example is given for thickness determination. Given the problem, the design for thickness is the same as previous sections. The use of this technique allows a wider joint spacing, therefore, fewer lineal feet of joint per job. It is also a crack control technique. However, shrinkage-compensating concrete demands the use of reinforcing steel, properly sized and properly placed. The best reference is ACI 223 (*Reference 13*) which gives procedures for determining the expansion and shrinkage strains and for selecting the proper percentages of steel. Since steel is mandatory, it is also feasible to consider the use of structurally active reinforcement as discussed in Section 1.5.1 and illustrated in Section 6.7.

For example, once the slab thickness is set based on loadings or construction, the use of shrinkage-compensating concrete requires additional values to assure that compensation will in fact occur. These are:

- a selection of reinforcing steel
- the expected concrete prism expansion for the mix
- expected member (slab) expansion in the field

The concrete prism expansion is measured by a testing lab following ASTM C 878. The expected expansion of the slab in the field, which should equal or exceed the concrete's anticipated drying shrinkage, is needed for joint design. The reinforcing steel is required to maintain the integrity of the slab and restrain the concrete's expansion and contraction.

The procedure is as follows:

- Slab thickness has been set at 6 inches.
- Joint spacing is 50 feet.
- Reinforcing steel is selected as #4 bars at 16 inches each way, based on shrinkage compensating crack control.
- Calculate the steel percentage. Area of steel is 0.20 square inches per 16 inches of slab width, or 0.15 square inches per foot of slab width.

  steel percentage $p = 0.15/(6 \times 12) = 0.208$

  Establish a diagonal on the chart of *Figure 39* (from *Reference 13*) to represent this steel percentage (shown as a dashed line).
- Calculate *V/SA*, the ratio of slab volume to surface area.

  $V/SA = (6 \times 12 \times 12)/(12 \times 12) = 6$
- Find the intersection of *V/SA* = 6 curve with the diagonal representing steel percentage. This is Point *X* in *Figure 39*.
- From Point *X* draw a vertical line down to find the maximum prism expansion of 0.036%. The concrete mix, with shrinkage compensating additive or Type K cement, should be specified to have a prism expansion of 0.036%, as measured according to ASM C 878. The concrete must expand more than it subsequently shrinks.
- To find the expected field expansion of this concrete, draw a horizontal line from Point X to the left, and read 0.030%. The actual shrinkage of this slab is expected to be less than this amount.

Commentary:

*Panels between joints should be as nearly square as possible. Distance between joints can be as much as 80 to 100 feet (*References 8 *and* 13*); however, the authors recommend restricting the spacing to 50 or 60 feet. The first pour will serve as a test. Subsequent pours allow corrections if needed. Performance specifications for this type of floor construction may be preferred.*

Reinforcing bars are selected on the basis of judgment and experience. Bars are commonly #3, #4, or #5 with spacings of 14 inches or more. The decision is arbitrary, but the resulting percentage must be usable in Figure 39 *or similar designs.*

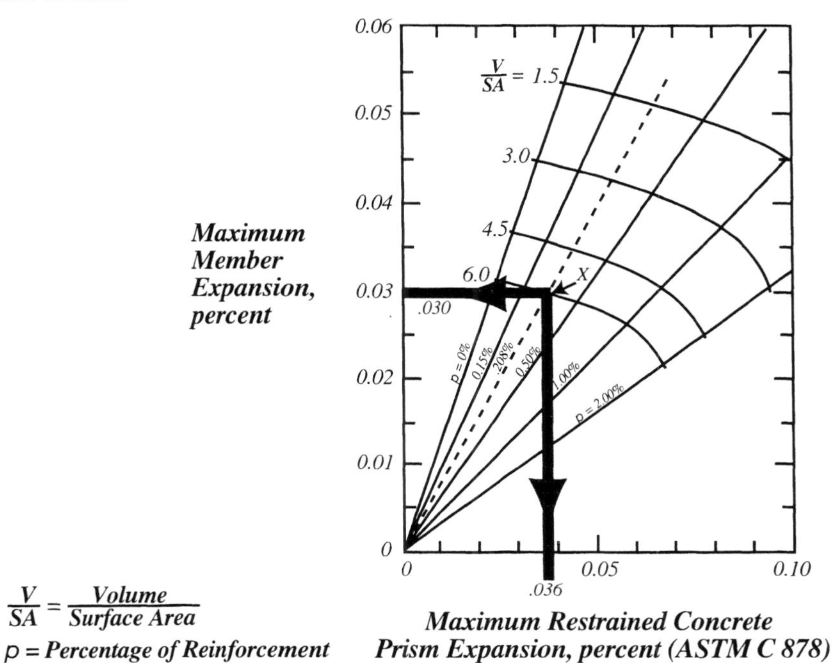

Figure 39 *Graph relating reinforcing steel percentage, member (slab) expansion and restrained concrete prism expansion, all in percents.*

- Use the expected field expansion of 0.030% to calculate joint filler thickness. Where the length of the slab is 50 feet, all motion is assumed to occur at one of the free ends, even though it is far more likely to be divided between opposite ends.

  Motion = 50 × 12 × 0.0003 = 0.18 inches (3/16 inches)
- Thickness of joint filler material = 2 × 3/16 inches = 3/8 inches, made of a material that will readily compress to 50% of its thickness (*Reference 13*).

6.6 — Using post-tensioning tendons for slabs with distributed uniform loading on both sides of an aisle

Prestressing the slab on grade by means of post-tensioned tendons serves two purposes. In this design example, it acts as a positive crack control. The prestress level is variable, but commonly produces approximately 100 psi of long-term (after prestress losses due to anchorage effects and tendon creep) compressive stress in the concrete. In addition to preventing the formation of shrinkage cracks, this stress increases the effective modulus of rupture.

The example in Section 6.3, which followed the WRI design charts for thickness determination, is solved here again, using the increased flexural stress due to the prestress of 100 psi applied to the slab. (See Section 3.7 for another prestressed slab example.) The original (Section 6.3) problem variables and intermediate results were:

- E_c = 3,605,000 psi
- k = 300 pci
- D/k = 11 (from *Figure 37*)
- w = 2500 psf
- Aisle width = 90 inches
- Moment = 5700 foot-pounds per foot (applied moment from *Figure 38*)

Commentary:
The use of prestressing tendons has two other advantages. The spacing of tendons, usually in the range of 30 to 40 inches, allows field workers to walk between the tendons (Reference 23). *Further, experience has shown that the* MOR *is actually increased beyond the calculated level shown. This is because the prestress prevents the propagation of microcracking, which is part of the phenomenon that decreases the tensile strength* (MOR) *of the concrete.*

Due to the prestress, the new effective allowable stress is

$$335 + 100 = 435 \text{ psi}$$

This can be used in *Figure 38* to obtain a graphical solution.

Two results can be achieved from the applied moment (5700 foot-pounds per foot) and the higher allowable bending stress (435 psi).

- The first result is a higher permissible uniform load. Although no separate figure is shown with this solution, the right side of *Figure 38* would indicate a new load of 3200 psf instead of the original 2500 psf.
- The second result is a thinner slab. Use the moment of 5700 foot-pounds per foot (not exact due to the change in D/k), the new allowable stress of 435 psi, and the relationship

$$M = \frac{bt^2}{6} \times f_c$$

Solving for t gives a theoretical slab thickness of 8.9 inches.

6.7 — Using structurally reinforced concrete for slabs loaded uniformly on both sides of an aisle

Steel reinforcement is placed in a slab for one or more reasons. Regardless of the reason, however, the steel must be accurately positioned in the slab. The four reasons are:

1. To act as crack control. Here, the steel is commonly selected by use of the subgrade drag equation. This is discussed in Section 1.5.1.3.

2. To act as the required steel with shrinkage-compensating concrete slabs. In this case the steel is essential for good slab performance since joints are usually at a much wider spacing than normally used. The steel is selected following ACI 223 *(Reference 13)*, and the procedure is discussed in Section 6.5.

3. To provide reserve load capacity. For this purpose, the slab remains at the thickness based on the uncracked and unreinforced slab section. The steel area is increased so that the moment capacity of the reinforced cracked section is greater than the allowable moment for the slab *(Reference 11)*. This is illustrated below in *Example A*.

4. To allow use of a thinner slab. Here, the safety factor provided for load capacity is due to a substantially larger area of steel (properly located) than that indicated by the subgrade drag equation *(Reference 11)*. This slab is expected to develop hairline cracks due to loading. The process is illustrated in *Example B*.

The two examples start with the original selection of a 10-inch slab, retaining the *MOR* of 570 psi and the allowable stress of 335 psi. *Example A* follows reason 3 above, providing reserve strength. *Example B* will follow reason 4, reducing the thickness and providing load support with steel reinforcement. All moment capacities are calculated by conventional concrete and reinforced concrete equations; however, tabulated solutions of these equations are given in the appendix.

***Example A*:** This slab is to be left 10 inches thick and is to be reinforced for reserve strength. The applied moment is 5700 foot-pounds per foot from the WRI charts.

We will design for 25 percent more strength than necessary to provide for the applied moment of 5700 foot-pounds per foot. This is 1.25 times 5700 or 7125 foot-pounds per foot.

- Select reinforcement placed at middepth to supply 7125 foot-pounds per foot moment resistance.
- From *Table 15*, we see that #6 bars at 12-inch spacing provide 8.91 foot-kips per foot resistance in a 10-inch slab. More tables like this one are reproduced in the appendix.
- Required spacing = $(8.91 \div 7.125) \times 12$
  = 15.0 inches
- Therefore, #6 rebars (both ways at middepth), 15 inches on center, provide the reserve strength requested.

Commentary:

To be effective, steel reinforcement must be in the proper location after the concrete hardens. That position depends on why the steel has been specified; however single layers must be at or above the slab's middepth. The authors recommend that the steel be supported during construction by slab bolsters or other positive devices. Either deformed rebars or WWF sheets are acceptable.

First in Example A, *the calculated value for the moment is 5583 foot-pounds per foot. This differs from the graphical value by 2 percent, which is both acceptable and negligible. Second, this design can be done as it is here using the actual moment. It also can be done using the cracking strength; that is, providing a reserve moment or load capacity beyond the cracking moment instead of beyond the applied moment.*

Since the cracking moment of the 10-inch slab is 9500 foot-pounds per foot, this reserve strength is available only in the case of unexpected full-depth cracks in the slab, such as might occur due to excess drying shrinkage and restraint to it.

Slab Moment Capacities For One or Two Layers of Rebar at 12 - Inch Spacing

Slab Thickness		8-in.						10-in.					
		#3	#4	#5	#6	#7	#8	#3	#4	#5	#6	#7	#8
Number of Layers	A_s (in.²)	0.11	0.20	0.31	0.44	0.60	0.79	0.11	0.20	0.31	0.44	0.60	0.79
	d_b (in.)	0.38	0.50	0.63	0.75	0.88	1.00	0.38	0.50	0.63	0.75	0.88	1.00
ONE	*d* (in.)	4.00	4.00	4.00	4.00	4.00	4.00	5.00	5.00	5.00	5.00	5.00	5.00
	resulting cover (in.)	3.63	3.50	3.38	3.25	3.13	3.00	4.63	4.50	4.38	4.25	4.13	4.00
	M_u (ft-k/ft)	1.78	3.24	5.02	7.13	9.72	12.80	2.23	4.05	6.28	8.91	12.15	16.00
	p (%)	0.11	0.21	0.32	0.46	0.62	0.82	0.09	0.17	0.26	0.37	0.50	0.66
	wgt (psf)	0.75	1.37	2.12	3.01	4.10	5.40	0.75	1.37	2.12	3.01	4.10	5.40
TWO	*d* (in.)	6.38	6.25	6.13	6.00	5.88	5.75	8.38	8.25	8.13	8.00	7.88	7.75
	M_u (ft-k/ft)	2.84	5.06	7.69	10.69	14.28	18.40	3.73	6.68	10.20	14.26	19.14	24.80
	p (%)	0.23	0.42	0.65	0.92	1.25	1.65	0.18	0.33	0.52	0.73	1.00	1.32
	wgt (psf)	1.50	2.73	4.24	6.01	8.20	10.79	1.50	2.73	4.24	6.01	8.20	10.79

NOTES: 1. $f'_c = 4000$ *psi;* $f_y = 60{,}000$ *psi; cover = 1.25 in; b = 12 in;* $\phi = 0.90$

2. Design assumptions made in table:

	One Layer	Two Layers
d	$t/2$	$t - (cover + d_b)$
$j_u d$	$0.9d$	
M_u	$\phi A_s f_y (j_u d)$	

3. Percentage of reinforcement based on gross section, $b \times t$

Table 15 *Slab moment capacities (resistance) in foot-kips per foot of slab width, using either one or two layers of reinforcing bars, from* Reference 11.

***Example B*:** The original 10-inch slab is to be reduced to 8 inches. Strength will be provided with reinforcing steel selected to provide a safety factor (load support) of 2.

- The actual moment applied is 5700 foot-pounds per foot, as before.
- The cracking moment of the 8-inch slab is 6080 foot-pounds per foot, as determined by calculation using

$$M_{cr} = \frac{bt^2}{6} \times MOR$$

and then dividing by 12 to adjust units to foot-pounds per foot. Or use Table A.35 in the appendix, which has precalculated cracking moment values.

- This means the safety factor against cracking is 6080/5700 = 1.066 which is too low.
- Select a safety factor (or load factor) of 2.
- Moment capacity (to be provided by steel) is:

2 (5700) = 11,400 foot-pounds per foot

- Two possibilities exist, both using *Table 15 (Reference 11)* for the 8-inch slab:
  #6 bars (one layer) at 7.5 inches.
  Spacing 12(7.13/11.4) = 7.505 inches
  #5 bars (two layers) at 8 inches.
  Spacing 12(7.69/11.4) = 8.09 inches
- Use of #6 bars in one layer at 7.5 inches both ways is recommended to provide the needed strength.

Commentary:
A safety factor this low (1.066) is absolutely unacceptable without any reinforcement. Reserve strength must be provided.

If this slab's loading stays below the design value of 2500 psf, it is unlikely to crack due to applied loads. However, if the actual load reaches or exceeds the design value, then hairline cracks in the slab will occur. These are load cracks that will stay tight due to the presence of the reinforcing steel.

6.8 — Use of fibers for concrete enhancement

Although many kinds of fibers, including glass and natural material, have been used to improve the properties of concrete, the important categories for fiber enhanced concrete for slabs on ground are steel fiber and synthetic (polymeric) fiber.

It is recommended that the actual modulus of rupture used for design be measured by beam tests, following ASTM C 1018, for the trial mix including the intended dosage of steel fibers.

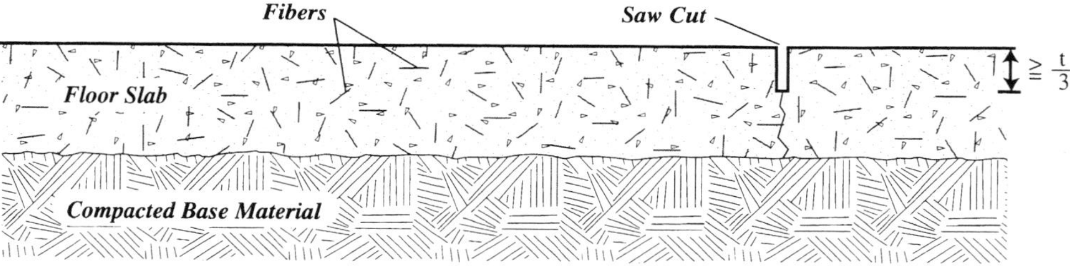

Figure 40 *Slab on grade reinforced with steel fibers for crack control.*

These are the types of fiber enhanced concretes that are discussed in this section. Caution is recommended in referring to the use of fibers as reinforcement because dosages are seldom sufficient to critically alter shear and moment capacity of a given concrete section. The concrete properties that are most positively affected by the presence of fibers are toughness and fatigue. Other improvements include control of plastic shrinkage cracking and abrasion resistance.

Since fiber enhanced concrete is still in the research and development stage, it would be advisable for the designer to keep abreast of developments and specifications. Specifically, reference to the publications of ACI Committee 544 (Fiber Reinforced Concrete) (*Reference 24*) is encouraged.

Some form of cracking must occur in the concrete before the steel fibers become effective. The designer must select the appropriate value of MOR, *taking into account any differences between the* MOR *at first visible crack and the* MOR *at final rupture of the reinforced beam when tested.*

An important physical characteristic of fibers in fiber reinforced concrete is aspect ratio, the relationship of the diameter or thickness of a fiber to its length. This ratio varies from an approximate low of 30 for some steel fibers to 300 for certain synthetic fibers.

6.8.1— Steel fiber

Steel fibers used in concrete are discontinuous and range in length from 0.25 inches to 3.0 inches. Today most steel fibers are deformed, hooked, or crimped. They are made of either carbon or stainless steel, and their tensile strength ranges from 50 to 200 ksi. The aspect ratio (length-diameter ratio) of steel fibers ranges from 30 to 150 or more. *Figure 41* shows various shapes of steel fibers.

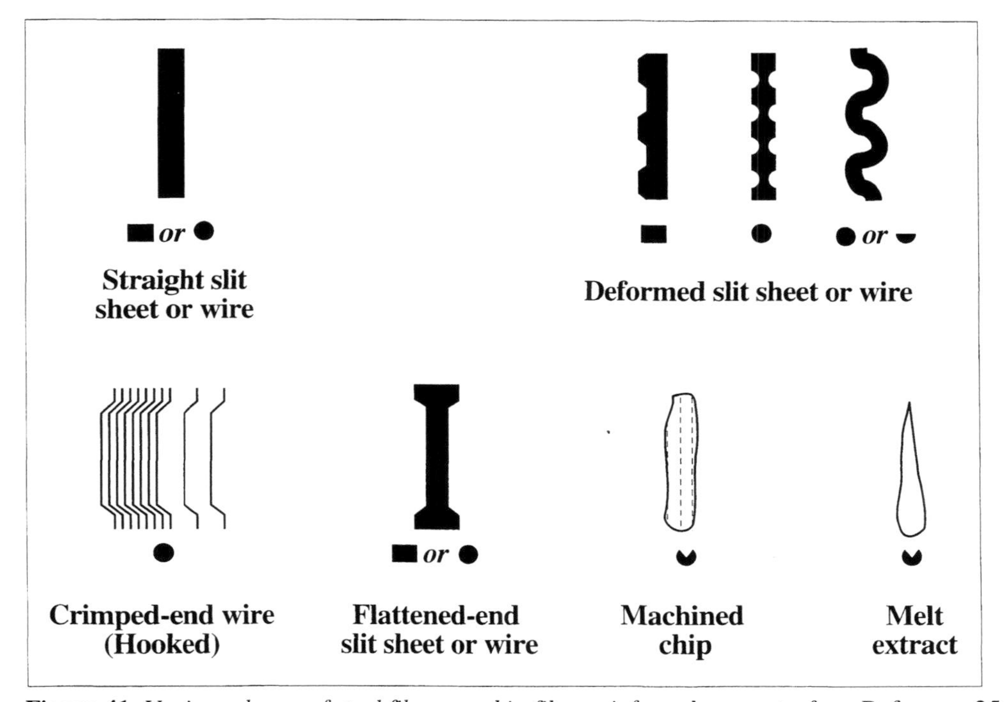

Figure 41 *Various shapes of steel fibers used in fiber reinforced concrete, from* Reference 25.

Fiber concentrations in concrete mixes generally range from 0.1% to 1% by volume. These concentrations, which are generally regarded as low fiber concentrations from a design standpoint, are used to improve shrinkage control and enhance dynamic loading ability.

Slab thickness is determined by the same methods as previously illustrated. The PCA, WRI, and COE charts, as well as cited equations, all work the same way with steel fiber enhanced concrete and synthetic fiber enhanced concrete.

6.8.2— Synthetic (polymeric) fibers

Synthetic (polymeric) fibers, which are the result of research and development in the textile and petrochemical industries, have been steadily increasing in use for concrete slabs on ground. Most of them have very high tensile strength but relatively low modulus of elasticity. Hence their most positive effect on concrete is during the plastic formative stages of concrete hardening. High aspect ratio is a unique quality of polymeric fibers that helps in these early stages. Diameters are in the micrometers, resulting in aspect ratios of 300 and upwards. Table 16 summarizes the physical properties of the most common fibers.

Fiber Type	Eff. dia. $\times 10^{-3}$ in. (10^{-3}mm)	Specific gravity	Tensile strength, ksi (MPa)	Elastic modulus ksi (GPa)	Ultimate elongation
Acrylic	0.5 – 4.1 (13 – 104)	1.17	30 – 145 (207 – 1,000)	2,000 – 2,800 (14.6 – 19.6)	7.5% – 50.0%
Aramid I	0.47 (12)	1.44	525 (3,620)	9,000 (62)	4.4%
Aramid II (high modulus)	0.40 (10)	1.44	525 (3,620)	17,000 (117)	2.5%
Nylon		1.16	140 (965)	750 (5.17)	20.0%
Polyester		1.34 – 1.39	130 – 160 (896 – 1,100)	2,500 (17.5)	
Polyethylene	1.0 – 40.0 (25 – 1,020)	0.96	29 – 35 (200 – 300)	725 (5.0)	3.0%
Polypropylene		0.90 – 0.91	45 – 110 (310 – 760)	500 – 700 (3.5 – 4.9)	15.0%

Table 16 *Physical properties of polymeric fibers, from* Reference 25.

The primary synthetic fibers used in concrete slabs on ground are polypropylene and nylon. (Polypropylene may be the most widely used polymeric fiber for slab enhancement in the United States.) The fibers are either fibrillated or monofilament. Fibrillated fibers are extruded polypropylene sheets that are stretched and slit, producing interconnected fiber strands that open during concrete mixing. The fibrillated fibers are generally cut into lengths ranging from ¼ to 2¼ inches.

Monofilament fibers are smooth and have a relatively smaller surface area than fibrillated fibers, therefore their mechanical anchoring is not as strong as that of fibrillated fibers. This limits their benefits in plastic shrinkage control.

Nylon fibers are made of Nylon 6 and are monofilament. The distinct difference of nylon from both polypropylene and polyester fibers is that it is hydrophilic. Both polypropylene and polyester fibers are hydrophobic. This means that nylon has both a chemical and mechanical bond to the concrete, while polypropylene and polyester have mechanical bonds only. Whether this hydrophilic nature has a long-term negative effect on the mechanical bond is not currently known.

A typical dosage of synthetic fibers is 1.5 pounds per cubic yard. Fibers generally come in 1.5 or 1.6 pound bags that can be charged directly into the hopper of a ready mix truck.

Synthetic fibers primarily control plastic shrinkage cracking, and they aid in enhancing the properties of ready mix concrete. Therefore, the fiber volume can be kept around 1.5 to 3 pounds per cubic yard (0.1% to 0.2%) and still be of use. Synthetic fibers should not be considered as a source of primary reinforcement.

6.8.3—Joint treatment

For dosages of steel fibers under 130 lb per cubic yard (1%), joint spacing may be slightly greater than that of equivalent conventionally reinforced concrete. Higher dosages will not eliminate jointing or cracking but will instead hold cracks tighter. If joints are not provided in steel fiber enhanced slabs having dosages as high as 65 to 125 pounds per cubic yard, a major crack may appear approximately every 150 feet. In sawing joints, saw cuts should be made to a depth of $t/3$ in lieu of the more conventional $t/4$ (*see Figure 40*). If load transfer is necessary at joints, use dowels, transfer plates, or adequately designed tongue and groove details at construction joints.

Some think that contraction joint spacing can be extended from 50% to 100% when using 1.5 to 3 pounds per cubic yard (0.1% to 0.2% of synthetic fibers by volume). A designer would be prudent to treat this consideration conservatively. An increase (up to 40 times the slab depth) may be considered, but tight restraints should remain on the water-to-cement ratio.

6.8.4— Compatibility

Both steel fibers and synthetic fibers have been quite helpful in controlling shrinkage cracking in post-tensioned floors prior to tensioning of the tendons. The slab benefits from the toughness associated with the multi-directional fibers.

One of the prime advantages of fiber enhanced concrete is the improved performance under dynamic loading. Although shrinkage can be controlled, shrinkage cracks cannot be totally eliminated. The advantages of combining fibers with other elements, such as conventional reinforcing, post-tensioning, and shrinkage compensating concrete can be seen.

Improved shrinkage crack control and increased dynamic loading capabilities are the distinctive features of fiber enhanced concrete. These factors do not necessarily affect the thickness of a slab or the use of recognized thickness design procedures.

Both steel and synthetic fibers can be used to enhance the qualities of ready mix concrete, but they must be taken for what they are: An additional tool available to the designer to help fulfill the needs of a client when used properly.

Design example: Use the following values as presented in *Example A* of Section 6.7. In that example, a 10-inch slab was selected for an applied moment of 5700 foot-pounds per foot, caused by a uniform load of 2500 psf and an aisle width of 90 inches. The modulus of rupture (*MOR*) was 570 psi and the safety factor, 1.70.

- For this steel fiber example, assume the beam tests show a *MOR* of 850 psi.
- Using the same factor of 1.70, the allowable bending stress is 850/1.70 = 500 psi.
- The required section modulus of the slab is then (5700 × 12)/500 = 136.8 cubic inches per foot of slab.
- $t = \sqrt{136.8/2} = 8.3$ inches
- The required slab thickness calculates to 8.3 inches instead of the previous 10 inches. A design thickness of 8 1/2 inches would be recommended for use.

Commentary:
Some form of cracking must occur in the concrete before the steel fibers become effective. The designer must select the appropriate value of MOR, *taking into account any differences between the* MOR *at first visible crack and the* MOR *at final rupture of the reinforced beam when tested.*

CHAPTER 7
DESIGN OF SLABS SUPPORTED ON PLASTIC CLAYS

7.1 — Introduction

In most slab-on-grade designs the soil supports the floor and resists slab movement, thereby supporting the loads on the slab. Typically the soil does not initiate movement and produce loading on the slab. However, in plastic soils, changes in moisture content can cause volume changes in the soil sufficient to make the soil itself move. These moisture changes cause a loading on the slab resulting from the soil either pushing on or receding from the slab. Since this is an overall motion caused by a large volume of soil, it affects the entire slab, not just a localized area. Therefore it is necessary to design the slab as an integral unit, including the stiffening beams that will most likely be required.

The process is mathematically complicated because of the climatic factors and soil composition that influence the volume change of the soil. Solution of the soil motion part of the problem is empirical and complex. The structural part of the design is conventional and not difficult.

To better understand the design example, study the two flow charts provided on pages 72 and 73. *Flow Chart 1* goes through the necessary steps in slab design. *Flow Chart 2* is a subroutine to Step 4 of *Flow Chart 1*. You should use it when information from the soils specialist is incomplete. The design equations (from *Reference 10*) supporting *Flow Chart 1* are listed sequentially in the appendix, page 232. The equations taken together with the flow chart steps can be used as the basis for a computer solution to this design problem.

7.2 — Plastic clay conditions

Plastic clays are fine-grained soils that have a high potential to shrink and swell. They generally have a liquid limit, *LL*, of 20% and a plasticity index, *PI*, of 10% or greater. When working with soils having properties in this range, it is quite possible that problem plastic clays may be involved. Further information should be sought from a qualified geotechnical engineer.

For slabs supported on clay, be wary of k values (modulus of subgrade reaction) on the order of 100 or greater unless specifically supported by the geotechnical engineer. Long term loading of a clay that is 3 feet deep or deeper could result in an actual value of k that is not reflected in a plate test because of the clay's plastic nature.

If the clay shows a potential for volume change of 4 inches or more vertically, serious consideration should be given to either:

- A pile- or pier-supported structural slab built over collapsible voids
- Changing the base material through lime stabilization
- Removal of the problem soil

With lower potential volume changes, a stiffened ribbed slab like the one in the following example is a reasonable solution.

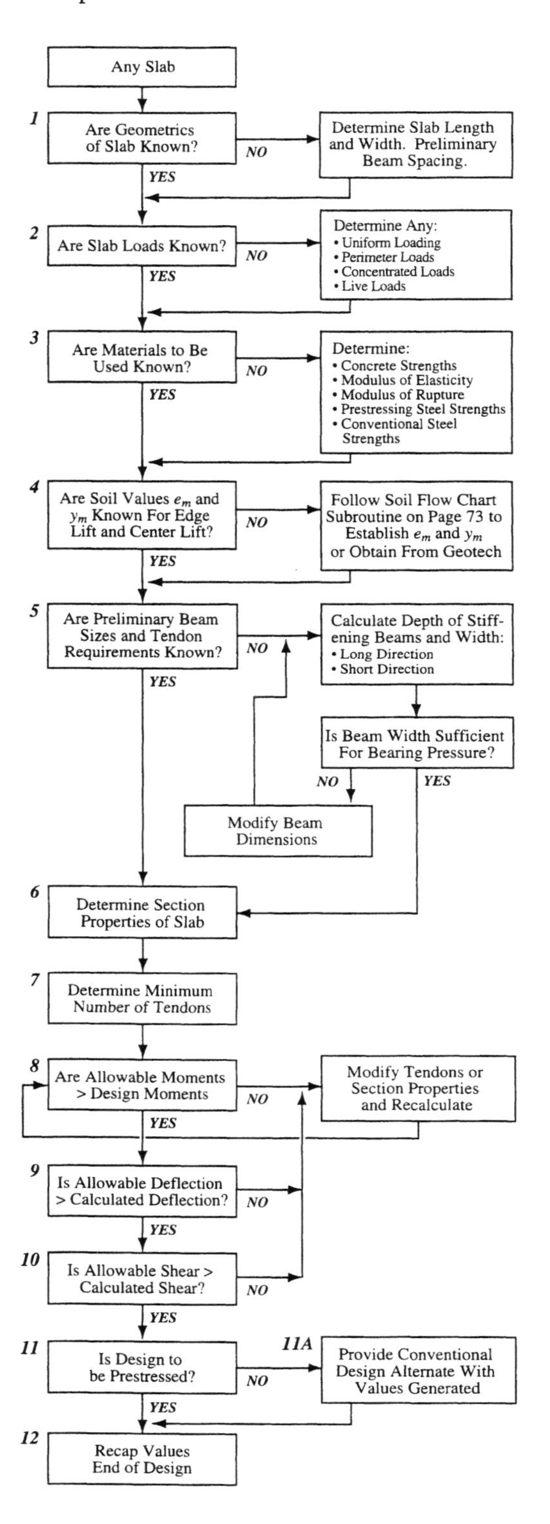

Flow Chart 1 *Procedure for design of slabs on plastic clays.*

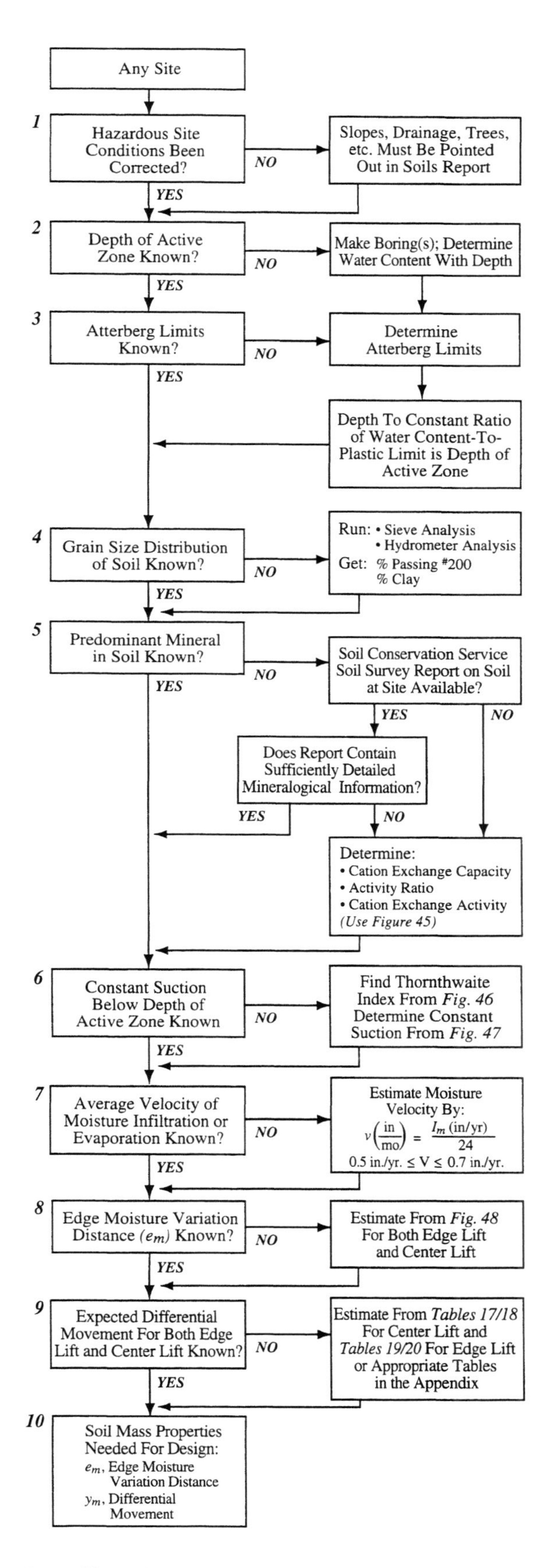

Flow Chart 2 *Subroutine for determination of soil properties, from the PTI Manual* (Reference 10).

7.3 — The PTI method for slab design on plastic clay

The Post-Tensioning Institute (PTI) sponsored research at Texas A & M University, resulting in a book, *Design and Construction of Post-Tensioned Slabs-on-Ground (Reference 10)*. This publication, which the authors refer to as the PTI Manual, sets forth a procedure for stiffening a slab with interior grade beams in order to control differential deflection and movement associated with shrink-swell potential. The PTI method is followed in the example given in this chapter.

Due to the enhanced T-beam action that is characteristic of stiffened or ribbed post-tensioned slabs on ground, the most common approach to the problem has been a prestressed post-tensioned design. It can be seen in the design example that the entire slab is analyzed as a single monolithic unit. This is valid since the slab is post-tensioned in both directions. This results in modeling the slab as a large T-beam, which gives a relatively high neutral axis.

Post-tensioning is not the only solution that can be obtained following the PTI procedure. With sufficient reinforcing steel in the slab and stiffening beams, a conventionally reinforced stiffened slab is a viable solution. Fibers have also provided the sole reinforcement in some stiffened slab-on-grade applications.

Because of the way the PTI Manual is set up, examples here follow the path of least resistance, first doing a post-tensioned slab design, then converting to an equivalent conventionally reinforced counterpart.

7.3.1 — Analysis of slab loads

For design purposes the slab on ground on plastic clay can generally be considered as uniformly loaded overall, with an added perimeter wall load. Concentrated loads or point loads can be treated by separate analysis. Uniform loading is generally determined by summing all live and dead loads and considering them applied uniformly over the entire slab. This is reasonable since there are stiffening beams throughout which permit even distribution. The stiffening beams are generally located on a grid, with spacing not to exceed 20 feet both ways *(Figure 42)*.

Commentary:
In contrast to other loading analyses presented in previous examples, the design of slabs on plastic clay treats the entire slab, not just an isolated section. This is because the stresses that occur affect the whole unit, not a localized area. Primary stresses in the slab result from soil shrinkage and swelling that affect the entire slab as an integral structure, and this is why stiffening ribs are needed.

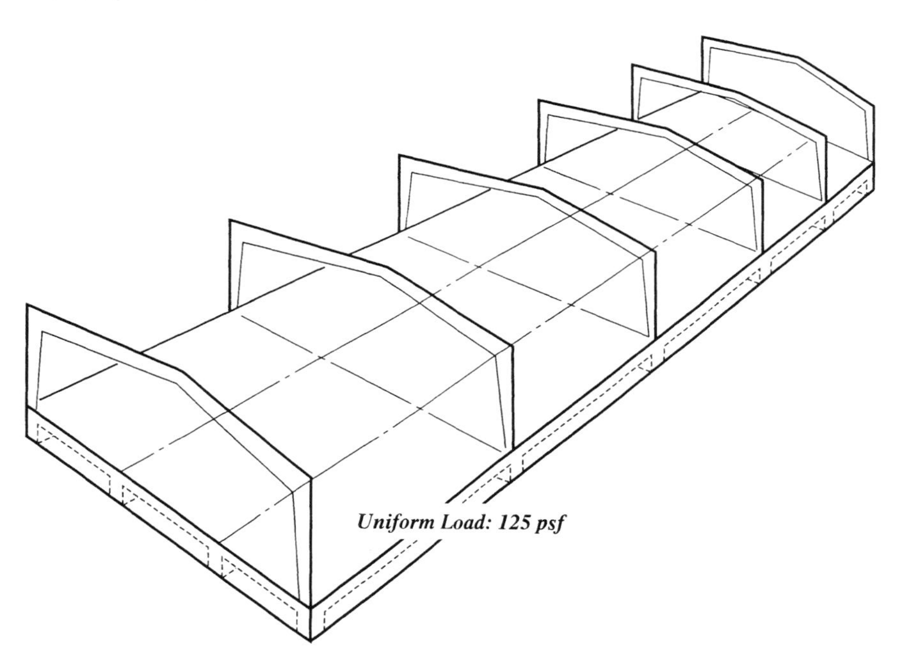

Figure 42 *Stiffening beams not farther than 20 feet apart each way distribute loads and cause slab to work as an overall unit. The rigid frame or other building structure places a perimeter load on the slab.*

7.3.2 — Values needed to solve the problem

In dealing with a ribbed and stiffened slab, four primary areas must be considered:

- Slab geometry, including preliminary beam spacing
- Loads on the slab
- Materials to be used in the slab
- Soil conditions

These are shown as Items 1 through 4, *Flow Chart 1* (page 72). If soil values e_m (edge moisture variation distance) and y_m (differential movement) are not known, then several steps in *Flow Chart 2* must also be followed.

Commentary:
Since clay is a fairly elastic material whose properties vary widely depending on moisture content and clay content of the soil, significantly more soils information is required to define the problem.

7.3.3 — Design objectives

The design presented in the following example checks the ability of the slab to resist both positive and negative bending moments associated with the shrink-swell potential of the soil. The slab can take either a domed or dished shape *(Figure 43)*, depending on the relative shrinkage or swelling experienced across the edge moisture variation distance e_m.

Of equal if not greater importance in design is the determination of adequate stiffness. Field experience indicates that sufficient stiffness in the slab design cannot be overemphasized.

The edge moisture variation distance is determined from the basic input data. It is the distance measured inward from the slab edge, over which the soil moisture content varies. Climate is the most important factor affecting this distance. The two soil- and climate-dependent values needed to proceed with slab design are e_m *and* y_m*. When these values are supplied by the soils specialist, it greatly simplifies the problem for the slab designer.*

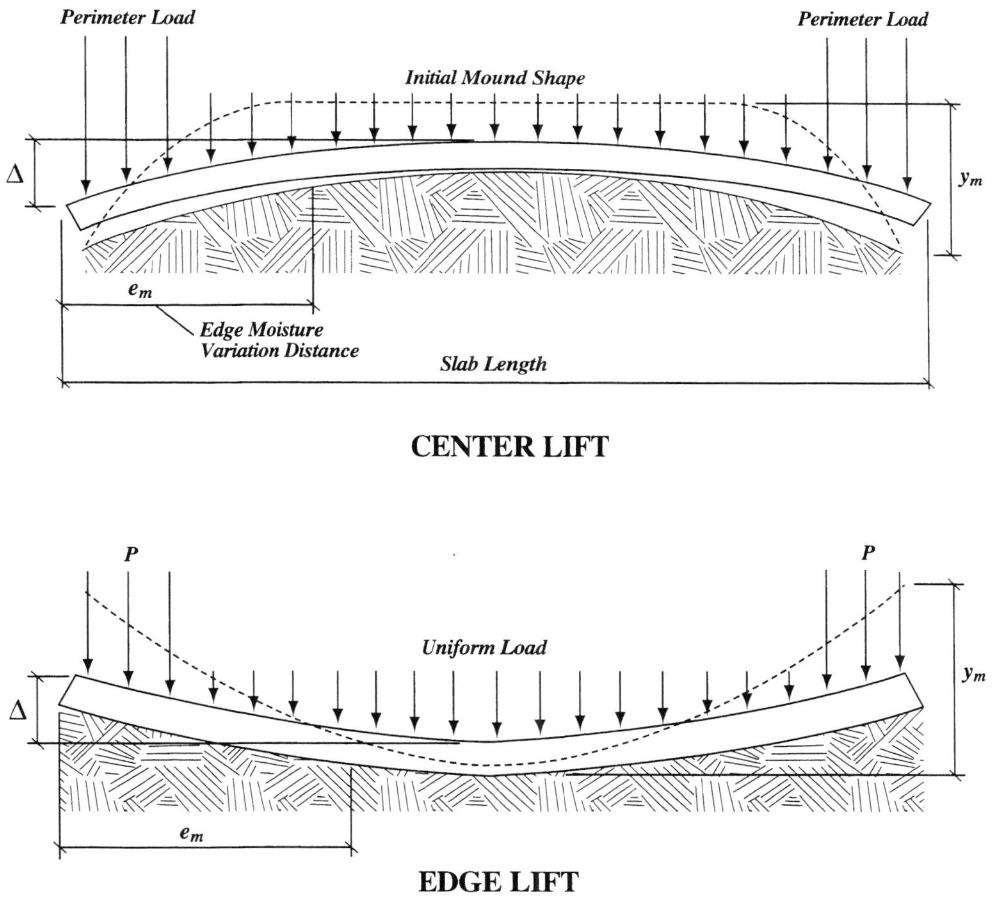

Figure 43 *Soil-structure interaction models, from the PTI Manual* (Reference 10). *The domed shape is called "center lift condition," and the dished shape is the "edge lift condition."*

The slab design example presented here follows the flow chart sequence, and references by number the applicable steps in the charts. For convenience of the user, all PTI formulas and charts required for this design case are presented in the appendix *(page 232)*. Only those needed for the sample problem are given in the text of the example. Charts in the appendix are generally larger and therefore should be used for repeated reference in solving other problems. The smaller chart versions in the example are for quick reference in understanding the step-by-step solution.

7.3.4 — Computer solutions

A number of vendors, including the Post-Tensioning Institute, have developed computer programs for producing these PTI solutions. The flow chart for this chapter provides a good outline for developing a computer program for the plastic clay solution. The detailed problem that follows provides all formulas needed to develop either simplifying subroutines for hand-held calculators or a full-scale computer program. If this slab design procedure is used on a regular basis, purchase of one of the existing programs or the preparation of either a program or spread sheet is highly recommended.

7.4 — DESIGN EXAMPLE: Rectangular post-tensioned slab on plastic clay with uniform and perimeter loads

7.4.1—Symbols and notation

A area of gross concrete cross section, square inches

A_o a co-dependent variable used in factoring center lift moment design, dependent on the physical properties of the slab, loading conditions, and soil properties

Ac activity ratio of clay

b width of an individual stiffening beam, inches

B a nondimensional constant used in factoring center lift design, dependent on soil properties

C a nondimensional constant used in factoring center lift design, dependent on loading condition and soil properties

$CEAc$ cation exchange activity of soil

d depth of stiffening beam measured from top surface of slab to bottom of beam, inches

e eccentricity of post-tensioning force, inches

e_m edge moisture variation distance, feet

E_c long-term or creep modulus of elasticity of concrete, psi

E_s modulus of elasticity of soil, psi

f_c allowable compressive stress in concrete, psi

f_c' 28-day compressive strength of concrete, psi

f_{ps} permissible stress in prestressing tendon, psi

f_{pu} ultimate stress in prestressing tendon, psi

f_t allowable tensile stress in concrete, psi

I gross moment of inertia, in.^{4}

I_m Thornthwaite index, moisture velocity in inches per year

L total slab length in the direction being considered, feet

LL liquid limit
M_ℓ design moment in long direction, foot-kips per foot
M_s design moment in the short direction, foot-kips per foot
${}_nM_t$ ${}_nM_c$ negative and positive bending moments including
${}_pM_t$ ${}_pM_c$ tension or compression in the extreme fibers, foot-kips per foot
n number of beams in a cross section
N modular ratio (modulus of elasticity of steel to modulus of elasticity of concrete)
N_t number of tendons
pF constant soil suction value
P perimeter loading on the slab, pounds per foot
P_r prestressing force, kips
P_re moment due to post-tensioning eccentricity, inch-kips
P/A prestress force resulting from tendon load divided by gross concrete area, psi
PI plasticity index
PL plastic limit
q_{allow} allowable soil bearing pressure
q_u unconfined compressive strength of soil, psf
S beam spacing, feet
S_B section modulus with respect to the bottom fiber, in.^{3}
S_T section modulus with respect to the top fiber, in.^{3}
t slab thickness, inches
v design shear stress, psi
v_c allowable concrete shear stress, psi
V design shear force, kips per foot
V_s expected shear force in short direction, kips per foot
V_ℓ expected shear force in long direction, kips per foot
w soil bearing pressure, kips per square foot
W slab width, feet
W_{slab} slab weight, pounds
$\bar{y}$ neutral axis location of stiffened cross section, inches
y_m maximum differential soil movement, inches
Z depth to constant suction, feet
β relative stiffness length, feet
Δ expected differential deflection under service load, inches
Δ_{allow} allowable differential deflection of slab, inches
μ coefficient of friction

7.4.2 — The problem and initial assumptions; materials data

A single-story rigid frame metal building in Lafayette, Louisiana *(Figure 44)*, has a perimeter wall load of

900 pounds per linear foot

The concentrated load from the rigid frame has been included in this perimeter load value. Uniform floor loading on the slab is

125 pounds per square foot

The slab measures 100 x 50 feet, and the assumed spacing of stiffening beams is 20 feet in one direction, 16 feet 8 inches in the other. (NOTE: These are values required by Steps 1 and 2 of *Flow Chart 1.*)

Commentary:
You can run an analysis of the transformed section of the edge beam to show whether the concentrated load of the rigid frame can be satisfactorily distributed along its length from point to point. A uniform load is a reasonable assumption in many instances. If concentrated loads are questionable, they should be checked and accommodated with a modified section.

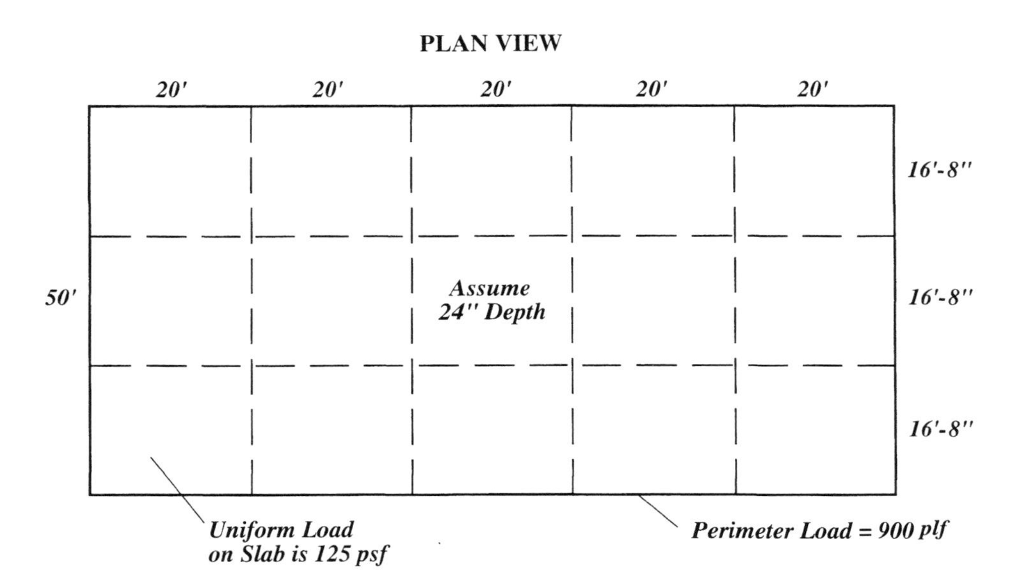

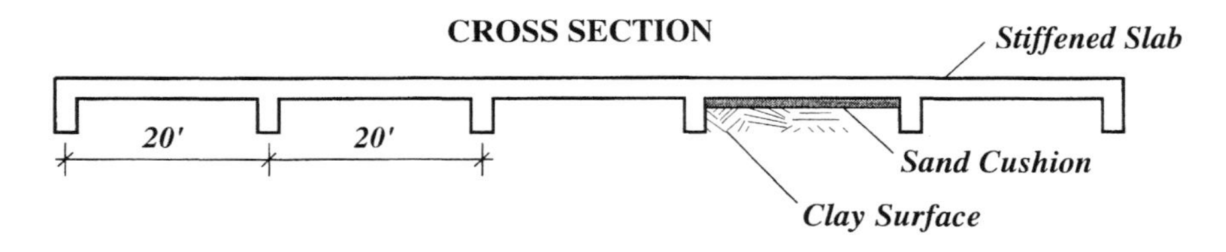

Figure 44 *Plan and cross section of the floor slab design example.*

If a soil is extremely stable and does not necessarily require a plastic design analysis, a stiffened design may still be desirable. In such instances, stiffening beams may be placed to accommodate rigid frames, bearing partitions, or other elements which may be spaced more than 20 feet apart.

The beam spacing should not exceed 20 feet as a matter of standard practice. This has governed the spacing selected here where rigid frames are 20 feet apart in the long direction of the slab. The 16 foot 8 inch dimension is chosen as the largest equal division of 50 feet that comes within the 20-foot limit. Where rigid frame bay spacing exceeds 20 feet, an intermediate beam is recommended in most instances except for extremely stable soil conditions. Then it is sometimes a reasonable judgment call to permit extended beam spacing.

The materials to be used (Step 3 of *Flow Chart 1*) are as follows:

Concrete compressive strength $f_c' = 3000$ psi
Concrete creep modulus of elasticity $E_c = 1{,}500{,}000$ psi
Prestressing steel: 270 k, 1/2-inch-diameter 7-wire strand

7.4.3 — Soils investigation

The design procedure requires determination of the amount of climate-controlled differential movement of the expansive soils, Step 4, *Flow Chart 1*. We have some of the soils information, as follows:

Atterberg limits:
Plastic limit $PL = 30$
Liquid limit $LL = 70$
Plasticity index $PI = 40$
Clay content = 65%
Unconfined compressive strength $q_u = 3000$ psf

Soil modulus of elasticity E_s = 1000 psi
Depth to constant suction Z = 5 feet
Location: Lafayette, Louisiana

There is not enough information to complete Step 4, *Flow Chart 1*, so we must go to the soils subroutine in *Flow Chart 2* (page 73). Assuming there are no hazardous site conditions, there is enough information to satisfy Steps 2, 3, and 4, *Flow Chart 2*. Then proceed with additional steps outlined in *Flow Chart 2*.

7.4.3.1 — Determine the predominant clay mineral in the soil

This is Step 5, *Flow Chart 2*. Using the known values of plastic limit and percent of clay, determine the cation exchange activity, *CEAc*.

$CEAc = PL^{1.17} \div$ percent clay
$CEAc = 30^{1.17} \div 65$
$CEAc = 53.48 \div 65 = 0.82$

Then find the clay activity ratio *Ac*, using known value of plasticity index, *PI*.

$Ac = PI \div$ percent clay
$Ac = 40 \div 65 = 0.62$

With these two values, we can enter the clay classification chart *(Figure 45)* to determine the predominant clay mineral in the soil. Enter the chart from the bottom, drawing a vertical line through $Ac = 0.62$. Draw a horizontal line through $CEAc = 0.82$. The two lines intersect in the area labeled "montmorillonite," indicating that montmorillonite is the principal clay mineral.

Commentary:
It has been observed that if the plastic limit (PL) *of the soil is above 25%, illite and kaolinite will be removed from consideration, and montmorillonite can be assumed for design.*

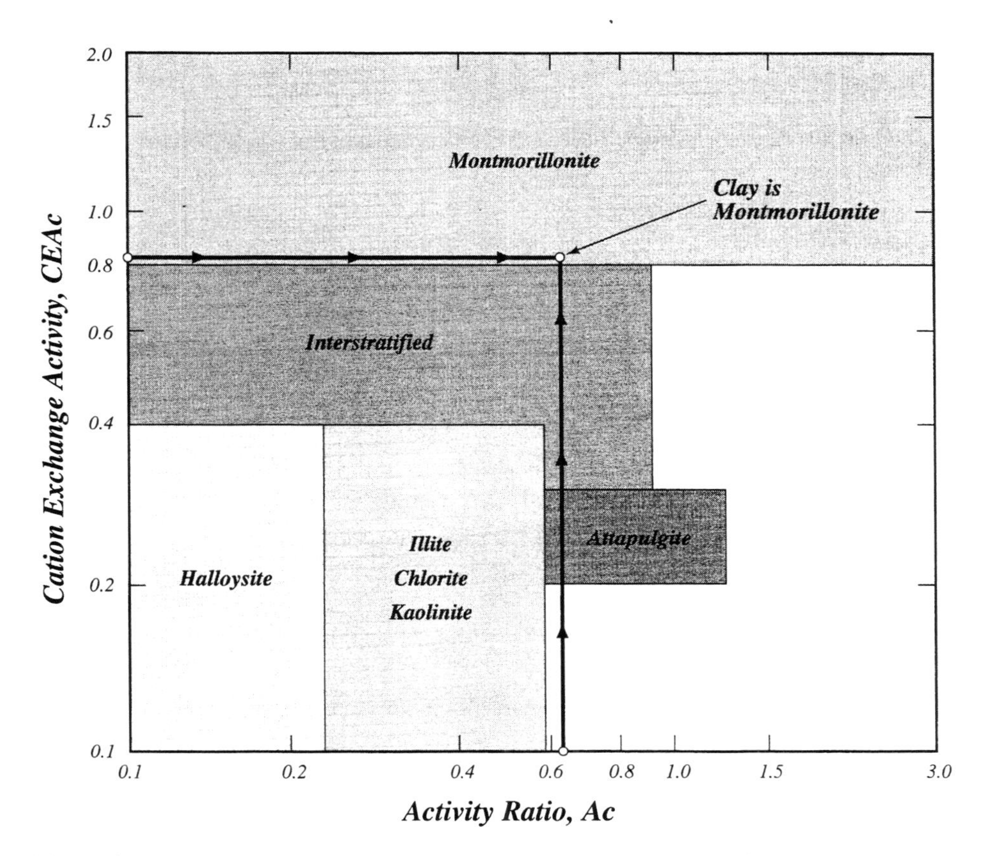

Figure 45 *Clay type classification related to cation exchange activity and clay activity ratio, from the PTI Manual* (Reference 10).

7.4.3.2 — Find the constant suction value for the soil

Commentary:
It is always more conservative to choose the lower number in selecting a Thornthwaite moisture index.

This is Step 6, *Flow Chart 2*. The constant suction value is needed for Step 9 below. First it's necessary to select a Thornthwaite moisture index from the map of *Figure 46*. By interpolation the moisture index $I_m = +30$.

Enter the chart of *Figure 47* at bottom with the Thornthwaite index value of +30. Draw a vertical line to the intersection with the curve, then move left to read the soil suction value pF as 3.2.

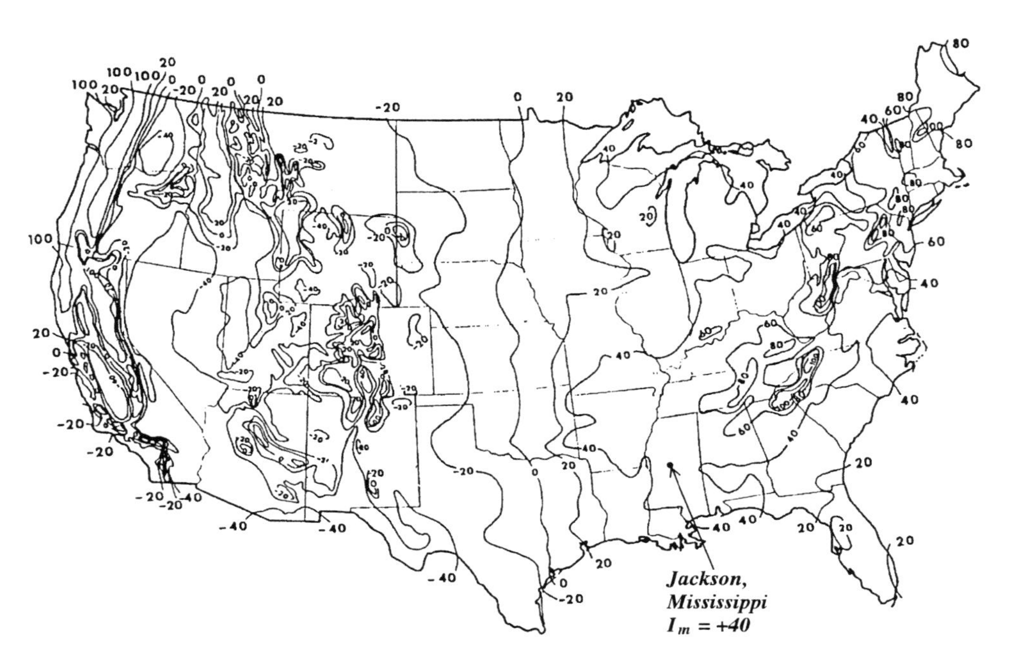

Figure 46 *Thornthwaite moisture index distribution in the United States, from the PTI Manual* (Reference 10).

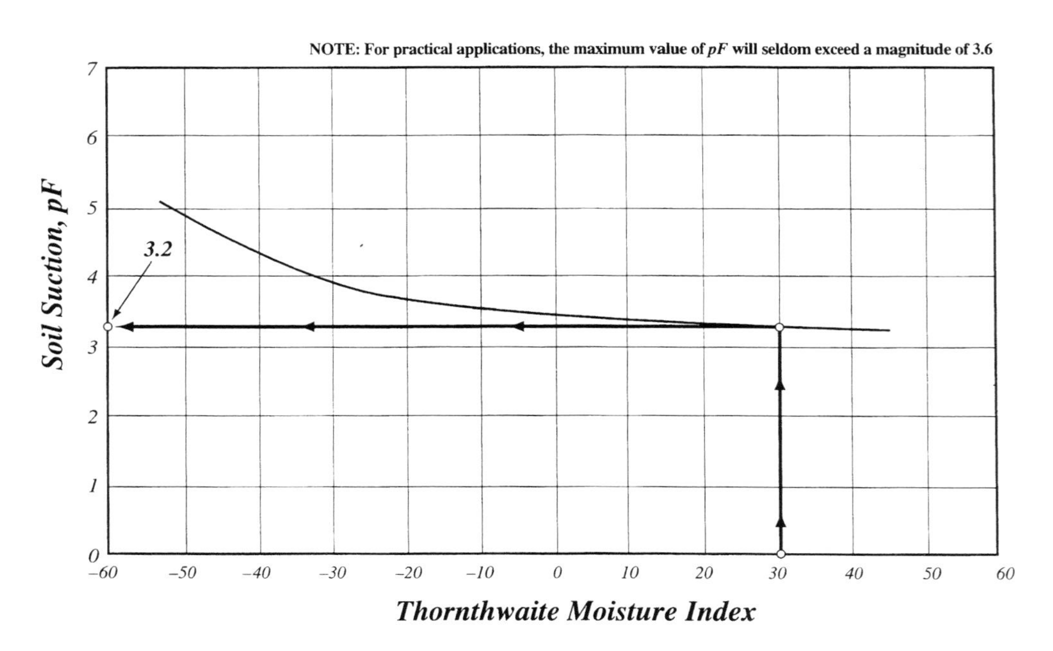

Figure 47 *Variation of constant soil suction with Thornthwaite moisture index, from the PTI Manual* (Reference 10).

7.4.3.3 — Determine the average moisture movement velocity

This is Step 7, *Flow Chart 2*. The estimated velocity of moisture flow is calculated using the Thornthwaite moisture index I_m of +30 obtained in the previous step.

$$\text{moisture velocity} = 0.5 \times I_m/12$$
$$= 0.5 \times 30/12 = 1.25 \text{ inches per month}$$

However, according to the PTI procedure *(Reference 10)*, the maximum moisture velocity shall be 0.7 inches per month. Therefore use 0.7 inches per month in this problem.

Commentary:
The example uses a value midway across the band (Figure 48), *but if the designer is extremely familiar with the local conditions, the chart bands allow some leeway for interpretation suited to site conditions. For example, if a particular vicinity is known to be a cause for concern, be more conservative by moving to values higher in either of the bands. To be less conservative, move to values closer to the bottom of the band.*

7.4.3.4 — Find edge moisture variation distance

With data given and developed in previous steps, we can now go to Step 8, *Flow Chart 2*. With the Thornthwaite moisture index of +30 determined in Section 7.4.3.2, enter the chart of *Figure 48* at the bottom, and draw a vertical line to the middle of the center lift band, then proceed horizontally to the left to read:

e_m, edge moisture variation distance = 3.8 feet (center lift)

Then continue the vertical line to the middle of the band for edge lift condition and again proceed horizontally to the left, reading:

e_m, edge moisture variation distance = 5.2 feet (edge lift)

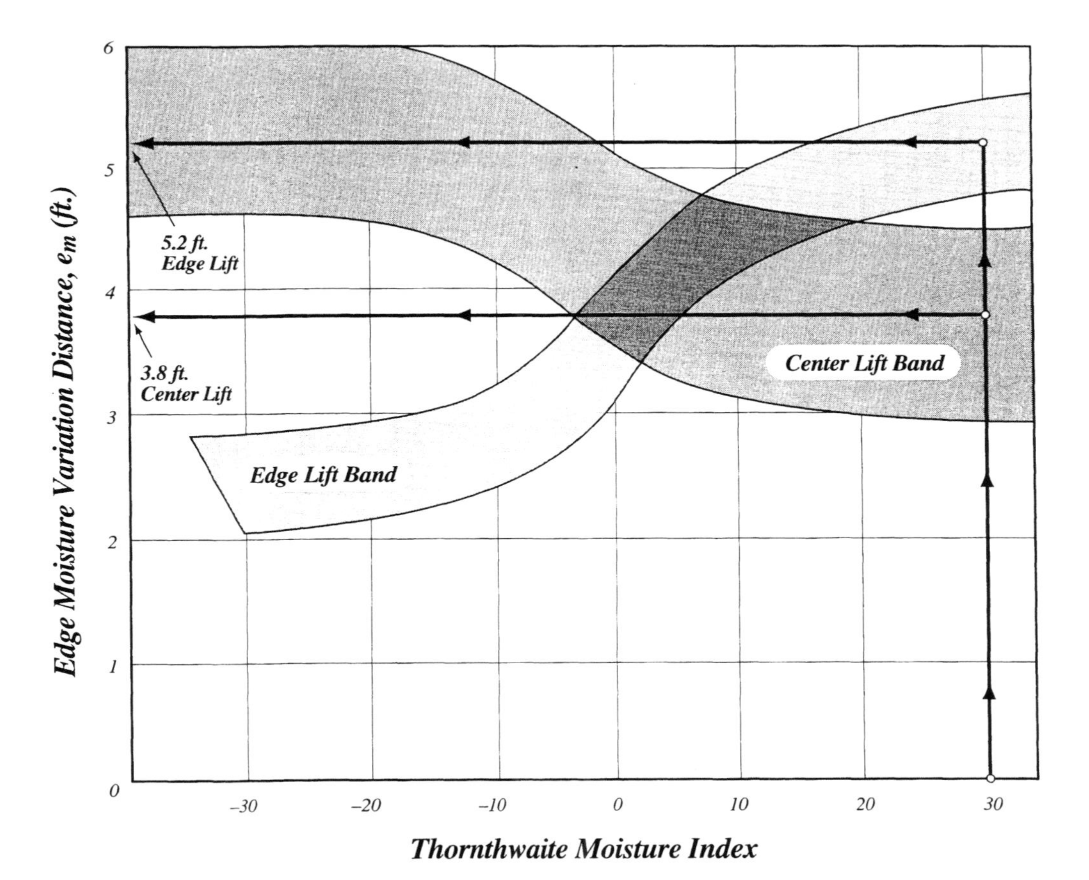

Figure 48 *Approximate relationship between Thornthwaite index and edge moisture variation distance. Note that extremely active clays may generate larger values of edge moisture variation than reflected by the above curves and related tables. Therefore these curves should be used only in conjunction with a site-specific soils investigation by knowledgeable geotechnical engineers (from* Reference 10).

7.4.3.5 — Determine expected differential swell for edge lift and center lift conditions

We now have all the supporting values to do Step 9, *Flow Chart 2*.

Center lift:

Since the soil is 65% clay, interpolate between *Table 17* for 60% clay and *Table 18* for 70% clay.

For $Z = 5$ feet, *moisture velocity* = 0.7 inches per month, $pF = 3.2$, and $e_m = 3.8$, we find:

$$y_m = 0.224 \text{ inches}$$

Edge lift:

Again interpolate between *Table 19* for 60% clay and *Table 20* for 70% clay.

For $Z = 5$ feet, *moisture velocity* = 0.7 inches per month, $pF = 3.2$, and $e_m = 5.2$, we find:

$$y_m = 0.23 \text{ inches}$$

Commentary:

As mentioned previously, it would be important for the slab designer to have e_m and y_m values supplied by an expert soils authority. Without this kind of site-specific information, calculations must be simplified in the conservative direction. Always assume that the clay mineral is montmorillonite. Use a conservative value of depth Z to constant suction. If the clay percentage varies, using the higher percentage is more conservative.

Percent Clay (%)	Depth to Constant Suction (ft.)	Constant Suction (pF)	Velocity of Moisture Flow (in./mo.)	Differential Swell (in.) — Edge Distance Penetration (ft.) 1 ft.	2 ft.	3 ft.	4 ft.	5 ft.	6 ft.	7 ft.	8 ft.
60	3	3.2	0.1	0.003	0.006	0.009	0.013	0.016	0.019	0.022	0.026
			0.3	0.009	0.019	0.028	0.038	0.048	0.059	0.069	0.080
			0.5	0.015	0.031	0.048	0.065	0.082	0.101	0.120	0.140
			0.7	0.022	0.044	0.068	0.092	0.119	0.147	0.176	0.209
		3.4	0.1	0.006	0.013	0.019	0.026	0.033	0.040	0.047	0.054
			0.3	0.019	0.039	0.060	0.082	0.105	0.130	0.157	0.186
			0.5	0.031	0.065	0.102	0.143	0.189	0.240	0.300	0.372
			0.7	0.044	0.093	0.147	0.211	0.286	0.380	0.503	0.683
		3.6	0.1	0.015	0.031	0.048	0.065	0.083	0.102	0.122	0.144
			0.3	0.044	0.095	0.152	0.219	0.300	0.403	0.543	0.758
			0.5	0.073	0.161	0.272	0.420	0.645	1.056	2.037	4.865
			0.7	0.101	0.229	0.407	0.689	1.246	2.689	6.912	—
	5	3.2	0.1	0.008	0.015	0.023	0.030	0.038	0.045	0.053	0.060
			0.3	0.022	0.044	0.067	0.090	0.113	0.138	0.162	0.187
			0.5	0.037	0.074	0.113	0.153	0.194	0.237	0.282	0.329
			0.7	0.050	0.103	0.158	0.217	0.278	0.343	0.412	0.487
		3.4	0.1	0.015	0.030	0.043	0.061	0.077	0.093	0.109	0.126
			0.3	0.045	0.091	0.140	0.191	0.245	0.302	0.363	0.426
			0.5	0.073	0.152	0.237	0.331	0.435	0.551	0.686	0.846
			0.7	0.102	0.214	0.341	0.485	0.655	0.865	1.140	1.541
		3.6	0.1	0.035	0.071	0.109	0.149	0.190	0.233	0.279	0.326
			0.3	0.103	0.219	0.349	0.499	0.678	0.901	1.200	1.652
			0.5	0.169	0.370	0.618	0.945	1.425	2.280	4.284	9.923
			0.7	0.234	0.526	0.922	1.528	2.686	5.574	—	—
	7	3.2	0.1	0.013	0.027	0.041	0.055	0.069	0.083	0.097	0.111
			0.3	0.041	0.082	0.125	0.168	0.212	0.256	0.302	0.349
			0.5	0.068	0.137	0.209	0.283	0.360	0.441	0.524	0.612
			0.7	0.093	0.191	0.294	0.402	0.516	0.637	0.767	0.907
		3.4	0.1	0.027	0.055	0.083	0.112	0.142	0.171	0.201	0.232
			0.3	0.082	0.167	0.256	0.351	0.449	0.555	0.666	0.786
			0.5	0.135	0.280	0.438	0.609	0.799	1.013	1.260	1.553
			0.7	0.188	0.395	0.627	0.892	1.204	1.587	2.092	2.840
		3.56	0.1	0.053	0.107	0.163	0.221	0.281	0.342	0.407	0.474
			0.3	0.156	0.326	0.512	0.720	0.957	1.232	1.566	1.994
			0.5	0.256	0.549	0.895	1.320	1.879	2.702	4.182	8.216
			0.7	0.354	0.779	1.317	2.059	3.247	5.677	—	—

Table 17 *Center lift condition, differential swell at the perimeter of a slab in predominantly montmorillonite clay soil (60 percent clay) from* Reference 10.

Percent Clay (%)	***Depth to Constant Suction (ft.)***	***Constant Suction (pF)***	***Velocity of Moisture Flow (in./mo.)***	***Differential Swell (in.)***							
				Edge Distance Penetration (ft.)							
				1 ft.	***2 ft.***	***3 ft.***	***4 ft.***	***5 ft.***	***6 ft.***	***7 ft.***	***8 ft.***
70	3	3.2	0.1	0.004	0.007	0.011	0.015	0.019	0.023	0.026	0.030
			0.3	0.011	0.022	0.034	0.046	0.057	0.070	0.082	0.095
			0.5	0.018	0.037	0.057	0.077	0.098	0.120	0.143	0.167
			0.7	0.026	0.052	0.082	0.110	0.141	0.174	0.210	0.248
		3.4	0.1	0.007	0.015	0.023	0.031	0.039	0.048	0.056	0.065
			0.3	0.023	0.047	0.072	0.098	0.126	0.156	0.188	0.222
			0.5	0.038	0.078	0.122	0.171	0.225	0.287	0.358	0.443
			0.7	0.052	0.110	0.176	0.251	0.341	0.452	0.600	0.814
		3.6	0.1	0.018	0.037	0.056	0.077	0.098	0.121	0.145	0.171
			0.3	0.054	0.114	0.182	0.262	0.359	0.481	0.648	0.903
			0.5	0.088	0.192	0.324	0.502	0.769	1.258	2.428	5.796
			0.7	0.120	0.273	0.485	0.821	1.485	3.203	8.234	—
	5	3.2	0.1	0.009	0.017	0.026	0.035	0.044	0.053	0.062	0.071
			0.3	0.026	0.053	0.080	0.107	0.135	0.164	0.193	0.223
			0.5	0.044	0.088	0.134	0.182	0.231	0.283	0.336	0.392
			0.7	0.060	0.123	0.189	0.258	0.331	0.409	0.491	0.580
		3.4	0.1	0.018	0.036	0.055	0.073	0.092	0.111	0.131	0.151
			0.3	0.053	0.109	0.167	0.228	0.292	0.360	0.433	0.511
			0.5	0.088	0.182	0.284	0.395	0.519	0.658	0.818	1.008
			0.7	0.121	0.256	0.406	0.578	0.781	1.031	1.358	1.837
		3.6	0.1	0.042	0.086	0.131	0.178	0.227	0.278	0.332	0.389
			0.3	0.123	0.260	0.415	0.594	0.807	1.073	1.429	1.968
			0.5	0.202	0.441	0.737	1.126	1.698	2.717	5.104	11.822
			0.7	0.278	0.627	1.098	1.820	3.199	6.640	—	—
	7	3.2	0.1	0.016	0.032	0.049	0.065	0.082	0.098	0.115	0.132
			0.3	0.048	0.097	0.148	0.199	0.251	0.305	0.359	0.415
			0.5	0.081	0.163	0.249	0.338	0.429	0.525	0.624	0.729
			0.7	0.112	0.229	0.351	0.480	0.616	0.759	0.915	1.081
		3.4	0.1	0.032	0.066	0.099	0.134	0.168	0.204	0.240	0.276
			0.3	0.098	0.199	0.306	0.418	0.536	0.661	0.794	0.937
			0.5	0.162	0.334	0.522	0.727	0.952	1.207	1.501	1.851
			0.7	0.224	0.470	0.747	1.063	1.435	1.891	2.492	3.383
		3.56	0.1	0.063	0.128	0.194	0.263	0.334	0.407	0.484	0.563
			0.3	0.185	0.387	0.609	0.857	1.139	1.468	1.865	2.376
			0.5	0.305	0.655	1.067	1.573	2.239	3.219	4.983	9.162
			0.7	0.421	0.928	1.569	2.453	3.868	6.763	—	—

Table 18 *Center lift condition, differential swell at the perimeter of a slab in predominantly montmorillonite clay soil (70 percent clay) from* Reference 10.

Percent Clay (%)	Depth to Constant Suction (ft.)	Constant Suction (pF)	Velocity of Moisture Flow (in./mo.)	Differential Swell (in.) Edge Distance Penetration (ft.)							
				1 ft.	2 ft.	3 ft.	4 ft.	5 ft.	6 ft.	7 ft.	8 ft.
60	3	3.2	0.1	0.003	0.006	0.008	0.011	0.014	0.016	0.019	0.022
			0.3	0.008	0.016	0.025	0.033	0.040	0.048	0.056	0.064
			0.5	0.014	0.027	0.041	0.054	0.066	0.079	0.091	0.104
			0.7	0.019	0.038	0.056	0.074	0.091	0.109	0.125	0.142
		3.4	0.1	0.006	0.012	0.017	0.023	0.029	0.035	0.040	0.046
			0.3	0.018	0.035	0.051	0.068	0.084	0.099	0.115	0.130
			0.5	0.029	0.057	0.084	0.110	0.135	0.160	0.183	0.206
			0.7	0.041	0.079	0.116	0.151	0.184	0.216	0.247	0.277
		3.6	0.1	0.014	0.029	0.043	0.056	0.070	0.083	0.096	0.109
			0.3	0.043	0.083	0.122	0.159	0.194	0.228	0.260	0.291
			0.5	0.071	0.136	0.195	0.251	0.303	0.352	0.399	0.433
			0.7	0.098	0.185	0.264	0.336	0.402	0.463	0.521	0.575
		3.8	0.1	0.035	0.069	0.102	0.133	0.163	0.191	0.219	0.246
			0.3	0.104	0.195	0.277	0.352	0.421	0.484	0.544	0.599
			0.5	0.169	0.309	0.428	0.533	0.627	0.712	0.790	0.863
			0.7	0.233	0.413	0.562	0.690	0.802	0.903	0.994	1.077
	5	3.2	0.1	0.006	0.012	0.018	0.024	0.030	0.036	0.042	0.048
			0.3	0.018	0.036	0.054	0.071	0.089	0.106	0.123	0.140
			0.5	0.030	0.060	0.090	0.118	0.146	0.174	0.202	0.229
			0.7	0.042	0.083	0.124	0.163	0.202	0.240	0.278	0.314
		3.4	0.1	0.013	0.026	0.039	0.051	0.064	0.076	0.089	0.101
			0.3	0.039	0.077	0.114	0.151	0.187	0.222	0.256	0.291
			0.5	0.065	0.127	0.188	0.246	0.303	0.359	0.413	0.465
			0.7	0.090	0.177	0.259	0.339	0.415	0.488	0.559	0.628
		3.6	0.1	0.032	0.064	0.095	0.126	0.156	0.186	0.215	0.244
			0.3	0.096	0.188	0.276	0.360	0.440	0.518	0.593	0.665
			0.5	0.160	0.308	0.446	0.575	0.697	0.812	0.921	1.025
			0.7	0.224	0.425	0.607	0.775	0.931	1.077	1.214	1.343
		3.8	0.1	0.080	0.156	0.230	0.301	0.369	0.436	0.500	0.562
			0.3	0.238	0.450	0.642	0.817	0.980	1.132	1.274	1.407
			0.5	0.395	0.724	1.009	1.261	1.488	1.695	1.886	2.063
			0.7	0.551	0.984	1.345	1.657	1.933	2.181	2.407	2.614
	7	3.2	0.1	0.010	0.021	0.031	0.042	0.052	0.062	0.072	0.083
			0.3	0.031	0.062	0.093	0.124	0.154	0.184	0.214	0.243
			0.5	0.052	0.104	0.155	0.205	0.254	0.303	0.351	0.398
			0.7	0.073	0.145	0.215	0.284	0.352	0.419	0.484	0.548
		3.4	0.1	0.023	0.045	0.067	0.090	0.112	0.134	0.155	0.177
			0.3	0.068	0.135	0.200	0.264	0.328	0.390	0.451	0.511
			0.5	0.113	0.223	0.330	0.434	0.535	0.633	0.729	0.823
			0.7	0.159	0.311	0.458	0.598	0.734	0.865	0.992	1.115
		3.6	0.1	0.057	0.113	0.168	0.222	0.275	0.328	0.380	0.431
			0.3	0.171	0.334	0.490	0.640	0.785	0.924	1.058	1.188
			0.5	0.286	0.551	0.799	1.031	1.251	1.459	1.658	1.847
			0.7	0.402	0.764	1.095	1.400	1.684	1.950	2.200	2.437
		3.8	0.1	—	—	—	—	—	—	—	—

Table 19 *Edge lift condition, differential swell at the perimeter of a slab in a predominantly montmorillonite clay soil (60 percent clay) from* Reference 10.

Percent Clay (%)	Depth to Constant Suction (ft.)	Constant Suction (pF)	Velocity of Moisture Flow (in./mo.)	Differential Swell (in.) Edge Distance Penetration (ft.)							
				1 ft.	2 ft.	3 ft.	4 ft.	5 ft.	6 ft.	7 ft.	8 ft.
70	3	3.2	0.1	0.003	0.007	0.010	0.013	0.016	0.020	0.023	0.026
			0.3	0.010	0.020	0.029	0.039	0.048	0.058	0.067	0.076
			0.5	0.016	0.032	0.048	0.064	0.079	0.094	0.109	0.123
			0.7	0.023	0.045	0.067	0.088	0.109	0.129	0.149	0.169
		3.4	0.1	0.007	0.014	0.021	0.028	0.034	0.041	0.048	0.054
			0.3	0.021	0.041	0.061	0.081	0.100	0.118	0.137	0.155
			0.5	0.035	0.068	0.100	0.131	0.161	0.190	0.219	0.246
			0.7	0.048	0.094	0.138	0.179	0.219	0.258	0.294	0.330
		3.6	0.1	0.017	0.034	0.051	0.067	0.083	0.099	0.114	0.129
			0.3	0.051	0.099	0.145	0.189	0.231	0.271	0.310	0.347
			0.5	0.084	0.162	0.233	0.299	0.361	0.420	0.475	0.528
			0.7	0.117	0.221	0.314	0.400	0.479	0.552	0.620	0.684
		3.8	0.1	0.042	0.082	0.121	0.158	0.194	0.228	0.261	0.293
			0.3	0.124	0.233	0.330	0.419	0.501	0.577	0.648	0.714
			0.5	0.202	0.368	0.510	0.635	0.747	0.849	0.942	1.028
			0.7	0.277	0.492	0.669	0.822	0.955	1.075	1.184	1.284
	5	3.2	0.1	0.007	0.014	0.022	0.029	0.036	0.043	0.050	0.057
			0.3	0.022	0.043	0.064	0.085	0.106	0.126	0.147	0.167
			0.5	0.036	0.071	0.106	0.140	0.174	0.207	0.240	0.273
			0.7	0.050	0.099	0.147	0.195	0.241	0.286	0.331	0.374
		3.4	0.1	0.015	0.031	0.046	0.061	0.076	0.091	0.106	0.121
			0.3	0.046	0.092	0.136	0.179	0.222	0.264	0.306	0.346
			0.5	0.077	0.151	0.223	0.293	0.361	0.427	0.492	0.554
			0.7	0.108	0.210	0.309	0.403	0.494	0.582	0.666	0.748
		3.6	0.1	0.038	0.076	0.113	0.150	0.186	0.221	0.256	0.290
			0.3	0.115	0.224	0.329	0.429	0.525	0.617	0.706	0.792
			0.5	0.191	0.367	0.531	0.685	0.830	0.967	1.097	1.221
			0.7	0.267	0.506	0.724	0.924	1.110	1.283	1.446	1.600
		3.8	0.1	0.095	0.186	0.274	0.358	0.440	0.519	0.595	0.669
			0.3	0.283	0.536	0.764	0.974	1.168	1.348	1.517	1.677
			0.5	0.470	0.862	1.262	1.502	1.773	2.020	2.247	2.458
			0.7	0.656	1.172	1.603	1.974	2.303	2.598	2.867	3.114
	7	3.2	0.1	0.012	0.025	0.037	0.050	0.062	0.074	0.086	0.098
			0.3	0.037	0.074	0.111	0.147	0.183	0.219	0.255	0.290
			0.5	0.062	0.124	0.184	0.244	0.303	0.361	0.418	0.475
			0.7	0.087	0.173	0.256	0.339	0.419	0.499	0.577	0.653
		3.4	0.1	0.027	0.054	0.080	0.107	0.133	0.159	0.185	0.211
			0.3	0.081	0.160	0.238	0.315	0.390	0.464	0.537	0.609
			0.5	0.135	0.266	0.393	0.517	0.637	0.754	0.869	0.980
			0.7	0.189	0.371	0.545	0.713	0.875	1.031	1.182	1.329
		3.6	0.1	0.068	0.134	0.200	0.264	0.328	0.391	0.453	0.514
			0.3	0.204	0.398	0.584	0.763	0.935	1.101	1.261	1.415
			0.5	0.341	0.656	0.951	1.229	1.490	1.739	1.975	2.200
			0.7	0.479	0.910	1.304	1.668	2.006	2.323	2.621	2.903
		3.8	0.1	—	—	—	—	—	—	—	—

Table 20 *Edge lift condition, differential swell at the perimeter of a slab in a predominantly montmorillonite clay soil (70 percent clay) from* Reference 10.

7.4.4 — Check preliminary beam sizes and tendon requirements, Step 5 of Flow Chart 1

With the soil values e_m and y_m determined in Section 7.4.3 , we can now proceed to determine preliminary beam sizes and tendon requirements. The edge lift condition is checked first.

7.4.4.1 — Preliminary determination, stiffening beam depth d, edge lift condition

$$d = x^{1.176}$$

$$where\ x = \frac{L^{0.35} \times S^{0.88} \times e_m^{0.74} \times y_m^{0.76}}{12 \times \Delta_{allow} \times P^{0.01}}$$

Long direction:

Beam length $L = 100$ feet
Beam spacing $S = 16.67$ feet
Perimeter load $P = 900$ pounds per foot

Commentary:
Using L/1700 for allowable deflection is an empirical approach which provides a reasonable starting point for beam selection. Later on, allowable deflection is calculated using L/800.

β, the relative stiffness length according to the PTI procedure is assumed to be 10 feet. The maximum distance over which differential deflection will occur is L or 6β, whichever is smaller.

$6\beta = 60$ feet governs
$\Delta_{allow} = (12 \times 60)/1700 = 0.424$ inches

This value of Δ is substituted in the equation above, along with the given values, and soil properties determined in Section 7.4.3, to find x:

$$x_{long} = \frac{100^{0.35} \times 16.67^{0.88} \times 5.2^{0.74} \times 0.23^{0.76}}{12 \times 0.424 \times 900^{0.01}}$$

$$x_{long} = \frac{5.01 \times 11.89 \times 3.39 \times 0.33}{12 \times 0.424 \times 1.07}$$

$$x_{long} = \frac{66.64}{5.44} = 12.24$$

$$d_{long} = 12.24^{1.176} = 19.04 \text{ inches}$$

Therefore an initially assumed depth of 24 inches is reasonable.

Short direction:

Perimeter load remains the same, 900 pounds per foot.
Beam length $L = 50$ feet
Beam spacing $S = 20$ feet

Again assuming $\beta = 10$ feet, the maximum distance over which the differential deflection occurs is L (50 feet) since this is smaller than 6β.

$\Delta_{allow} = (12 \times 50)/1700 = 0.353$ inches

Substituting in the same equations used above for the long direction:

$$x_{short} = \frac{50^{0.35} \times 20^{0.88} \times 5.2^{0.74} \times 0.23^{0.76}}{12 \times 0.353 \times 900^{0.01}}$$

$$x_{short} = \frac{3.93 \times 13.96 \times 3.39 \times 0.33}{12 \times 0.353 \times 1.07}$$

$$x_{short} = \frac{61.37}{4.53} = 13.55$$

$$d_{short} = 13.55^{1.176} = 21.44 \text{ inches}$$

For beam depth selection the factors which have the most influence are beam spacing and perimeter loading for edge lift. Since perimeter loading tends to work against you in the center lift design, it is desirable to maintain close beam spacing in order to limit depth requirements.

The assumption of 24 inches is reasonable for the short direction as well as the long direction.

7.4.4.2 — Soil pressure under the beams (a subroutine for Step 5, Flow Chart 1)

The allowable soil pressure q_{allow} = 3000 psf is given in Section 7.4.3. The load on the ground consists of the weight of slabs and beams plus applied uniform load and the perimeter load. Since the slab weight calculation uses a 4-inch thickness over the total area, beam depths are adjusted to 24 – 4 or 20 inches (1.67 feet).

Applied loadings:

Weight of slab = 100 × 50 × 0.33 × 0.150 = 247.5 kips
Weight, long beams = 4 × 100 × 1.0 × 1.67 × 0.150 = 100.2 kips
Weight, short beams = 6 × 46 × 1.0 × 1.67 × 0.150 = 69.1 kips
Perimeter load = 900 plf × 300 ft = 270.0 kips
Uniform live load = 100 × 50 × 125 psf = 625.0 kips
Total load applied to soil 1311.8 kips

We assume that all of the load is transmitted through contact of the beam bottoms. Therefore calculate the contact area of the beams, based on the assumed spacing shown in *Figure 44*. The beam bottoms are 1 foot wide.

(94 × 4) + (50 × 6) = 676 feet beam length × 1 foot beam width
= 676 square feet of beam bottom contact area

The soil bearing pressure is then:

w = 1311.8/676 = 1.94 kips per square foot
1.94 < 3.0; therefore bearing pressure is OK.

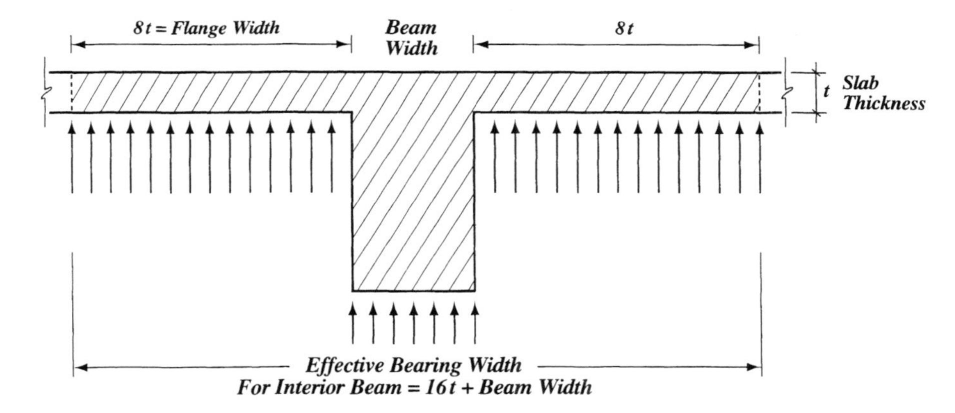

Figure 49A *Effective bearing width for interior grade beam, considering T-beam action.*

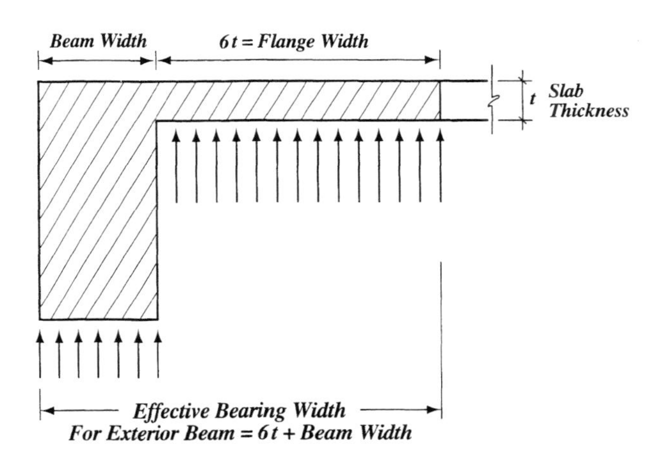

Figure 49B *Effective bearing width for exterior grade beam considering T-beam action.*

Commentary:
With a prestressed post-tensioned slab on ground, the enhanced T-beam action can be very helpful in distributing load. Observations indicate that using the T-beam section for load distribution is reasonable and effective. A distance of 8 times the slab thickness has been used successfully for computing the bearing area. For edge grade beams, 6 times the slab thickness on the one side having a flange would comply with the ACI Building Code. Figures 49A *and* 49B *show this condition. Shear should be reviewed at the interface when this procedure is followed. Should shear become critical, a thickened slab is a reasonable solution.*

7.4.5 — Determine section properties for full slab width, Step 6, Flow Chart 1

Commentary:
In calculating section properties, it can be seen that the transformed sections add significant stiffness to the system. Thus the desirability of an uncracked section can be easily recognized. It will also be seen that the greatest moments are negative moments, therefore the higher section modulus exists where it is desirable. These same section properties are useful in any conventionally reinforced slab using this technique as it will be seen that stresses can often fall within rupture modulus limits, thus simplifying a conventional design. This is due to the proximity of the neutral axis to the top of the slab.

Dimensions already determined (Figure 44)	*Long Direction*	*Short Direction*
Beam depth, d	24 in.	24 in.
Individual beam width	12 in.	12 in.
Number of beams	4	6
Total beam width	48 in.	72 in.
Slab thickness	4 in.	4 in.

Calculate section properties in the *long* direction:

Cross section	Area, in.^{2}	y, in.	Ay, in.^{3}
Slab = 46 × 12 × 4	2208	2	4,416
Beam = 12 × 24 × 4	1152	12	13,824
	3360		18,240

Distance from top of slab to neutral axis, $\bar{y}$ = 18,240/3360 = 5.43 inches

Moment of inertia (1/12 bh^3), beam and slab sections:

	Io	*A*	*d*	*Ad*^{2}	*TI*
Slab = 1/12 × 46 × 12 × 4^3	2944	2208	3.42	25,825	28,769
Beam = 1/12 × 4 × 12 × 24^3	55,296	1152	6.58	49,877	105,173
					133,942

S_T = 133,942/5.43 = 24,667 in.^{3}
S_B = 133,942/18.57 = 7213 in.^{3}

Calculate section properties in the *short* direction:

Cross section	Area, in.^{2}	y, in.	Ay, in.^{3}
Slab = 94 × 12 × 4	4512	2	9024
Beam = 12 × 24 × 6	1728	12	20,736
	6240		29,760

Distance from top of slab to neutral axis, $\bar{y}$ = 29,760 /6240 = 4.77 inches

Moment of inertia (1/12 bh^3), beam and slab sections:

	Io	*A*	*d*	*Ad*^{2}	*TI*
Slab = 1/12 × 94 × 12 × 4^3	6016	4512	2.77	34,370	40,386
Beam = 1/12 × 12 × 6 × 24^3	82,944	1728	7.24	90,577	173,521
					213,907

S_T = 213,907/4.77 = 44,844 in.^{3}
S_B = 213,907/19.23 = 11,123 in.^{3}

Summary of Section Properties

	Long direction	*Short direction*
Cross sectional area, A, sq. in.	3360	6240
Centroid of slab strands, inches from top fiber	−2.00	−2.00
Centroid of beam strands, inches from top fiber	−21.00	−21.00
Depth to neutral axis, inches from top fiber	5.43	4.77
Section modulus, S_T, in.^{3}	24,667	44,844
Section modulus, S_B, in.^{3}	7,213	11,123
Allowable concrete tensile stress, $f_t = 6\sqrt{3000}$ = 329 psi	0.329 ksi	0.329 ksi
Allowable concrete compressive stress, f_c = 0.45 (3000) = 1350	1.350 ksi	1.350 ksi

7.4.6 — Calculate minimum number of tendons required, Step 7, Flow Chart 1

Stress permitted per tendon, $f_{ps} = 0.7 \times f_{pu} = 0.7 \times 270 = 189$ ksi

Stress in tendon after losses: $f_{ps} = 189 - 30 = 159$ ksi

Force P_r per tendon:

Area per 1/2-inch-diameter tendon = 0.153 square inches

$P_r = 0.153 \times 159 = 24.33$ kips per tendon

Sufficient tendons must be installed to overcome slab-subgrade friction as well as to keep the minimum average prestress at 50 psi. Determine the number of tendons necessary for a minimum average prestress of 50 psi in the concrete.

$$\text{Number of tendons} = \frac{\text{concrete area} \times \text{average prestress}}{\text{force per tendon}}$$

$$N_t\,(long) = \frac{50 \text{ psi} \times 3360 \text{ sq.in.}}{24{,}330} = 6.90$$

$$N_t\,(short) = \frac{50 \text{ psi} \times 6240 \text{ sq.in.}}{24{,}330} = 12.82$$

Number of tendons to overcome slab-subgrade friction on polyethylene sheeting:

Weight of beams and slab = 416.8 kips (from Section 7.4.4.2)

$N_t = 0.5\,(\mu \times W_{slab})/24.330$

$= 0.5\,(0.75 \times 416.8)/24.33$

$= 6.42$ strands in each direction

Total number of tendons needed is the sum of those required to maintain minimum prestress and those needed to overcome subgrade friction.

$N_t\,(long) = 6.90 + 6.42 = 13.32$ use 14 tendons

$N_t\,(short) = 12.82 + 6.42 = 19.24$ 20 tendons acceptable

Recheck minimum number of tendons in each direction, following the guidelines of the commentary:

$N_t\,(long) = 50 \text{ ft}/5$ + number of beams

$= 10 + 4 = 14$ tendons — OK

$N_t\,(short) = 100 \text{ ft}/5$ + number of beams

$= 20 + 6 = 26$ tendons > 20.

Therefore use 26 tendons in the short direction.

Design prestress forces

Since maximum moments occur near the slab perimeter, friction losses will be minimal at points of maximum moments. Therefore, assume total prestressing force effective for structural calculations:

Long direction: $P_r = 14 \times 24.3 = 340.2$ kips

Short direction: $P_r = 26 \times 24.3 = 631.8$ kips

Commentary:

The value of 0.7 f_{pu} is assumed as a reasonable beginning approximation of f_{ps}, just as the losses of 30 ksi are an estimate. The number of tendons for this design procedure is generally low enough to encourage a conservative design approach.

These figures are based on stress-relieved 270k strand. Low-relaxation strand will permit a higher value. As the difference in number of tendons required is relatively minor, it is prudent to consider stress-relieved strand in calculations and permit low-relaxation as a substitute rather than vice versa.

The coefficient of friction used for polyethylene sheeting is 0.75 with a ribbed and stiffened section. This has been determined in the field to be not only realistic, but conservative. Care in subgrade preparation or a double layer of polyethylene can result in lowering this value.

Although 20 tendons are acceptable and meet all the criteria of the PTI Manual, the authors' experience is that tendon spacing greater than 5 feet on center is neither practical nor prudent. Also it is desirable to have at least one tendon in each stiffening beam in addition to holding to a maximum spacing of 5 feet on center.

7.4.7 — Check design moments against allowable moments for edge lift condition, Step 8, Flow Chart 1

Design moment, *long* direction, edge lift condition

$$M_\ell = \frac{S^{0.10} (d\, e_m)^{0.78}\, y_m^{0.66}}{7.2 \times L^{0.0065}\, P^{0.04}}$$

$$M_\ell = \frac{16.67^{0.10} (24 \times 5.2)^{0.78}\, 0.23^{0.66}}{7.2 \times 100^{0.0065} \times 1150^{0.04}}$$

$$M_\ell = \frac{132 \times 43.15 \times 0.38}{7.42 \times 1.32}$$

$$M_\ell = \frac{21.64}{9.79} = 2.21 \text{ foot-kips per foot}$$

Commentary:
P is 1150 because the perimeter load is 900 plf and the 24-inch-deep exterior beam weighs 250 plf: (24 inches – 4 inches)/ 12 × 150 = 250 pounds.

Design moment, *short* direction, edge lift condition

$$M_s = d^{0.35} \times \frac{19 + e_m}{57.75} \times M_\ell$$

$$M_s = 24^{0.35} \times \frac{19 + 5.2}{57.75} \times 2.21$$

$$M_s = 3.04 \times 0.42 \times 2.21$$

$$M_s = 2.82 \text{ foot-kips per foot}$$

Allowable service moments, *long* direction, tension in bottom fiber, edge lift condition

$$(12 \times 50)\ _pM_t = S_B (P_r/A + f_t) - P_r e$$

Note that the long direction has a 50-foot-wide cross section; hence the design moment is multiplied by 50 feet. It is also multiplied by 12 inches per foot to make units of the left side of the equation compatible with the right.

The quantity $P_r e$ accounts for the moment associated with the eccentricity of the post-tensioning tendons. It has been standard practice to drape beam tendons to within 3 inches of the bottom of stiffening beams in order to maximize positive moment capacity. This is because negative moment capacity is almost always more than ample due to the significantly higher section modulus. With this consideration, this eccentricity factor can therefore be either positive or negative as in Figure 50.

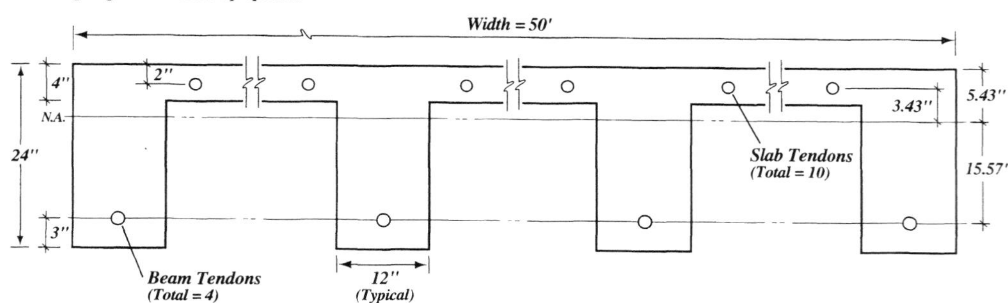

Figure 50 *Cross section of slab taken through the short direction, indicating long direction section properties. Location of neutral axis and eccentricities of beam and slab tendons are shown.*

$$\begin{aligned} P_r e &= \big[[N_t\,(\text{top}) \times 3.43] - [N_t\,(\text{beam}) \times 15.57]\big] \times 24.3 \\ &= [\,10 \times 3.43 - 4 \times 15.57\,] \times 24.3 \\ &= [\,34.3 - 62.28\,] \times 24.3 \\ &= -27.98 \times 24.3 \end{aligned}$$

$$P_r e = -679.91 \text{ inch-kips}$$

Now substitute this value for P_re in the equation above for tension in bottom fiber.

$$(12 \times 50)\ {}_pM_t = 7213\ [\ (340.2/3360) + 0.329] - (-679.91)$$
$$600\ {}_pM_t = 7213\ [0.101 + 0.329] + 679.91$$
$$600\ {}_pM_t = 7213\ (0.430) + 679.91$$
$${}_pM_t = (3102 + 680)/600$$
$$= 6.30 \text{ foot-kips per foot}$$
$$6.30 > 2.21 \text{ — OK}$$

Allowable service moments, *long* direction, compression in top fiber, edge lift condition

$$(12 \times 50)\ {}_pM_c = S_T\ [f_c - (P_r/A)] - P_re$$
$$600\ {}_pM_c = 24{,}667\ [1.350 - 340.2/3360] - P_re$$
$$600\ {}_pM_c = 24{,}667\ [1.350 - 0.101] - (-679.91)$$
$${}_pM_c = [30{,}809 + 680]/600$$
$$= 52.48 \text{ foot-kips per foot}$$
$$52.48 > 2.21 \text{ — OK}$$

Allowable service moment, *short* direction, tension in bottom fiber, edge lift condition

Again the quantity P_re must be calculated to account for the prestressing moment associated with the eccentricity of the post-tensioning tendons. *Figure 51* shows this cross section. Note that the neutral axis is slightly different for the two cross sections. Since the calculation of P_re depends on the location of the neutral axis, there are two sets of values *(Figures 50* and *51).*

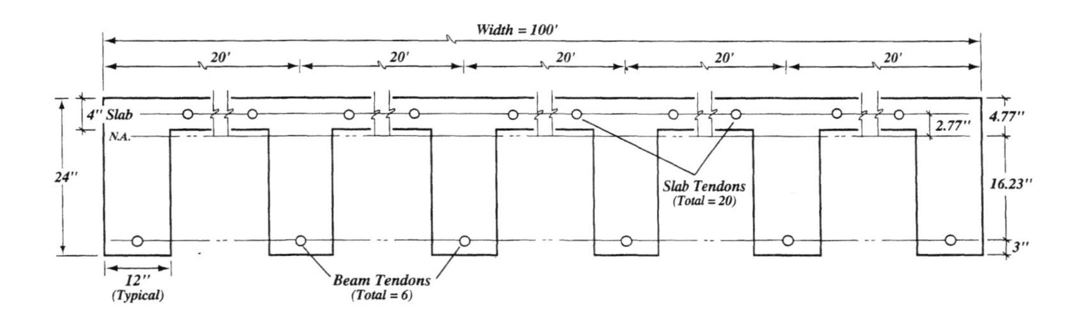

Figure 51 *Cross section of slab taken through the long direction, indicating short direction section properties. Neutral axis and eccentricities of slab and beam tendons are shown.*

$$P_re = \big[[N_t\,(\text{top}) \times 2.77] - [N_t\,(\text{beam}) \times 16.23]\big] \times 24.3$$
$$= [20 \times 2.77 - 6 \times 16.23] \times 24.3$$
$$= [55.4 - 97.38] \times 24.3$$
$$= -41.98 \times 24.3$$
$$P_re = -1020.11 \text{ inch-kips}$$

$$(12 \times 100)\ {}_pM_t = 11{,}123\ [631.8/6240 + 0.329] - (-1020.11)$$
$${}_pM_t = [(11{,}123 \times 0.430) + 1020.11]/1200$$
$${}_pM_t = 5803/1200$$
$$= 4.84 \text{ foot-kips per foot}$$
$$4.84 > 2.82 \text{ — OK}$$

Allowable service moments, *short* direction, compression in top fiber, edge lift condition

$$(12 \times 100)\ {}_pM_c = 44{,}844\ [1.350 - 631.8/6240] - (-1020.11)$$
$${}_pM_c = [(44{,}844 \times 1.249) + 1020.11]/1200$$
$$= 57{,}030/1200 = 47.52 \text{ foot-kips per foot}$$

Since 47.52 is greater than 2.82, section is OK.

7.4.8 — Deflection calculations, edge lift condition, Step 9, Flow Chart 1

Allowable differential deflection, *long* direction, edge lift

$$\beta = \frac{1}{12}\sqrt[4]{\frac{E_c I}{E_s}}$$

$$\beta = \frac{1}{12}\sqrt[4]{\frac{1{,}500{,}000 \times 133{,}942}{1000}}$$

$$\beta = \frac{1}{12}\sqrt[4]{200{,}913{,}000}$$

$$\beta = \frac{1}{12} \times 119.05$$

$$\beta = 9.92 \text{ feet}$$

$6\beta = 59.52$ feet < 100 feet so 6β governs

$\Delta_{allow} = [12 \times 59.52]/800 = 0.89$ inches

Commentary:
Proper deflection control is absolutely imperative. It appears to be of even greater importance when 6ß governs the design instead of actual slab length, L. Since the beam depth directly affects deflection, it is always advisable to be conservative in depth selection and unyielding in jobsite inspections. Residential construction is particularly sensitive to deflection control, because wall finishes such as stucco or masonry can be unmerciful in revealing cracks due to deflection in an otherwise sound design.

Expected differential deflection, *long* direction, edge lift

$$\Delta = \frac{L^{0.35} S^{0.88} e_m^{\ 0.74} y_m^{\ 0.76}}{15.90\ d^{0.85} P^{0.01}}$$

$$\Delta = \frac{100^{0.35}\ 16.67^{0.88}\ 5.2^{0.74}\ 0.23^{0.76}}{15.90 \times 24^{0.85} \times 1150^{0.01}}$$

$$\Delta = \frac{5.011 \times 11.89 \times 3.38 \times 0.32}{15.90 \times 14.89 \times 1.07}$$

$$\Delta = \frac{64.44}{253.32} = 0.25 \text{ inches}$$

The expected differential deflection of 0.25 inches < 0.89 inches allowable deflection — OK

Allowable deflection, *short* direction, edge lift

First determine β.

$$\beta = \frac{1}{12}\sqrt[4]{\frac{1{,}500{,}000 \times 213{,}907}{1000}}$$

$$\beta = \frac{1}{12} \times 133.84$$

$$\beta = 11.15 \text{ feet}$$

$$6\beta = 66.91 \text{ feet} > 50 \text{ feet}$$

Therefore 50 feet governs and

$\Delta_{allow} = (12 \times 50)/800 = 0.75$ inches

Expected differential deflection, short direction, edge lift

$$\Delta = \frac{50^{0.35}\,20^{0.88}\,5.2^{0.74}\,0.23^{0.76}}{15.90 \times 24^{0.85} \times 1150^{0.01}}$$

$$\Delta = \frac{3.93 \times 13.96 \times 3.38 \times 0.32}{15.90 \times 14.89 \times 1.07}$$

$$\Delta = \frac{59.34}{253.32} = 0.23 \text{ inches}$$

Since the expected deflection of 0.23 inches is less than the allowable 0.75 inches, cross section is OK. This completes the check of deflections for edge lift bending and all are less than allowable for both long and short directions.

7.4.9 — Shear calculations for edge lift condition, Step 10, Flow Chart 1

Expected shear force, *long* direction, edge lift condition

$$V_s \text{ or } V_\ell = \frac{L^{0.07}\,d^{0.40}\,P^{0.03}\,e_m^{\,0.16}\,y_m^{\,0.67}}{3 \times S^{0.015}}$$

$$V_\ell = \frac{100^{0.07}\,24^{0.40}\,1150^{0.03}\,5.2^{0.16}\,0.23^{0.67}}{3.0 \times 16.67^{0.015}}$$

$$V_\ell = \frac{1.38 \times 3.56 \times 1.23 \times 1.30 \times 0.37}{3 \times 1.04}$$

$$V_\ell = \frac{2.90}{3.12} = 0.93 \text{ kips per foot}$$

Commentary:
Using only the beams to resist shear is a conservative approach and in keeping with the PTI Manual. A strong case can be made for including a portion of the T-beam for shear resistance also.

Allowable shear stress

$$v_c = 1.5\sqrt{f_c'} = 1.5\sqrt{3000} = 82.2 \text{ psi}$$

Only the beams are considered to resist shear.

Total design shear stress *v*, *long* direction, edge lift

$$v = \frac{V_\ell \times W}{ndb}$$

$$v = \frac{0.93 \times 1000 \times 50}{4 \times 12 \times 24}$$

$$v = \frac{46{,}500}{1152}$$

$v = 40.36$ psi

40.36 psi is less than 82.2 psi allowable, so shear is OK in long direction

Although allowable shear is $1.5\sqrt{f_c'}$ according to the PTI Manual, a strong case has been made to permit a total allowable of $1.5\sqrt{f_c'} + P/A$. Some available computer programs also use this enhanced allowable. This only applies to prestressed slabs.

Expected shear force, *short* direction, edge lift condition

$$V_s = \frac{50^{0.07}\,24^{0.40}\,1150^{0.03}\,5.2^{0.16}\,0.23^{0.67}}{3.0 \times 20^{0.015}}$$

$$V_s = \frac{1.31 \times 3.56 \times 1.23 \times 1.30 \times 0.37}{3 \times 1.05}$$

$$V_s = \frac{2.76}{3.15} = 0.88 \text{ kips per foot}$$

Total design shear stress, *short* direction, edge lift

$$v = \frac{0.88 \times 1000 \times 100}{6 \times 12 \times 24}$$

$$v = \frac{88{,}000}{1728}$$

$$v = 50.93 \text{ psi}$$

The design shear stress of 50.93 psi is less than the allowable shear stress of 82.2 psi, so shear is OK in the short direction as well as the long direction. This completes the checks necessary for the edge lift condition.

7.4.10 — Center lift design

A check of bending, deflection, and shear must be made for the center lift condition just as has been done for the edge lift condition completed above. Since soil conditions and preliminary design are the same, we can begin at Step 8 of *Flow Chart 1*.

Design moments for *long* direction, center lift condition (Step 8, *Flow Chart 1*)

$$M_\ell = A_o\left[B \times e_m^{1.238} + C\right]$$

$$\text{where } A_o = \frac{1}{727}\left[L^{0.013}\, S^{0.306}\, d^{0.688}\, P^{0.534}\, y_m^{0.193}\right]$$

$$A_o = \frac{1}{727}\left[100^{0.013}\, 16.67^{0.306}\, 24^{0.688}\, 1150^{0.534}\, 0.224^{0.193}\right]$$

$$A_o = \frac{1}{727}\left[1.06 \times 2.36 \times 8.90 \times 43.09 \times 0.75\right]$$

$$A_o = \frac{719.52}{727} = 0.989$$

From the soils data, page 81, e_m, edge moisture variation distance is 3.8 feet for the center lift condition. According to the PTI procedure, for $e_m \leq 5$, $B = 1.0$ and $C = 0$. Using these values, determine M_ℓ:

$$M_\ell = 0.989\left[1 \times 3.8^{1.238} + 0\right]$$

$$M_\ell = 0.989 \times 5.22$$

$$M_\ell = 5.16 \text{ foot-kips per foot}$$

Design moments, *short* direction, center lift

$$M_s = \left[\frac{58 + e_m}{60}\right] \times M_\ell$$

$$M_s = \left[\frac{58 + 3.8}{60}\right] \times 5.16$$

$$M_s = 5.31 \text{ foot-kips per foot}$$

Allowable moments, *long* direction, center lift

Allowable moments must be calculated and compared with design moments. First, calculate negative bending moments.

• Tension in top fiber. NOTE: From page 90, $P_r e = -679.91$ inch-kips

$$(12 \times 50)\ {}_nM_t = S_T\left[\frac{P_r}{A} + f_t\right] + P_r e$$

$$600\ {}_nM_t = 24{,}667\left[\frac{340.2}{3360} + 0.329\right] + (-679.91)$$

$${}_nM_t = \frac{[24{,}667 \times 0.430] - 679.91}{600}$$

$$= 16.54 \text{ foot-kips per foot}$$

Since the allowable moment of 16.54 foot-kips per foot is greater than the design moment of 5.16, section is OK for tension in the top fiber.

• Compression in bottom fiber

$$(12 \times 50)\ {}_nM_c = S_B\left[f_c - \frac{P_r}{A}\right] + P_r e$$

$$600\ {}_nM_c = 7213\left[1.350 - \frac{340.2}{3360}\right] + (-679.91)$$

$${}_nM_c = \frac{7213\,(1.249) - 679.91}{600}$$

$$= 13.88 \text{ foot-kips per foot}$$

Since 13.88 is greater than 5.16 design moment, section is OK in long direction for compression in bottom fiber.

Allowable moments in *short* direction, center lift design

• Tension in top fiber NOTE: From page 91, $P_r e = -1020.11$ inch-kips

$$(12 \times 100)\ {}_nM_t = S_T\left[\frac{P_r}{A} + f_t\right] + P_r e$$

$$1200\ {}_nM_t = 44{,}844\left[\frac{631.8}{6240} + 0.329\right] + (-1020.11)$$

$${}_nM_t = \frac{44{,}844\,(0.101 + 0.329) - 1020.11}{1200}$$

$${}_nM_t = \frac{19{,}283 - 1020.11}{1200}$$

$$= 15.22 \text{ foot-kips per foot}$$

Since the allowable 15.22 foot-kips per foot is greater than the design moment of 5.31, section checks OK.

• Compression in bottom fiber

$$(12 \times 100)\ {}_nM_c = S_B\left[f_c - \frac{P_r}{A}\right] + P_r e$$

$${}_nM_c = \frac{11{,}123\ (1.350 - 0.101) + (-1020.11)}{1200}$$

$${}_nM_c = \frac{12{,}872.52}{1200}$$

$${}_nM_c = 10.73 \text{ ft.-kips per ft.}$$

Since the allowable moment, 10.73 foot-kips per foot is greater than the design moment of 5.21, section is OK. Moment capacities exceed expected service moments for center lift in both long and short directions.

Center lift deflection calculations, *long* direction (Step 9, *Flow Chart 1*)

• Allowable differential deflection

$$\Delta_{allow} = 12(L \text{ or } 6\beta)/360$$

Use 6β, since it is less than L (= 100 ft). β, the relative stiffness length, is 9.92 feet as calculated on page 92.

$$\Delta_{allow} = 12(59.52)/360$$
$$= 1.98 \text{ inches}$$

• Expected differential deflection

$$\Delta = \frac{[y_m L]^{0.205}\, S^{1.059}\, P^{0.523}\, e_m{}^{1.296}}{380 \times d^{1.214}}$$

$$\Delta = \frac{[0.224 \times 100]^{0.205}\, 16.67^{1.059}\, 1150^{0.523}\, 3.8^{1.296}}{380 \times 24^{1.214}}$$

$$\Delta = \frac{1.89 \times 19.68 \times 39.88 \times 5.64}{380 \times 47.37}$$

$$\Delta = \frac{8366}{18{,}004} = 0.46 \text{ inches}$$

The expected differential deflection, 0.46 inches is less than the allowable of 1.98 inches so section is OK for deflection in the long direction.

Center lift deflection in the *short* direction

• Allowable differential deflection

$$\Delta_{allow} = 12(L \text{ or } 6\beta)/360$$

Here L of 50 feet is smaller than 6β so it is used in the calculation of the allowable:

$$\Delta_{allow} = 12(50)/360$$
$$= 1.66 \text{ inches}$$

• Expected differential deflection

Substitute in the same expression used above for expected deflection, noting that only the values of S and L are different:

$$\Delta = \frac{[0.224 \times 50]^{0.205}\, 20^{1.059}\, 1150^{0.523}\, 3.8^{1.296}}{380 \times 24^{1.214}}$$

$$\Delta = \frac{1.64 \times 23.87 \times 39.88 \times 5.64}{380 \times 47.37}$$

$$\Delta = \frac{8805}{18{,}004} = 0.49 \text{ inches}$$

Since 0.49 inches is less than the allowable of 1.66 inches, section is OK for deflection. Deflections in both long and short directions are much less than the allowable deflection for center lift loading.

Shear calculations, *short* direction, center lift condition (Step 10, *Flow Chart 1*)

• Design shear force

$$V_s = \frac{1}{1350}\,[L^{0.19}\, S^{0.45}\, d^{0.20}\, P^{0.54}\, y_m^{0.04}\, e_m^{0.97}]$$

$$V_s = \frac{1}{1350}\,[50^{0.19}\, 20^{0.45}\, 24^{0.20}\, 1150^{0.54}\, 0.224^{0.04}\, 3.8^{0.97}]$$

$$V_s = \frac{1}{1350}\,[2.10 \times 3.85 \times 1.89 \times 44.95 \times 0.94 \times 3.65]$$

$$V_s = \frac{2357}{1350} = 1.75 \text{ kips per foot}$$

• Allowable shear stress (according to the PTI Manual)

$$v_c = 1.5\sqrt{f_c'} = 1.5\sqrt{3000} = 82.2 \text{ psi}$$

• Design shear stress

$$v = \frac{V \times W}{ndb}$$

$$v = \frac{1.75 \times 100 \times 1000}{6 \times 12 \times 24}$$

$$v = \frac{175{,}000}{1728}$$

$$v = 101.3 \text{ psi}$$

101.3 > 82.2 psi allowable shear stress;
this does not meet the PTI recommendation.

However, as explained earlier in these calculations, a strong case has been made for use of a shear value of

$$1.5\sqrt{f_c'} + P/A$$

If this is were applied here, the allowable shear would increase from 82.2 psi to 183.52 psi, well above the calculated 101.3 psi shear stress.

A second consideration is that only the stem of the beam has been taken into consideration, and not the flange. By observation, consideration of the flange would provide a unit shear force of 26.92 psi, a value well under the allowable.

A third possibility—if it is desired to follow the PTI recommendation rigorously—would be to alter the cross section to comply with the more rigid shear standards. This can be done by one of the following:

1. Increase beam width
2. Increase beam depth
3. Decrease beam spacing
4. Use web reinforcement over a distance of 1.5β at each end of the beam
5. A combination of 1, 2, and 3.

Increasing the short direction beam width to 14 inches would reduce the stress to 83.3 psi when calculated by the most conservative method shown above. This is close enough to the allowable of 82.2 psi to represent satisfactory compliance with the PTI recommendations.

However, the authors' choice is to keep the short direction beam width at 12 inches for simplicity of construction, relying on their experience that shear is rarely if ever critical for this type of construction.

Shear calculation in the *long* direction, center lift condition

• Design shear force and stress

$$V_\ell = \frac{1}{1940}\,[L^{0.09}\,S^{0.71}\,d^{0.43}\,P^{0.44}\,y_m^{\ 0.16}\,e_m^{\ 0.93}]$$

$$V_\ell = \frac{1}{1940}\,[\,100^{0.09}\,16.67^{0.71}\,24^{0.43}\,1150^{0.44}\,0.23^{0.16}\,3.8^{0.93}\,]$$

$$V_\ell = \frac{1}{1940}\,[\,1.51 \times 7.37 \times 3.92 \times 22.22 \times 0.79 \times 3.46\,]$$

$$V_\ell = \frac{2650}{1940} = 1.37 \text{ kips per ft.}$$

$$v = \frac{1.37 \times 50 \times 1000}{4 \times 12 \times 24} = \frac{68{,}500}{1152} = 59.46 \text{ psi}$$

The stress of 59.46 is less than the allowable 82.2. Section OK.

This completes the problem for center lift. Edge lift and center lift design results are summarized in *Table 21* (Step 12, *Flow Chart 1*).

Edge lift design	***Design***	***Allowable***	**Center lift design**	***Design***	***Allowable***
Moment, ft.-kips per ft.			Moment, ft.-kips per ft.		
Long Direction			*Long Direction*		
Tensile	2.21	6.30	Tensile	5.16	16.54
Compressive	2.21	52.48	Compressive	5.16	13.88
Short direction			*Short direction*		
Tensile	2.82	4.84	Tensile	5.31	15.22
Compressive	2.82	47.52	Compressive	5.31	10.73
Differential deflection, in.			Differential deflection, in.		
Long Direction	0.25	0.89	*Long Direction*	0.46	1.98
Short direction	0.23	0.75	*Short direction*	0.49	1.66
Shear stress, psi			Shear stress, psi		
Long Direction	40.36	82.2	*Long Direction*	59.46	82.2
Short direction	50.93	82.2	*Short direction**	101.30	82.2
			** NOTE: See comments regarding shear in Sections 7.4.9 and 7.4.10 of this example.*		

Tendons and beam requirements *(see Figure 52)*

Long direction

Fourteen 1/2-inch-diameter 270k strands total. Ten tendons in the slab, 2 inches below the top, beginning 30 inches from each end, then 5 feet on center. One tendon in each beam, 3 inches from the bottom fiber. Four beams 12 inches wide, 24 inches deep, evenly spaced at 16 feet 8 inches on center.

Short direction

Twenty-six 1/2-inch-diameter 270k strands total. Twenty in the slab, beginning 30 inches from each end then 5 feet on center. One tendon in each of the 6 beams, which are 12 inches wide, 24 inches deep, and evenly spaced at 20 feet on center.

Table 21 *Design Summary for the Post-Tensioned Slab on Plastic Clay.*

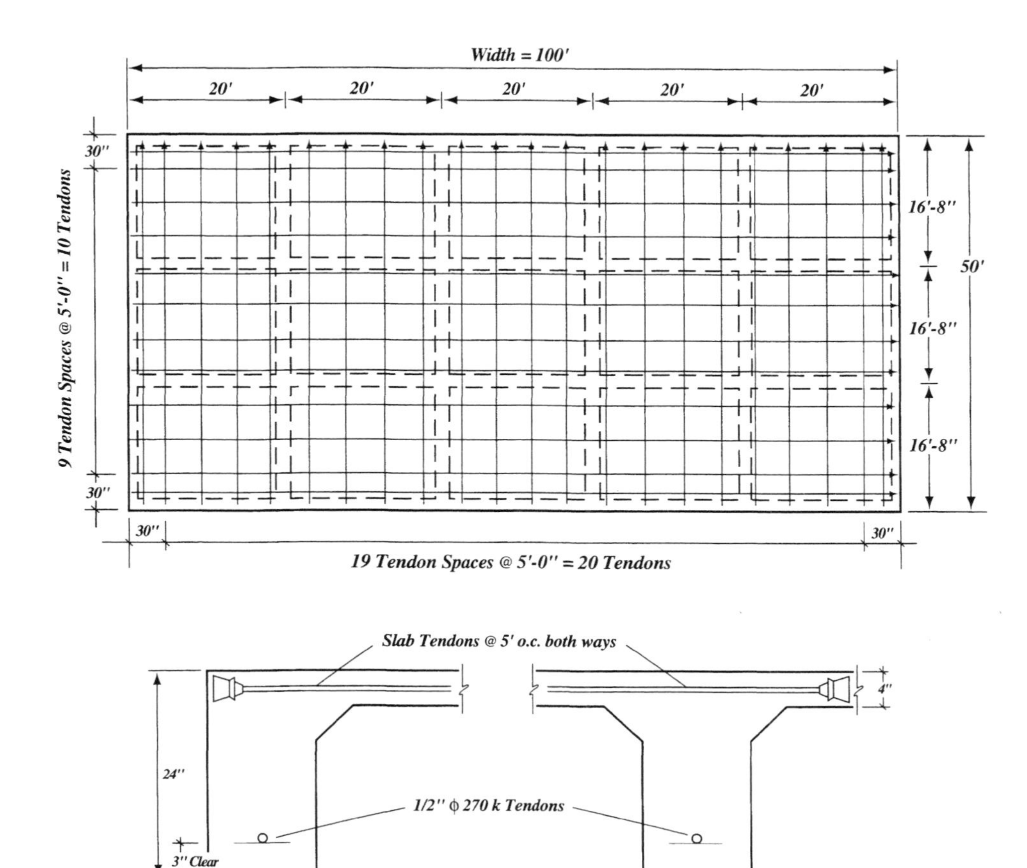

Figure 52 *Beam and tendon locations for the post-tensioned slab on plastic clay.*

7.5— DESIGN EXAMPLE: Simple rectangle, uniform thickness post-tensioned slab on plastic clay with uniform and perimeter load

7.5.1 — Introduction

The state-of-the-art Post-Tensioning Institute solution for slabs on plastic clays is the ribbed slab design presented in Section 7.4. However, revisions in progress to the PTI design manual (*Reference 10*) address the concept of a foundation of uniform thickness. The uniform thickness solution can be developed from a ribbed foundation design that satisfies all moment, shear, and differential deflection requirements of the established ribbed slab design procedures.

The new Post-Tensioning Institute procedure permits uniform thickness design for foundations at least 6 inches thick. If there is a perimeter beam that is at least 12 inches deep, the uniform thickness may be as small as 4 inches. The uniform thickness design is applicable for slabs on plastic clay as in this example, but it may also be useful for slabs on relatively stable soils that do not manifest large moment, shear, and deflection requirements.

In many areas, soil stability or construction preferences dictate a desire for uniformly thick foundations. The following example provides an equivalent uniform thickness foundation based on the design example found in Section 7.4.

7.5.2—Modeling the problem

The easiest way to design a uniformly thick post-tensioned slab on plastic clay is to prepare a customary ribbed foundation design, as explained in Section 7.4, and then derive from it an equivalent uniform thickness slab. Since Section 7.4 works out the ribbed foundation in detail for a 50x100-foot slab, we can start with a summary of results from that earlier example, provided in Table 22. As indicated on page 77, the slab has a uniform loading of 125 pounds per square foot and a perimeter wall load of 900 pounds per linear foot.

Edge lift design	***Design***	***Allowable***	**Center lift design**	***Design***	***Allowable***
Moment, ft.-kips per ft.			Moment, ft.-kips per ft.		
Long Direction			*Long Direction*		
Tensile	2.21	6.30	Tensile	5.16	16.54
Compressive	2.21	52.48	Compressive	5.16	13.88
Short direction			*Short direction*		
Tensile	2.82	4.84	Tensile	5.31	15.22
Compressive	2.82	47.52	Compressive	5.31	10.73
Differential deflection, in.			Differential deflection, in.		
Long Direction	0.25	0.89	*Long Direction*	0.46	1.98
Short direction	0.23	0.75	*Short direction*	0.49	1.66
Shear stress, psi			Shear stress, psi		
Long Direction	40.36	82.2	*Long Direction*	59.46	82.2
Short direction	50.93	82.2	*Short direction**	101.30	82.2
			** NOTE: See comments regarding shear in Sections 7.4.9 and 7.4.10 of this example.*		

Tendons and beam requirements *(see Figure 52)*
Long direction
Fourteen 1/2-inch-diameter 270k strands total. Ten tendons in the slab, 2 inches below the top, beginning 30 inches from each end, then 5 feet on center. One tendon in each beam, 3 inches from the bottom fiber. Four beams 12 inches wide, 24 inches deep, evenly spaced at 16 feet 8 inches on center.

Short direction
Twenty-six 1/2-inch-diameter 270k strands total. Twenty in the slab, beginning 30 inches from each end then 5 feet on center. One tendon in each of the 6 beams, which are 12 inches wide, 24 inches deep, and evenly spaced at 20 feet on center.

Table 22 *Summary of design data for ribbed slab on plastic clay, as developed in the example of Section 7.4.*

Since the allowable moment and shear will probably not be critical in the analysis of a uniformly thick slab, this leaves us with an analysis of differential deflection requirements. Use the ratio of expected differential deflection Δ to the allowable differential deflection Δ_{allow} in determining the needed thickness. Select the largest ratio in each direction.

$$\text{Long direction, edge lift design: } \frac{\Delta}{\Delta_{allow}} = \frac{0.25}{0.89} = 0.28$$

$$\text{Long direction, center lift design: } \frac{\Delta}{\Delta_{allow}} = \frac{0.46}{1.98} = 0.23 \quad \text{USE } 0.28$$

$$\text{Short direction, edge lift design: } \frac{\Delta}{\Delta_{allow}} = \frac{0.23}{0.75} = 0.31$$

$$\text{Short direction, center lift design: } \frac{\Delta}{\Delta_{allow}} = \frac{0.49}{1.66} = 0.30 \quad \text{USE } 0.31$$

7.5.3— Determination of equivalent thickness

Long direction: The moment of inertia I previously determined for the combined beam and slab in the long direction is 133,942 in.^{4} (see page 88). Apply the deflection ratio calculated in Section 7.5.2 to find the minimum acceptable I for an equivalent uniform thickness slab:

$$0.28 \times 133,942 \text{ or } 37,503 \text{ in.}^4$$

Calculate the depth d required for a uniform thickness slab 50 feet wide, having this same moment of inertia.

$$I = 1/12\, bd^3$$
$$12\, I = bd^3$$
$$d^3 = \frac{12\, I}{b}$$

Using the I value determined for the 50-foot width of the long direction of this slab

$$d^3 = \frac{12\, I}{12 \times 50}$$

$$d = \sqrt[3]{\frac{I}{50}} = \sqrt[3]{\frac{37{,}503}{50}} = \sqrt[3]{750}$$

$$d = 9.08 \text{ inches}$$

Short direction: The moment of inertia I previously determined for the short direction is 213,907 in.^{4} (page 88). Adjusting by the deflection ratio previously determined, the minimum acceptable I for a uniform thickness slab is

$$0.31 \times 213,907 \text{ or } 66,311 \text{ in.}^4$$

For a uniformly thick slab 100 feet wide, the thickness required to provide this moment of inertia is

$$d = \sqrt[3]{\frac{I}{100}} = \sqrt[3]{\frac{66{,}311}{100}} = \sqrt[3]{663} = 8.72 \text{ inches}$$

The larger of the two values is 9.08 inches. Therefore it is recommended that the equivalent uniform thickness slab be 9¼ inches thick. This can readily be formed using nominal 2x10 lumber, which has an actual depth of 9¼ inches.

7.5.4— Determine number of tendons required

Stress permitted per tendon: $f_{ps} = 0.7 \times f_{pu} = 0.7 \times 270 = 189$ ksi
Stress in tendon after losses: $f_{ps} = 189 - 30 = 159$ ksi
Force P_r per ½-inch-diameter tendon = area of tendon × net stress in tendon
= 0.153 square inches × 159 ksi
= 24.33 kips per tendon

Sufficient tendons must be installed to overcome friction between slab and subgrade, as well as to keep the minimum average prestress at 50 psi. Since there are no ribs in the slab, there are no supplemental beam tendons. Although a net prestress of 50 psi is permissible, it is the authors' experience that an average prestress on the order of 75 psi is more effective in

Commentary:
The value of 0.7 p_u is assumed as a reasonable amount of permitted stress per tendon. The 30 ksi loss of prestress is estimated. The small number of tendons for this design procedure encourages a conservative approach.

controlling temperature and shrinkage cracks. Therefore this higher prestress value of 75 psi will be applied to determine the number of tendons required.

$$\text{Number of tendons} = \frac{\text{concrete area} \times \text{average prestress}}{\text{force per tendon}}$$

$$N_t\,(long) = \frac{75\text{ psi} \times 50 \times 12 \times 9.25}{24{,}330} = 17.1$$

$$N_t\,(short) = \frac{75\text{ psi} \times 100 \times 12 \times 9.25}{24{,}330} = 34.2$$

Commentary:
The coefficient of friction used for polyethelyne sheeting is 0.5 with a uniform thickness slab. Care in subgrade preparation and a double layer of polyethelyne can lower this value.

These figures are based on stress relieved 270k strand. Low-relaxation strand has a net force P_r *of 27 kips for 1/2-inch diameter strand. If one is confident of obtaining a low relaxation 1/2-inch diameter strand, the number of tendons in the long direction may be reduced to 24.33/27 × 23 or 21 tendons, and in the short direction to 37 tendons.*

Tendons must also be installed to overcome slab-subgrade friction on polyethelene sheeting.

$$\text{Weight of slab} = \frac{9.25 \times 50 \times 100 \times 150}{12} = 578\text{ kips}$$

$$N_t = \frac{0.5\,(\mu \times W_{slab})}{24{,}330/1000}$$

$$= \frac{0.5\,(0.5 \times 578)}{24.330}$$

= 5.9 tendons, each direction

Total number of tendons needed is the sum of those required to maintain minimum prestress and those needed to overcome subgrade friction.

$N_t\,(long) = 17.1 + 5.9 = 23$ tendons

$N_t\,(short) = 34.2 + 5.9 = 40.1$ — use 41 tendons

7.5.5— Recheck design moments against allowable moment capacity

Using neutral axis prestressing, it is unnecessary to perform separate checks for edge lift and center lift conditions. Simply take the maximum moments in the long and short directions from *Table 22*. Since the maximum moments occur near the perimeter, friction losses will be minimal at points of maximum moment. Therefore, we can assume the total prestressing force is effective for structural calculations.

P_r total prestressing force = number of tendons × force per tendon
P_r long direction = 23 × 24.3 = 558.9 kips
P_r short direction = 41 × 24.3 = 996.3 kips

Allowable tensile moment capacity, long direction: Unit prestress applied to concrete in the long direction is

$$\frac{\text{total force}}{\text{area of section}} = \frac{558.9}{50 \times 12 \times 9.25} = 101\text{ psi}$$

The allowable tensile stress in the concrete without prestressing is

$$6\sqrt{f_c'} = 6\sqrt{3000} = 329 \text{ psi}$$

Combining this allowable with the prestressing force gives a total tensile capacity in the long direction of

$$329 + 101 = 430 \text{ psi}$$

Use this allowable stress and apply the flexure formula to find total tensile moment capacity in the long direction

Commentary:
Use the selected thickness, 9 1/4 inches, to find the section modulus per foot of slab width.
$S = 1/6\ bd^2$
$= 1/6 \times 12 \times 9.25^2$
$= 171.13 \text{ in.}^3$

$$M = fS = 430 \times 171.13 = 73{,}586 \text{ inch-pounds per foot}$$
$$= 6.13 \text{ foot-kips per foot}$$

Since the design moment of 5.16 foot-kips per foot is less than this allowable capacity, the section is OK in the long direction.

Allowable tensile moment capacity, short direction: In the short direction, the tendons apply a prestress of

$$\frac{996.3}{100 \times 12 \times 9.25} = \frac{996.3}{11{,}100} = 0.090 \text{ ksi or } 90 \text{ psi}$$

Adding this to the 329 psi tension capacity of the concrete before prestressing gives a total capacity of 329 + 90 or 419 psi in the short direction. We then have an allowable moment of

$$M = fS = 419 \times 171.13 = 71{,}703 \text{ inch-pounds per foot}$$
$$= 5.97 \text{ foot-kips per foot}$$

Since this allowable is more than the design moment of 5.31 foot-kips per foot in the short direction (*Table* 22), the section is OK.

Allowable compressive moment capacity in the long direction: The allowable compressive stress in the concrete (before prestressing) is

$$0.45 \times f_c' = 0.45 \times 3000 = 1350 \text{ psi}$$

To get the net allowable compression, this must be reduced by the prestress of 101 psi applied in the long direction, giving

$$1350 - 101 = 1249 \text{ psi}$$

The allowable moment then is

$$M = fS = 1249 \times 171.13 = 213{,}741 \text{ inch-pounds per foot}$$
$$= 17.81 \text{ foot-kips per foot}$$

This substantially exceeds the compressive design moment in the long direction, 5.16 foot-kips per foot, so the section is OK for compressive moment in the long direction.

Allowable compressive moment in the short direction: The allowable concrete compression (before prestressing) is 1350 psi as determined in the previous step. It must be reduced by 90 psi applied prestress in the short direction to get the net allowable compression of

1350 − 90 = 1260 psi.

Using this allowable compressive stress, the allowable moment is calculated as

$M = fS = 1260 \times 171.13 = 215{,}623$ inch-pounds per foot
= 17.97 foot-kips per foot

Since this allowable is larger than the maximum design moment (compression, short direction) of 5.31, the section is OK.

7.5.6 — Check of shear capacity

Determine the allowable shear stress v_c in the concrete

$$v_c = 1.5\sqrt{f_c'} = 1.5\sqrt{3000} = 82.2 \text{ psi}$$

Using this allowable shear stress, find the allowable shear force per foot of slab width.
Allowable shear force per foot of slab width = area $\times$ v_c
= 12 × 9.25 × 82.2
= 9124 pounds per foot

Commentary:
One advantage of the uniformly thick slab is the ability to easily meet shear requirements. This is due to the uniform stiffness of the entire slab cross section.

Compare this allowable shear force with the expected shear force determined in the original design. From the example solved in Section 7.4, we find the following expected shear forces:

Edge lift
- Long direction — 930 pounds per foot (page 93)
- Short direction — 880 pounds per foot (page 93)

Center lift
- Long direction — 1370 pounds per foot (page 98)
- Short direction — 1750 pounds per foot (page 97)

The largest expected shear force of 1750 pounds per foot is well below the allowable of 9124 pounds per foot of slab width. Therefore shear capacity of the section is satisfactory.

7.5.7— Summary of results

Table 23 summarizes the results of shear and moment checks for the uniform thickness slab on plastic clay, comparing design values with allowables for both edge lift and center lift conditions. Since differential deflection was the governing factor in selecting the uniform thickness, the resulting differential design deflections will be close to, but within, the original allowables for the Section 7.4 slab on which this one was modeled. A formal deflection check is not required.

Edge lift design	*Design*	*Allowable*	**Center lift design**	*Design*	*Allowable*
Moment, ft.-kips per ft.			Moment, ft.-kips per ft.		
Long Direction			*Long Direction*		
Tensile	2.21	6.13	Tensile	5.16	6.13
Compressive	2.21	17.81	Compressive	5.16	17.81
Short direction			*Short direction*		
Tensile	2.82	5.97	Tensile	5.31	5.97
Compressive	2.82	17.97	Compressive	5.31	17.97
Shear capacity, pounds per foot of slab width			Shear capacity, pounds per foot of slab width		
Long Direction	930	9124	*Long Direction*	1370	9124
Short direction	880	9124	*Short direction*	1750	9124

Tendon requirements *(as shown in Figure 53)*

Long direction

Twenty-three ½-inch-diameter 270-k strands located at the neutral axis (4⅝ inches from slab top)

Short direction

Forty-one ½-inch-diameter 270-k strands located at the neutral axis (4⅝ inches from slab top)

Table 23 *Design Summary for Uniform Thickness Post-Tensioned Slab on Plastic Clay.*

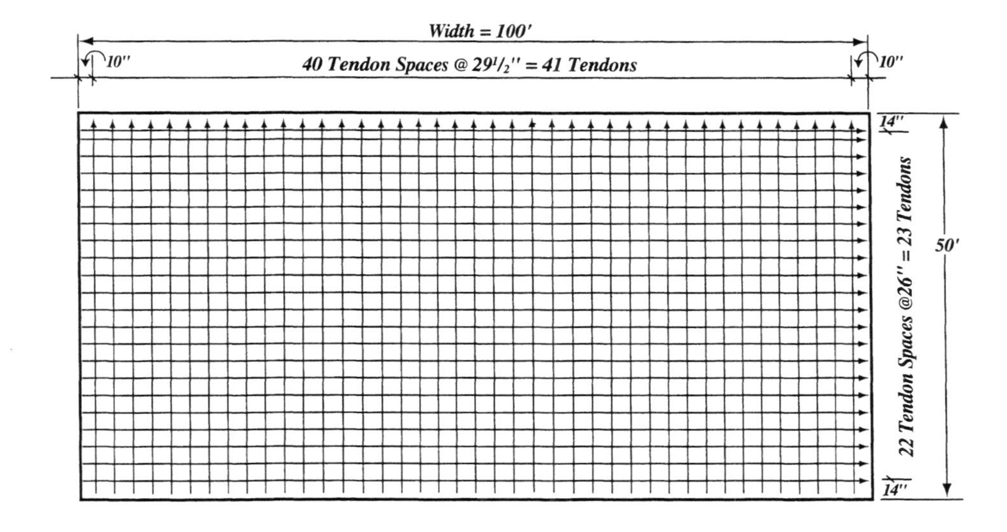

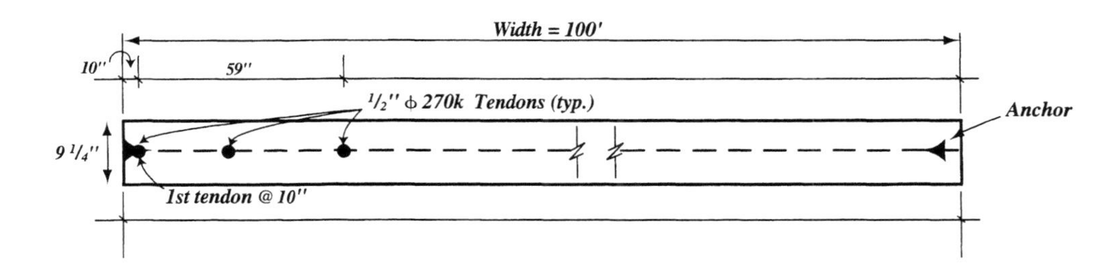

Figure 53 *Tendon locations for uniform thickness post-tensioned slab on plastic clay.*

7.6—DESIGN EXAMPLE: Simple rectangle, conventionally reinforced slab on plastic clay with uniform and perimeter load

7.6.1 — Introduction

The PTI design procedure shown in Section 7.4 is the state-of-the-art solution for slabs on plastic clays. The authors' choice for slabs on plastic clays is to use post-tensioning tendons for two distinct reasons:

1. The available design procedure is initially geared to post-tensioning; and
2. In many regions, post-tensioning is the most economical and most forgiving solution.

However, this is does not rule out the use of conventional reinforcement with this design procedure. Reinforcing bars can be readily adapted to the solution of the problem, as we will show here. Familiarity with the previous example of this chapter is necessary for the solution because the refinement necessary for a conventional reinforcement alternative is based on all previous work.

The primary difference between the two solutions is that the post-tensioned slab is analyzed as an uncracked section, and stiffness is based on the gross moment of inertia. With conventional reinforcement, a cracked section will have to be considered—at least for positive moment capacity.

In negative moment considerations, stresses will often be low enough for the gross uncracked moment of inertia to be considered.

7.6.2 — Modeling the problem, long direction

The easiest way to provide an equivalent conventionally reinforced solution is to provide a slab which has both an equivalent strength, and equally important, an equivalent stiffness. *Figure 54* shows a model of the post-tensioned section of the previous problem in the long direction.

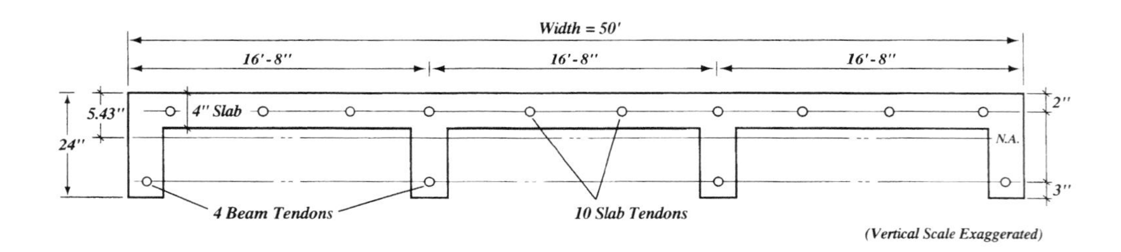

Figure 54 *Model of post-tensioned section obtained in solution of problem in Section 7.4. This section is cut across the short direction for use in analysis of the stiffened slab in the long direction.*

Cross section properties previously obtained are:

I = 133,942 in.^{4}
A = 3360 square inches
Depth to neutral axis (NA) = 5.43 inches from top
S_B = 7213 in.^{3}
S_T = 24,667 in.^{3}

As is the case with other slab solutions, the rupture modulus is often one of the primary considerations in design strength. This can also be considered in a conventional design when looking at the negative moment requirements. This is because of a top section modulus which is on the order of three times as large as the bottom section modulus.

For positive moment capacity, the slab is likely to crack when conventional reinforcing is implemented. The cross section of a conventionally reinforced slab will look something like the one shown in *Figure 55*.

Commentary:
Center lift produces a negative moment. Since the neutral axis of the slab is quite high, the top fiber stress rarely exceeds the concrete rupture modulus. In addition, the conventionally reinforced design almost invariably results in a greater beam depth, thereby providing a greater section modulus. However, the authors always recommend conventional deformed reinforcement or heavy flat-sheet welded wire fabric in the slab to insure its integrity for negative moments.

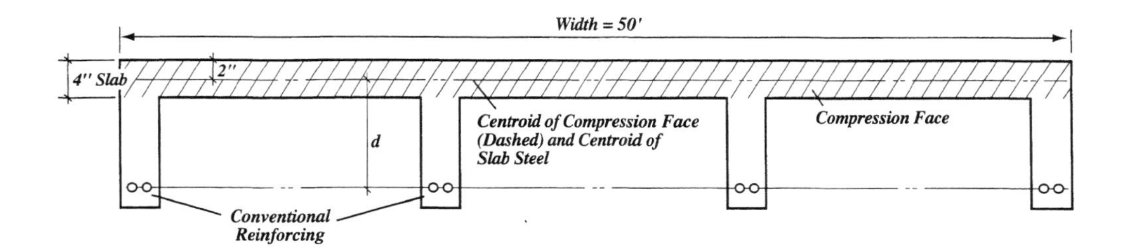

Figure 55 *Conventionally reinforced slab cross section (long direction analysis).*

For a simple, practical solution to the problem, assume the center of the 4-inch slab as the center of the compression face. Then use a modular ratio $N = 9$ in computing the effective transformed area for conventional reinforcing steel.

The authors have found that with this design method the critical design parameter appears to be deflection. This simplifies the conventional design procedure to determining a cracked section stiffness that satisfies the deflection requirements of the edge lift design condition.

7.6.3 — Values needed to solve the problem with conventional reinforcement

We need two values from the previous post-tensioned design (Section 7.4) to determine deflection requirements:

- Allowable differential deflection
- Expected differential deflection

Most likely, the cracked moment of inertia for the conventional design will be less than the original uncracked post-tensioned section. The decreased moment of inertia, however, provides a lower allowable differential deflection. This creates a potential loop where a direct proportion determining the minimum acceptable moment of inertia is invalid.

Use the edge lift condition values from the post-tensioned slab design (summarized in *Table 21*):

Previous long direction allowable differential deflection = 0.89 inches
Previous long direction expected differential deflection = 0.25 inches

Therefore the apparent minimum acceptable moment of inertia is 0.25/0.89 times previous I of 133,942 in.^{4} :

$0.28 \times 133{,}942 = 37{,}624$ in.^{4}
Use $1.33 \times 37{,}624 = 50{,}040$ in.^{4}

Trial and error have indicated that a good place to start is with a cracked moment of inertia requirement 1.33 times the minimum now acceptable by direct proportion.

A cracked section with this moment of inertia in the long direction is the starting point for the reinforced concrete solution.

Commentary:
Selection of the 30-inch depth shown in Figure 56 *is based on the authors' experience. Several trial depths may be necessary; however, it is reasonable to select something greater than the post-tensioned depth, due to the lower stiffness found in the cracked section.*

7.6.4 — The reinforced concrete solution for slab on plastic clay

Long direction section:

Choose a section for the long direction *(Figure 56)* that would approximate the desired moment of inertia from the previous example. Try three #8 bars in the bottom of each of the four beams.

Total steel area = 3 × 4 × 0.78 = 9.36 square inches

Develop the idealized section shown in *Figure 57*. This assumes the 4-inch slab is the total compression flange and the tension flange is a transformed area equal to 9 × 9.36 = 84.24 square inches.

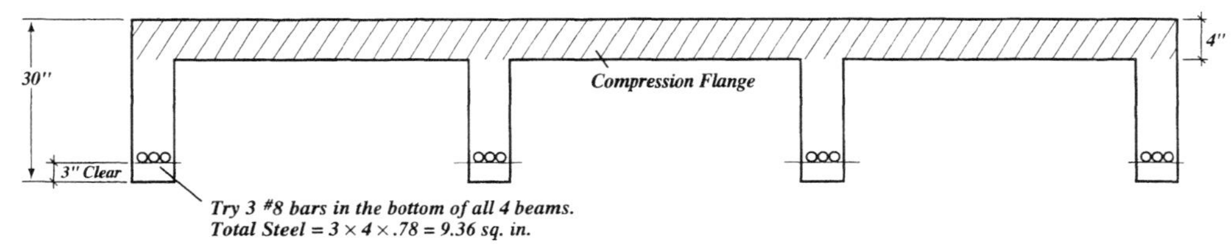

Figure 56 *Cross section selection approximating the desired moment of inertia.*

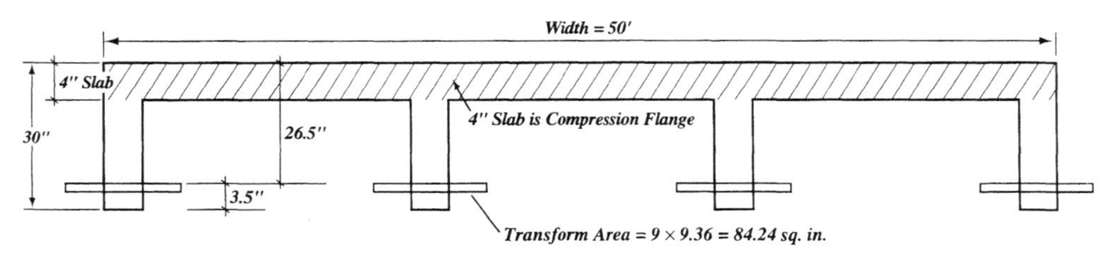

Figure 57 *Idealized section showing transformed steel area.*

Determine neutral axis and section properties:

	Area, in.^{2}	y, in.	Ay, in.^{3}
Concrete (4 × 50 × 12)	2400	2	4800
Steel *(transformed)* (9 × 9.36)	84.24	26.5	2232
TOTAL	2484.24		7032

$\bar{y}$ = 2.83 inches (distance from top fiber to neutral axis)

Determine moment of inertia:

	Io	*d*	Ad^2	*TI*
Concrete (1/12 bd^3)	3200	0.83	1653	4853
Steel *(transformed)*	0	23.67	47,197	47,197
			TOTAL	52,050

Check chosen section for allowable deflection:

$$\beta = \frac{1}{12}\sqrt[4]{\frac{E_C I}{E_S}}$$

$$\beta = \frac{1}{12}\sqrt[4]{\frac{1{,}500{,}000 \times 52{,}050}{1000}}$$

$$\beta = \frac{1}{12} \times 94.00 = 7.83 \text{ feet}$$

$$6\beta = 6 \times 7.83 = 47.00 \text{ feet}$$

$$\Delta_{allow} = \frac{12 \times 47}{800} = 0.70 \text{ inches}$$

Determine expected differential deflection (*long* direction analysis):

Check the reinforced section obtained by proportioning against post-tensioned section for determining expected differential deflection.

I of post-tensioned section = 133,942 $in.^4$
I of cracked conventional section = 52,050 $in.^4$

Previous expected differential deflection (post-tensioned solution) = 0.25 inches
Expected deflection of conventionally reinforced section = 133,942/52,050 × 0.25 = 0.64 inches
Allowable = 0.70 inches > 0.64 inches — OK

Select four beams 12 inches wide and 30 inches deep, each reinforced with 3 #8 bars, 3 inches from the bottom.

For slab steel, it is recommended that #3 deformed bars be used at mid-depth of the slab for crack control. Authors suggest #3 bars 12 inches on center, both ways.

7.6.5 — Short direction section with conventional reinforcement

From the previous post-tensioned design example solution summarized in *Table 21*:

Previous allowable deflection, short direction = 0.75 inches
Previous expected deflection, short direction = 0.23 inches
Previous *I*, short direction 213,907 $in.^4$

To start, determine the approximate acceptable moment of inertia from the above values:

0.23/0.75 × 213,907 = 65,598 $in.^4$

Assume a slightly larger section as was done in analyzing the long direction:

65,598 × 1.33 = 87,245 $in.^4$ (approximate value needed)

For simplicity of design use the same bars and beams determined for the long direction, that is:

12 x 30 inches, with three #8 bars per beam

As shown in *Figure 58*, there are six beams in the assumed section for the short direction.

Three #8 bars = 3 × 0.79 = 2.37 square inches steel per beam
2.37 × 6 = 14.22 square inches for the entire section

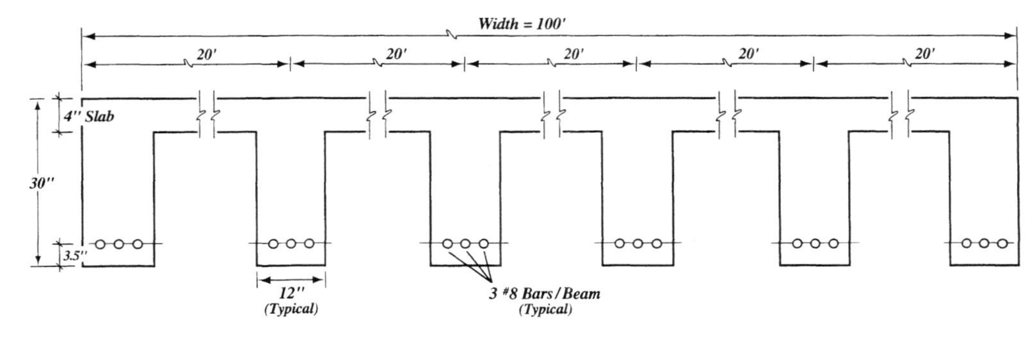

Figure 58 *Assumed section for short direction analysis. Cross section taken through the long direction.*

Determine the neutral axis and section properties of the assumed section:

	Area, in.^{2}	y, in.	Ay, in.^{3}
Concrete (4 × 100 × 12)	4800	2	9600
Steel (9 × 14.22)	128	26.5	3392
	4928		12,992

$\bar{y}$ = 12,992/4928 = 2.63 inches

Determine moment of inertia:

	Io	*d*	Ad^2	*TI*
Concrete (1/12 × 100 × 4^3)	6400	0.63	1905	8305
Steel	0	23.87	72,920	72,920
				81,255

It is assumed that 81,255 in.^{4} is sufficiently close to the 87,245 in.^{4} approximation to continue with the analysis.

Determine allowable deflection of the chosen section (*short* direction):

$$\beta = \frac{1}{12}\sqrt[4]{\frac{E_C I}{E_s}}$$

$$\beta = \frac{1}{12}\sqrt[4]{\frac{1{,}500{,}000 \times 81{,}225}{1000}}$$

$$\beta = 8.75 \text{ feet}$$

$$6\beta = 52.2 \text{ feet} > 50 \text{ so use } 50 \text{ feet}$$

$$\Delta_{allow} = \frac{12 \times 50}{800} = 0.75 \text{ inches}$$

Determine the expected differential deflection:

Check the reinforced section obtained as a proportion of the post-tensioned design section to determine the expected differential deflection:

I of the post-tensioned section (short direction) = 213,907 in.^{4}
I of the cracked conventional section = 81,225 in.^{4}

Previous expected differential deflection (for the post-tensioned design): = 0.23 inches

Expected deflection of the conventionally reinforced section in the short direction = 213,907/81,225 × 0.23 = 0.60 inches

The allowable deflection computed above for the short direction is 0.75 inches.
0.60 < 0.75, so section is OK for deflection.

The final design with conventional reinforcement has the same beam layout as the post-tensioned design (see *Figure 59*). The beams are 12 x 30 inches, reinforced with three #8 bars at the bottom. Stirrups are optional for construction purposes only. The slab should be reinforced with #3 bars at 12 inches on center both ways.

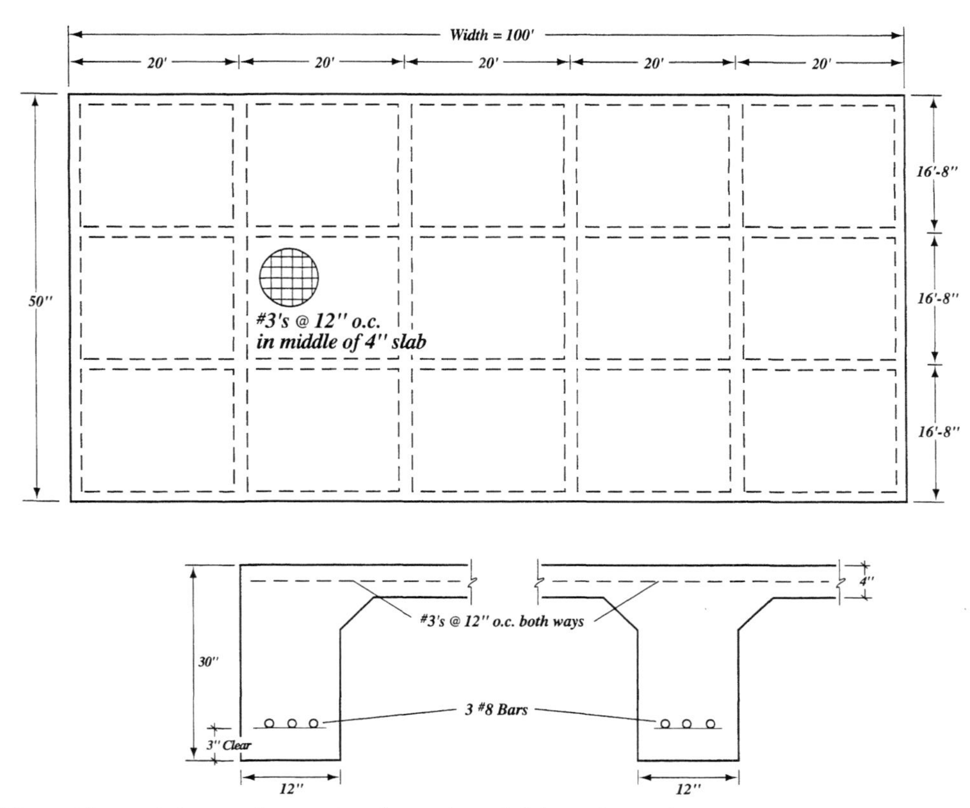

Figure 59 *Final design of conventionally reinforced slab on plastic clay.*

CHAPTER 8

DESIGN OF SLABS ON COMPRESSIBLE CLAYS

8.1 — Introduction

The more common slab on grade designs involve the generally local effect of the concrete slab due to loading in an isolated area. This may involve areas from one, two, or possibly three, floor panels ranging in area from about 100 square feet to 1000 square feet. The design generally assumes uniform pressure distribution and results in a simple static analysis.

By contrast, design of a slab on compressible clay must be done by including the footprint of the entire floor area. This is true because the footprint will tend to saucer (sink down) toward the middle of the slab due to long-term settlement. In addition, compressive clays undergo perimeter moisture changes that require the designer to address variations in perimeter support conditions.

This combination of factors calls for an empirical solution that is mathematically complicated. However, once moments, shears, and deflections are obtained, the structural part of the design is not difficult. *Flow Charts 3* and *4 (pages 114-115)*, along with the formulas given in sequence in the appendix *(page 234)*, should aid both in following the example and in solving future design problems.

8.2 — What is compressible clay?

Compressible clays are fine grained soils that have a tendency toward almost rapid change in height of a given soil mass due to applied loads. *Consolidation* refers to the slower change in height associated with water flowing out of a soil mass. The broader term *settlement* refers to change in height due to both compression and consolidation. However, it is common to refer to clays with a potential for settlement as compressible clays. It is always advisable to obtain further input from a qualified geotechnical engineer when such materials are expected.

The Building Research Advisory Board recommends that a soil be classified compressible when the ratio of the unconfined compressive strength q_u of the soil to the average total load W_{slab} on the slab is between 2.5 and 7.5.

Commentary:
W_{slab} *is the weight of the slab plus all applied loads divided by the area of the slab.*

For example, assume that the total slab load (its own weight plus applied loads), as distributed over its entire surface, is 300 psf. With this loading, a soil that would qualify for this analysis would probably be in the range of q_u = 750 to q_u = 2250. If the soil had a q_u = 1500, q_u/W_{slab} would equal 5, thereby establishing the structure and soil as a likely candidate for the design method presented in this chapter.

If the ratio were at the lower end of the range, the slab would more than likely be pile supported. For ratios above 7.5, the plastic clay formulas of *Chapter 7* may be more appropriate. The ultimate deciding factor is the soils expert's determination of the maximum anticipated settlement under load. If the anticipated differential settlement of the footprint exceeds 3 inches, or 1/200 times the slab length, the designer should consider alternate support mechanisms such as piling or piers founded in more stable material.

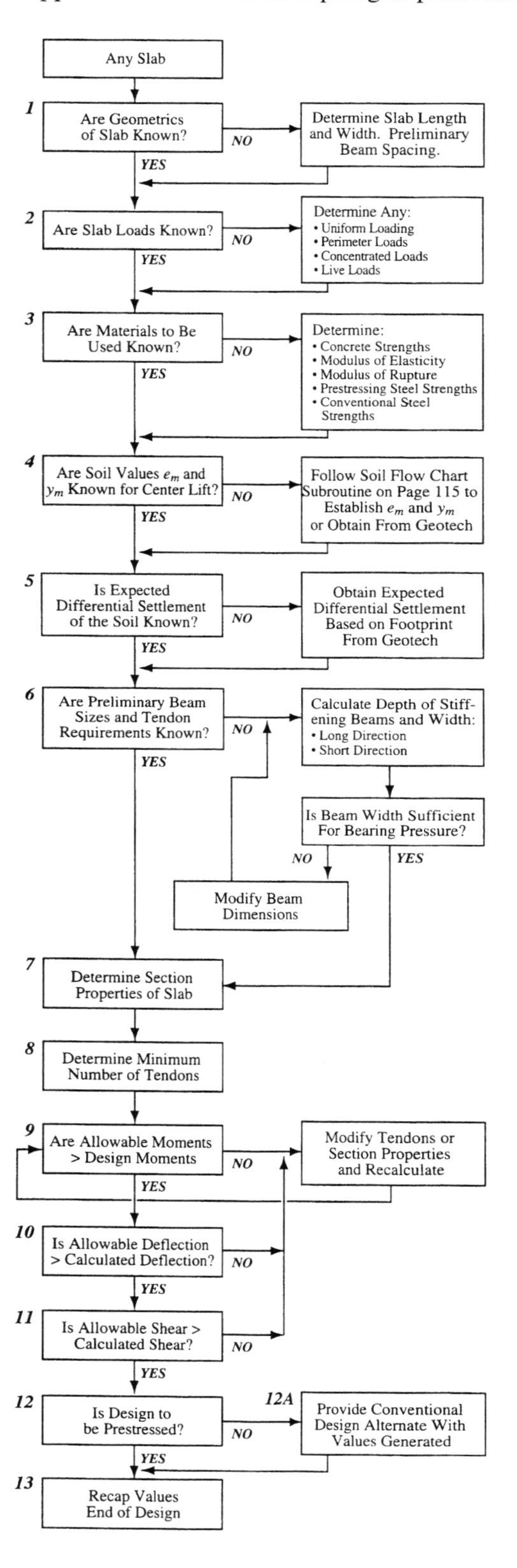

Flow Chart 3 *Procedure for design of slabs on compressible clays.*

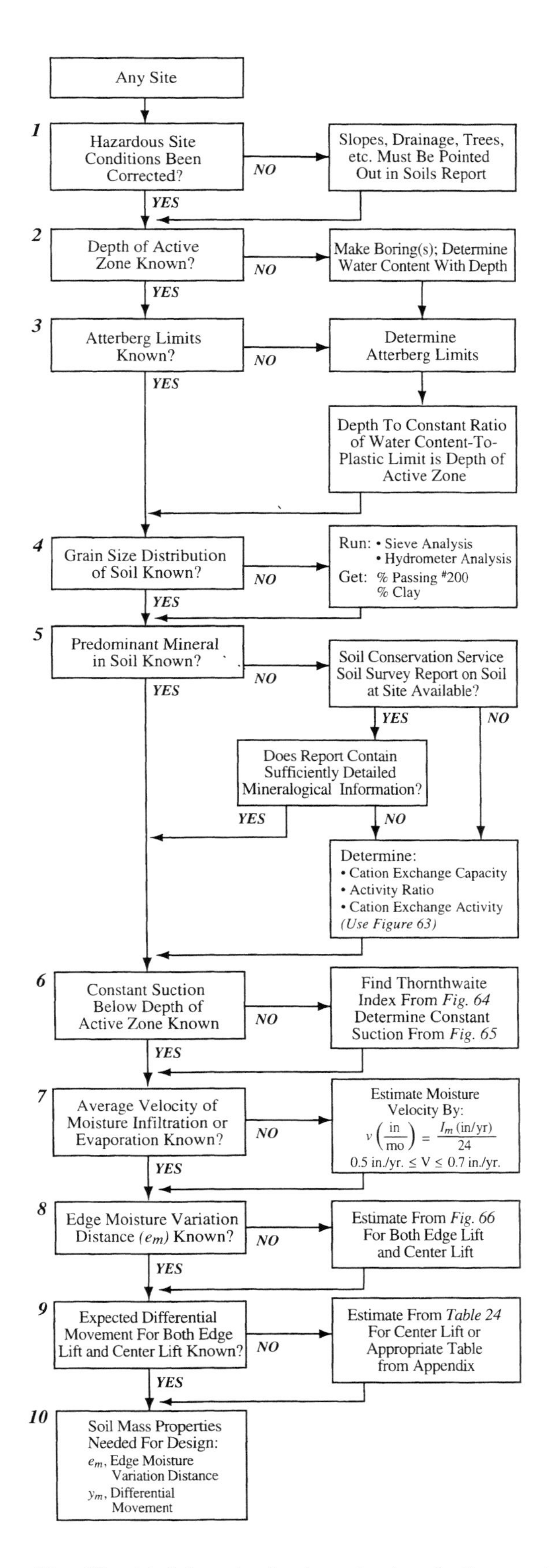

Flow Chart 4 *Subroutine for determination of soil properties, from the PTI Manual* (Reference 10).

8.3 — PTI method for designing slabs supported on compressible clays

The Post Tensioning Institute manual *(Reference 10)* gives a design procedure for ribbed and stiffened slabs supported on compressible soils. As with the procedure for slabs on plastic clays, the use of post-tensioning is not the only available solution. Since the PTI design procedure is geared to a post-tensioned solution, the authors present that option first, followed by a conversion to an equivalent conventionally reinforced slab.

8.3.1 — Analysis of slab loads

As indicated with the slab on ground on plastic clay, the slab supported on a compressible clay can be idealized as a uniform loading condition with perimeter wall loading. Concentrated loads or point loads can be treated by separate analysis.

Similar to the plastic clay example, loading consists of taking all of the live loads and dead loads of the slab and applying them uniformly over the whole slab. This procedure is valid since the stiffening beams in both direction permit even distribution. Stiffening beams are generally located on a grid not exceeding 20 feet on center *(Figure 60)*.

Commentary:
As with the analysis of slabs on plastic clays, this design procedure treats the entire slab as opposed to an isolated section. This is because the stresses encountered affect the whole unit, and not simply an isolated section. The primary stresses in the slab are the result of the saucering (dipping down) of the slab toward the center because of the soil compressibility. Stiffening ribs resist the tendency of the slab to saucer, minimizing differential settlement.

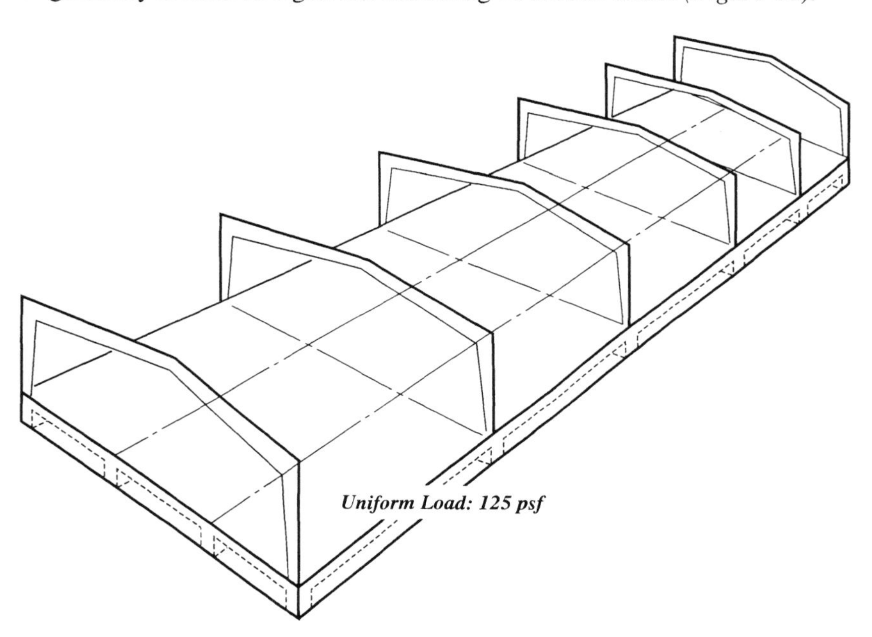

Figure 60 *Stiffening beams not farther than 20 feet apart both ways cause the slab to work as an overall unit to resist saucering due to compressible clays.*

8.3.2 — Values needed to solve the problem

Four primary areas of information are needed for design of a stiffened slab on compressible clay:

- Slab geometry, including preliminary beam spacing
- Loads on the slab
- Materials to be used in the slab
- Soil conditions

These are shown in Steps 1 through 4, *Flow Chart 3*. If the edge moisture variation distance e_m and differential movement y_m are not known, then several steps in the soils subroutine of

Flow Chart 4 must be followed. Since shrinking of the clay is also possible in the compressible mode, the moisture index and shrink potential data are also needed.

To simplify following the steps in the flow charts, all of the PTI formulas and charts associated with the intermediate answers are provided in the appendix *(page 234)*. The example presents a step-by-step solution for the design of slabs on compressible clays, using additional smaller copies of the design charts and tables at appropriate points where they are needed in the design.

8.3.3 — Design objectives

The design checks the ability of the slab to resist the positive moment created due to the saucering mode associated with soil settlement. Negative moments resulting from the condition referred to by PTI as *center lift* are also checked. This condition more appropriately should be identified as the negative moment resulting from perimeter shrinkage of the soil. This creates a condition emulating center lift. The doming of the slab is a result of the edge moisture variation distance, known as e_m. The dishing of the slab (saucering) is a function of the expected differential settlement. These conditions can be seen in *Figure 61*.

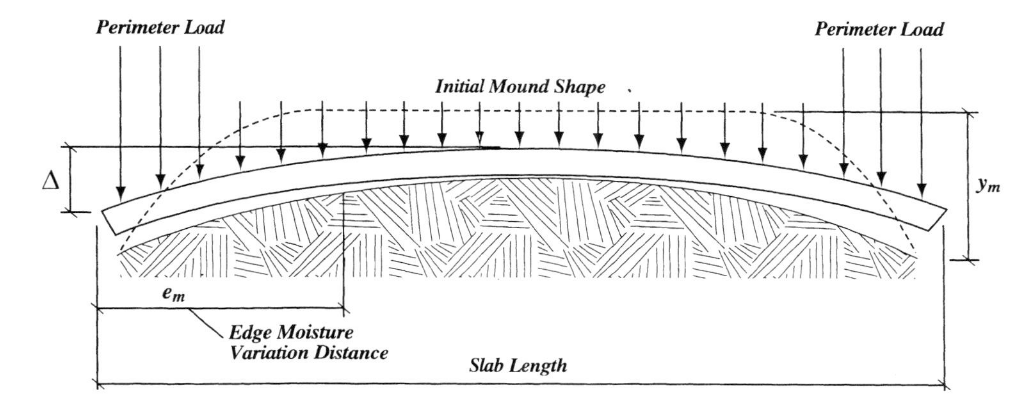

CENTER LIFT
(Edge Shrinkage)

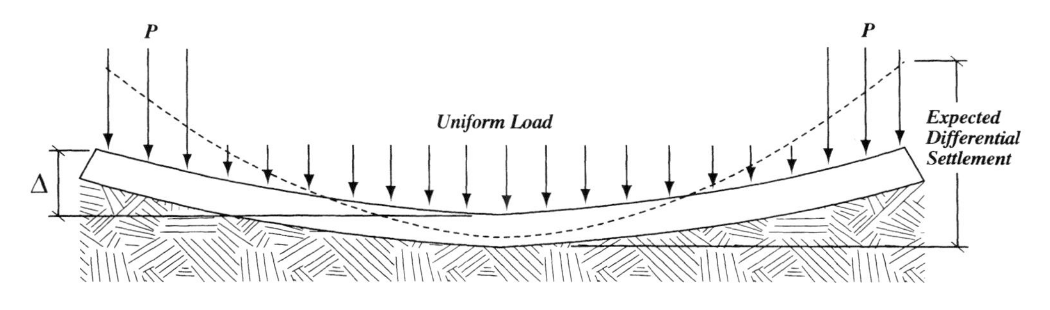

EDGE LIFT
(Dishing Due To Expected Differential Settlement)

Figure 61 *Soil structure interaction models, from the PTI Manual* (Reference 10).

Commentary:
For the center lift design, the edge moisture variation distance e_m *is determined from the basic input data. It is the distance, measured inward from the slab edge, over which the soil moisture content varies. Climate is the primary factor affecting this distance. The other soil- and climate-dependent value used for this analysis is* y_m*, the maximum differential soil movement. If these values could be supplied by the geotechnical engineer, it would greatly simplify the problem for the designer.*

8.3.4 — Computer solutions

A number of vendors, including the Post-Tensioning Institute have developed computer programs for producing these PTI solutions. The flow chart for this chapter provides a good outline for developing a computer program for the compressible clay solution. The detailed problem that follows provides all formulas needed to develop either simplifying subroutines for hand-held calculators or a full scale computer program. If this slab design procedure is used regularly, the authors strongly recommend purchase of one of the existing programs or the preparation of either a program or a spread sheet.

8.4 — DESIGN EXAMPLE: Rectangular post-tensioned slab on compressible clay with uniform and perimeter loads

8.4.1 — Symbols and notation

A area of gross concrete cross section, square inches

A_o a co-dependent variable used in factoring center lift moment design, dependent on the physical properties of the slab, loading conditions, and soil properties

Ac activity ratio of clay

b width of an individual stiffening beam, inches

B a nondimensional constant used in factoring center lift design, dependent on soil properties

C a nondimensional constant used in factoring center lift design, dependent on loading condition and soil properties

$CEAc$ cation exchange activity of soil

d depth of stiffening beam measured from top surface of slab to bottom of beam, inches

e eccentricity of post-tensioning force, inches

e_m edge moisture variation distance, feet

E_c long-term or creep modulus of elasticity of concrete, psi

E_s modulus of elasticity of soil, psi

f_c allowable compressive stress in concrete, psi

f_c' 28-day compressive strength of concrete, psi

f_{ps} permissible stress in prestressing tendon, psi

f_{pu} ultimate stress in prestressing tendon, psi

f_t allowable tensile stress in concrete, psi

I gross moment of inertia, in.^{4}

I_m Thornthwaite index, moisture velocity in inches per year

L total slab length in the direction being considered, feet

LL liquid limit

M_{cs} moment occurring as a result of constructing over compressible soil, foot-kips per foot

M_{ns} moment occurring in the "no-swell" condition, foot-kips per foot

M_ℓ design moment in long direction, foot-kips per foot

M_s design moment in the short direction, foot-kips per foot

${}_nM_t$ ${}_nM_c$ negative and positive bending moments including

${}_pM_t$ ${}_pM_c$ tension or compression in the extreme fibers, foot-kips per foot

n number of beams in a cross section

N modular ratio (modulus of elasticity of steel to modulus of elasticity of concrete)

N_t number of tendons
pF constant soil suction value
P perimeter loading on the slab, pounds per foot
P_r prestressing force, kips
P_re moment due to post-tensioning eccentricity, inch-kips
P/A prestress force resulting from tendon load divided by gross concrete area, psi
PI plasticity index
PL plastic limit
q_{allow} allowable soil bearing pressure
q_u unconfined compressive strength of soil, psf
S beam spacing, feet
S_B section modulus with respect to the bottom fiber, $in.^3$
S_T section modulus with respect to the top fiber, $in.^3$
t slab thickness, inches
v design shear stress, psi
v_c allowable concrete shear stress, psi
V design (service) shear force, kips per foot
V_s expected (service) shear force in short direction, kips per foot
V_ℓ expected (service) shear force in long direction, kips per foot
V_{ns} expected (service) shear, no-swell condition, kips per foot
V_{cs} expected (service) shear, compressible soil condition, kips per foot
w soil bearing pressure, kips per square foot
W slab width, feet
W_{slab} slab weight, pounds
$\bar{y}$ neutral axis location of stiffened cross section, inches
y_m maximum differential soil movement, inches
Z depth to constant suction, feet
exp Z natural antilog of a function Z, used to calculate expected differential deflection. Z in this instance depends on the physical properties of the slab and loading conditions. Designation is e^Z.
β relative stiffness length, feet
δ expected settlement occurring in compressible soil, inches
Δ expected differential deflection of slab under service load, inches
Δ_{allow} allowable differential deflection of slab, inches
Δ_{cs} differential deflection occurring as a result of constructing over compressible soil, inches
Δ_{ns} differential deflection occurring in the "no-swell" condition, inches
μ coefficient of friction

8.4.2 — The problem and initial assumptions; materials data

A single-story rigid frame metal building in Mobile, Alabama, has a perimeter wall load of

900 pounds per linear foot

The concentrated load from the rigid frame has been included in the perimeter load value. The uniform live load on the slab is

125 pounds per square foot

The slab *(Figure 62)* measures 100 feet x 50 feet and stiffening beams are assumed at spacings of 20 feet and 16 feet 8 inches. (NOTE: These are values required in Steps 1 and 2 of *Flow Chart 3*.)

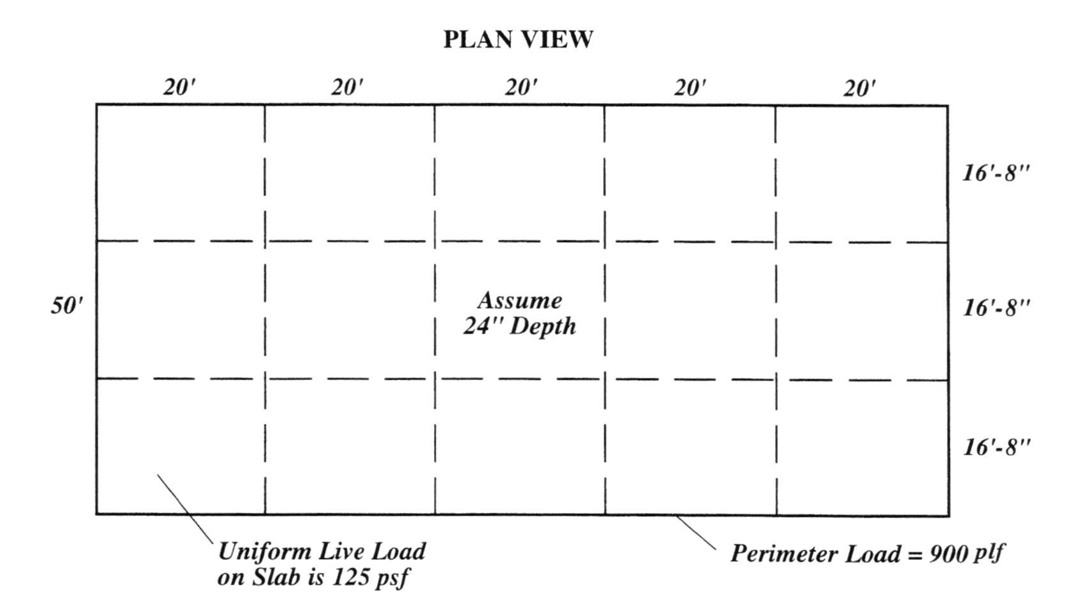

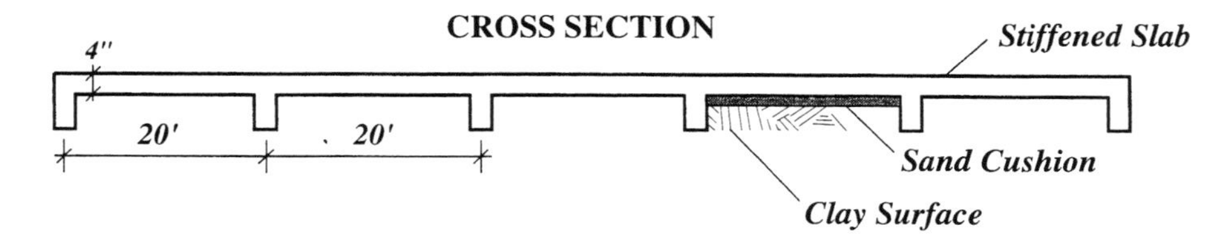

Figure 62 *Plan and cross section of the example floor slab.*

Commentary:
You can run an analysis of the transformed section of the edge grade beam to show whether the concentrated load of the rigid frame can be satisfactorily distributed along its length from point load to point load. A uniform load is a reasonable assumption in many instances. If concentrated loads are questionable, they should be checked and accommodated with a modified section.

The beam spacing should not exceed 20 feet on center as a matter of standard practice. This has governed the spacing selected here, where rigid frames happen to be 20 feet apart in the long direction of the slab. The 16-foot-8-inch dimension was chosen as the largest equal division of 50 feet that comes within the 20-foot limit. Where rigid frame bay spacing exceeds 20 feet, an intermediate beam is recommended in most instances except for extremely stable soil conditions. Then it is sometimes a reasonable judgment call to permit extended beam spacing.

The materials to be used [Step 3 of *Flow Chart 3*] are as follows:

Concrete compressive strength $f_c' = 3000$ psi
Concrete creep modulus of elasticity $E_c = 1{,}500{,}000$ psi
Prestressing steel: 270k, 1/2-inch-diameter 7-wire strand

8.4.3 — Soils investigation

The design requires knowledge of soil movement (Steps 4 and 5, *Flow Chart 3*). We have some information available regarding site soil conditions, as follows:

Atterberg limits:
Plastic limit $PL = 30$
Liquid limit $LL = 70$
Plasticity index $PI = 40$
Clay content = 70%

Unconfined compressive strength q_u = 1500 psf
Soil modulus of elasticity E_s = 1000 psi
Depth to constant suction Z = 7 feet
Expected differential settlement δ = 1.25 inches
Location: Mobile, Alabama

Some of this information must be used in the subroutine of *Flow Chart 4* to determine the soil values e_m and y_m required by Step 4, *Flow Chart 3*. Note that the soils specialist has given us the expected differential settlement δ needed for Step 5, *Flow Chart 3*.

8.4.3.1 — Determine the predominant clay mineral in the soil (Step 5, Flow Chart 4)

Since we have information covering the first four steps of *Flow Chart 4*, we can start at Step 5. There are two steps needed to determine the predominant clay mineral in the soil: Determine cation exchange activity, *CEAc*:

$CEAc = PL^{1.17} \div$ percent clay
$CEAc = 30^{1.17} \div 70$
$CEAc = 53.48 \div 70 = 0.764$

Determine the clay activity ratio, *Ac*:

$Ac = PI \div$ percent clay
$Ac = 40 \div 70 = 0.57$

With these two values, we can use the clay classification chart *(Figure 63)* to determine the predominant clay mineral in the soil. Enter the chart from the bottom with *Ac* and from the left with *CEAc*. Project lines upward and horizontally from these values. The area in which their intersection occurs determines that the principal clay mineral is interstratified. Therefore use the tables for montmorillonite in later steps.

Commentary:
It has been observed that if the plastic limit of the soil, PL, is above 25%, illite and kaolinite will be removed from consideration. Therefore, montmorillonite can be assumed for design purposes.

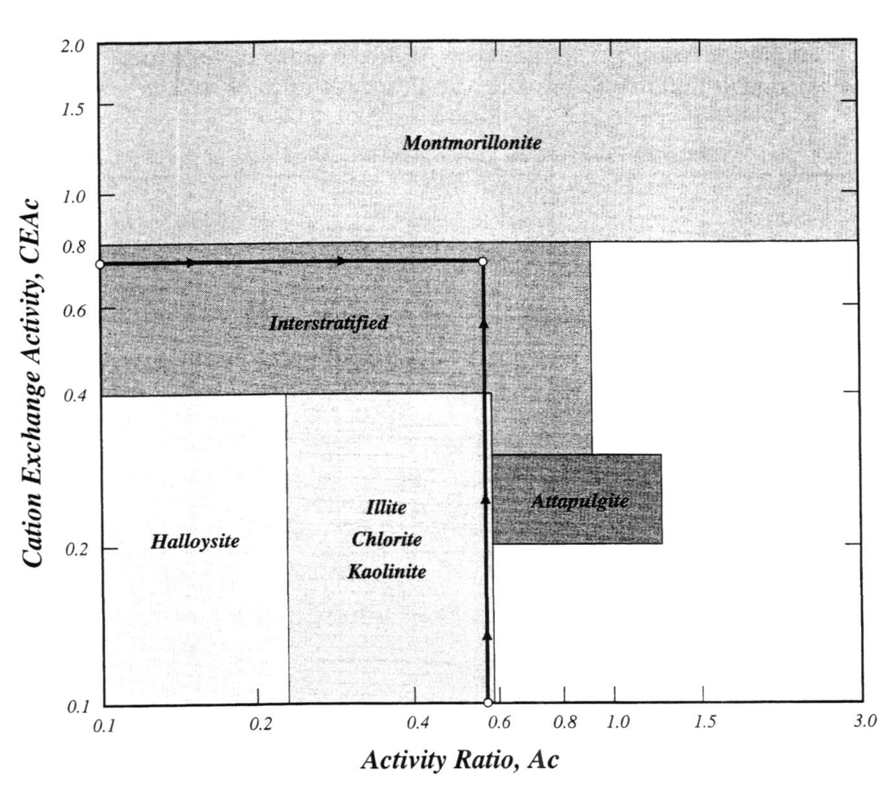

Figure 63 *Clay classification based on cation exchange and clay activity ratios (from the PTI Manual).*

8.4.3.2 — Constant soil suction value for soil

Commentary:
It is always more conservative to choose the lower number in selecting a Thornthwaite moisture index.

Step 6, *Flow Chart 4*, involves determination of the soil's constant suction value, pF. For this we need to find the Thornthwaite moisture index, I_m from the map of *Figure 64*. Interpolating a value for Mobile, Alabama, we read $I_m = +40$; that is, something greater than 40. For the design, we use a value of 40.

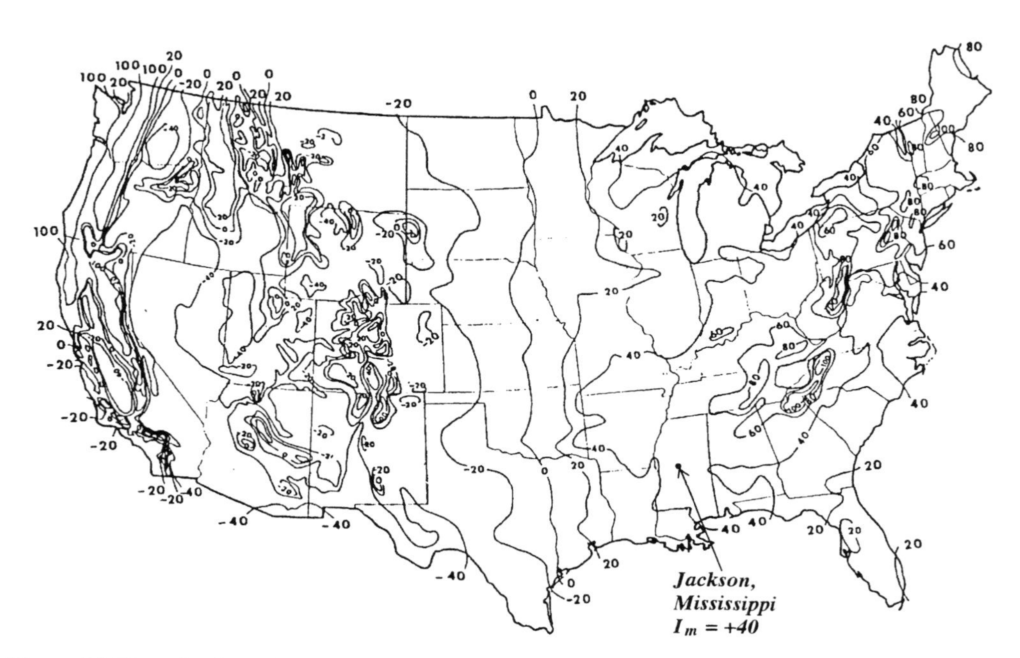

Figure 64 *Thornthwaite moisture index values for the United States, from the PTI Manual* (Reference 10).

With the Thornthwaite index, I_m, we can enter *Figure 65* at the bottom, move up to the curve, and then horizontally left to read the constant soil suction value pF as 3.2.

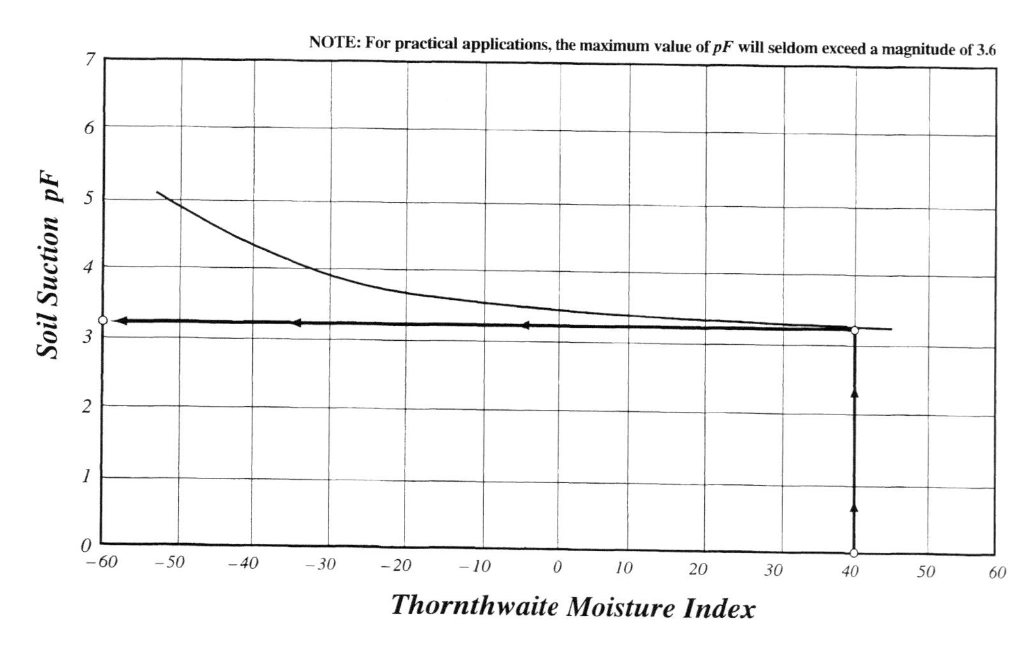

Figure 65 *Relationship of constant soil suction value pF to Thornthwaite moisture index, from the PTI Manual* (Reference 10).

8.4.3.3 — Velocity of moisture flow

Step 7, *Flow Chart 4*, calls for determination of the average moisture flow in the soil. This velocity is estimated using the Thornthwaite index I_m which is expressed in inches per year.

moisture velocity = $0.5 \times I_m/12$
$= 0.5 \times 40/12 = 1.67$ inches per month

According to the PTI Manual, the maximum moisture velocity is limited to 0.7 inches per month. Therefore use 0.7 inches per month instead of the calculated value.

8.4.3.4 — Determine edge moisture variation distance, e_m

Step 8, *Flow Chart 4*, calls for the edge moisture variation distance, which we will determine for center lift from *Figure 66*. Enter the chart at the bottom with the previously determined Thornthwaite moisture index I_m of +40. Proceed vertically upward to the center of the band for center lift, then horizontally to the left to read

e_m, edge moisture variation distance = 4.0 feet (center lift)

Center lift is the only mode considered because the soil compressibility accounts for the positive moment or edge lift condition. Since a compressible soil is still capable of volume change due to edge moisture variation, center lift (which could also be referred to as edge shrinkage) must be considered from a plastic clay perspective.

Commentary:

The example uses a value midway across the band (Figure 66), *but if the designer is extremely familiar with the local conditions, the chart bands allow some leeway for interpretation suited to site conditions. For example, if a particular vicinity is known to be a cause for concern, be more conservative by moving to values higher in either of the bands. To be less conservative, move to values closer to the bottom of the band.*

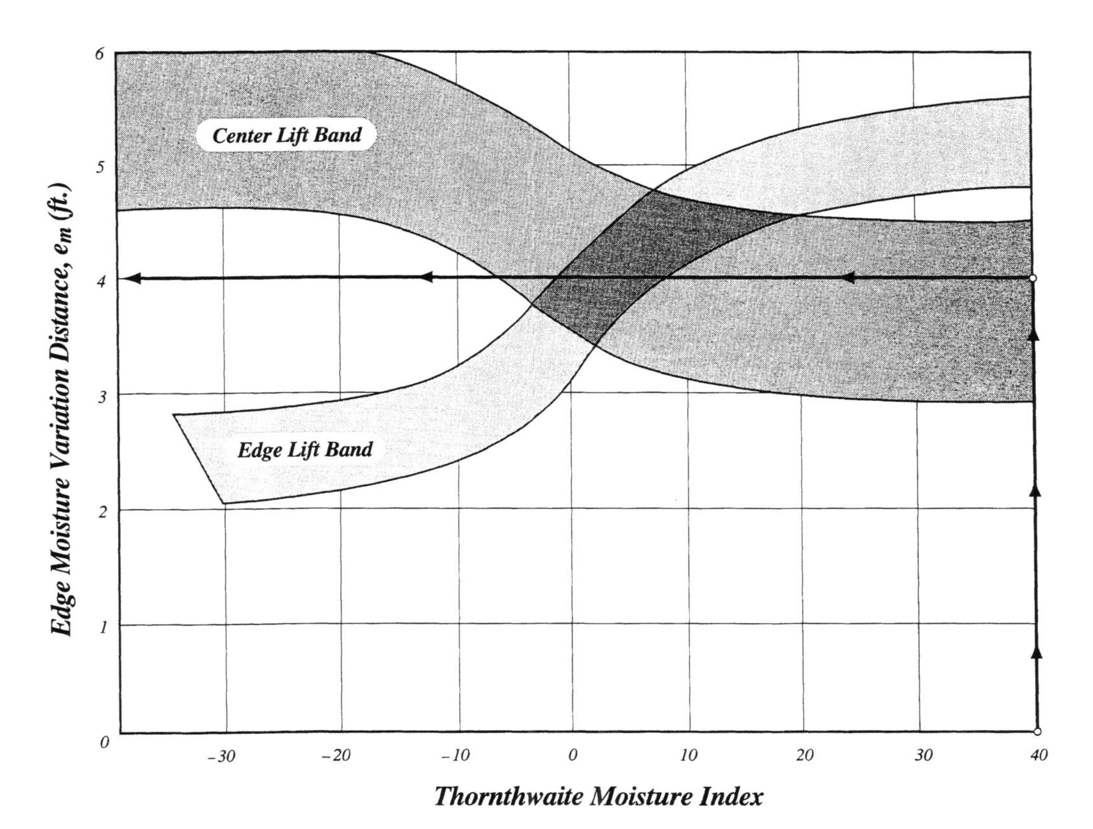

Figure 66 *Approximate relationship between Thornthwaite index and edge moisture variation distance, from the PTI Manual* (Reference 10). *Note that extremely active clays may generate larger values of edge moisture variation than reflected by the above curves and related tables. Therefore these curves should be used only in conjunction with a site-specific soils investigation by knowledgeable geotechnical engineers.*

8.4.3.5 — Estimated differential movement (swell)

We now have all the values needed to determine the expected differential movement (Step 9, *Flow Chart 4*), using *Table 24* for soil having 70 % clay:

Depth to constant suction = 7 feet
Constant suction value pF = 3.2
Velocity of moisture flow = 0.7 inches per month
Edge moisture variation distance e_m = 4.0 feet

From *Table 24*, read y_m = 0.48 inches for center lift conditions. As previously noted, we check only the center lift condition for the problem because the soil compressibility accounts for the edge lift condition.

Commentary:
To repeat, it is of special importance to the designer to have the soils expert supply the values of e_m *and* y_m *for center lift and edge lift if the soils are expansive. For compressible soils, the designer needs* e_m *and* y_m *for center lift, as well as the value of* δ*. Without this site-specific information, calculations must be simplified in the conservative direction; always assume the clay mineral to be montmorillonite and use a conservative value of Z, depth to constant suction. If the clay percentage varies, a higher percent is the more conservative choice.*

Percent Clay (%)	Depth to Constant Suction (ft.)	Constant Suction (pF)	Velocity of Moisture Flow (in./mo.)	Differential Swell (in.) Edge Distance Penetration (ft.)							
				1 ft.	2 ft.	3 ft.	4 ft.	5 ft.	6 ft.	7 ft.	8 ft.
70	3	3.2	0.1	0.004	0.007	0.011	0.015	0.019	0.023	0.026	0.030
			0.3	0.011	0.022	0.034	0.046	0.057	0.070	0.082	0.095
			0.5	0.018	0.037	0.057	0.077	0.098	0.120	0.143	0.167
			0.7	0.026	0.052	0.082	0.110	0.141	0.174	0.210	0.248
		3.4	0.1	0.007	0.015	0.023	0.031	0.039	0.048	0.056	0.065
			0.3	0.023	0.047	0.072	0.098	0.126	0.156	0.188	0.222
			0.5	0.038	0.078	0.122	0.171	0.225	0.287	0.358	0.443
			0.7	0.052	0.110	0.176	0.251	0.341	0.452	0.600	0.814
		3.6	0.1	0.018	0.037	0.056	0.077	0.098	0.121	0.145	0.171
			0.3	0.054	0.114	0.182	0.262	0.359	0.481	0.648	0.903
			0.5	0.088	0.192	0.324	0.502	0.769	1.258	2.428	5.796
			0.7	0.120	0.273	0.485	0.821	1.485	3.203	8.234	—
	5	3.2	0.1	0.009	0.017	0.026	0.035	0.044	0.053	0.062	0.071
			0.3	0.026	0.053	0.080	0.107	0.135	0.164	0.193	0.223
			0.5	0.044	0.088	0.134	0.182	0.231	0.283	0.336	0.392
			0.7	0.060	0.123	0.189	0.258	0.331	0.409	0.491	0.580
		3.4	0.1	0.018	0.036	0.055	0.073	0.092	0.111	0.131	0.151
			0.3	0.053	0.109	0.167	0.228	0.292	0.360	0.433	0.511
			0.5	0.088	0.182	0.284	0.395	0.519	0.658	0.818	1.008
			0.7	0.121	0.256	0.406	0.578	0.781	1.031	1.358	1.837
		3.6	0.1	0.042	0.086	0.131	0.178	0.227	0.278	0.332	0.389
			0.3	0.123	0.260	0.415	0.594	0.807	1.073	1.429	1.968
			0.5	0.202	0.441	0.737	1.126	1.698	2.717	5.104	11.822
			0.7	0.278	0.627	1.098	1.820	3.199	6.640	—	—
	7	3.2	0.1	0.016	0.032	0.049	0.065	0.082	0.098	0.115	0.132
			0.3	0.048	0.097	0.148	0.199	0.251	0.305	0.359	0.415
			0.5	0.081	0.163	0.249	0.338	0.429	0.525	0.624	0.729
			0.7	0.112	0.229	0.351	0.480	0.616	0.759	0.915	1.081
		3.4	0.1	0.032	0.066	0.099	0.134	0.168	0.204	0.240	0.276
			0.3	0.098	0.199	0.306	0.418	0.536	0.661	0.794	0.937
			0.5	0.162	0.334	0.522	0.727	0.952	1.207	1.501	1.851
			0.7	0.224	0.470	0.747	1.063	1.435	1.891	2.492	3.383
		3.56	0.1	0.063	0.128	0.194	0.263	0.334	0.407	0.484	0.563
			0.3	0.185	0.387	0.609	0.857	1.139	1.468	1.865	2.376
			0.5	0.305	0.655	1.067	1.573	2.239	3.219	4.983	9.162
			0.7	0.421	0.928	1.569	2.453	3.868	6.763	—	—

Table 24 *Differential swell at the perimeter of a slab for center lift swelling in predominantly montmorillonite clay soil (70% clay), from the PTI Manual* (Reference 10).

8.4.4 — Preliminary beam sizes, tendon requirements

With e_m and y_m determined in Section 8.4.3 above, and with the value of expected soil settlement δ (under load) given, we are ready to proceed with Step 6, *Flow Chart 3*. Since the soil is compressible, the distribution made and resulting stresses will emulate the edge lift condition as seen in plastic soils.

8.4.4.1 — Estimate the required depth d of stiffening beams.

$$d = x^{1.176}$$

$$\text{where } x = \frac{L^{0.35} \times S^{0.88} \times e_m^{0.74} \times y_m^{0.76}}{12 \times \Delta_{allow} \times P^{0.01}}$$

$$\text{and } \Delta_{allow} = \frac{12 \times (L \text{ or } 6\beta)}{1700}$$

$$= \frac{L \text{ or } 6\beta}{140}$$

Commentary: *Remember that this formula for estimating* d *is empirical in origin. For compressible clays, substituting a value of 0.67 for* e_m *appears to provide a reasonable result for estimating beam depth.*

Assume that $e_m = 0.67$ for purposes of estimating the initial beam depth only. Also assume that y_m, the differential deflection of the soil, is 1.25 inches. We do this because the compressible mode will dictate the greatest imposed loads on the slab. Therefore y_m becomes δ.

Long Direction:

Slab length $L = 100$ feet
Beam spacing $S = 16.676$ feet
Perimeter load $P = 900$ pounds per foot

Since this building is on compressible soil, use 6 feet as an initial approximation of β.

$6\beta = 36$ feet < 100 and so 6β governs
$\Delta_{allow} = 36/140 = 0.26$ inches

Use this allowable deflection and other given and calculated values to determine a value of x in the long direction, so that a beam depth d can be determined.

The PTI Manual uses an initial approximation of 4 feet for compressible soils. It has been the authors' experience that 6 feet more closely approximates the stiffness required.

$$x_{long} = \frac{100^{0.35} \times 16.67^{0.88} \times 0.67^{0.74} \times 1.25^{0.76}}{12 \times 0.26 \times 900^{0.01}}$$

$$x_{long} = \frac{5.01 \times 11.89 \times 0.74 \times 1.18}{3.12 \times 1.07}$$

$$x_{long} = \frac{52.01}{3.34} = 15.57$$

$$d_{long} = 15.57^{1.176} = 25.24 \text{ inches}$$

Short Direction:

Slab length $L = 50$ feet
Beam spacing $S = 20$ feet
Perimeter load $P = 900$ pounds per foot
$6\beta = 36$ feet < 50 so 6β again governs, and the allowable deflection is 36/140 or 0.26 inches.

$$x_{short} = \frac{50^{0.35} \times 20^{0.88} \times 0.67^{0.74} \times 1.25^{0.76}}{12 \times 0.26 \times 900^{0.01}}$$

$$x_{short} = \frac{3.93 \times 13.96 \times 0.74 \times 1.18}{3.34}$$

$$x_{short} = \frac{47.91}{3.34} = 14.34$$

$$d_{short} = 14.34^{1.176} = 22.91 \text{ inches}$$

Try 24-inch-deep beams in both directions.

8.4.4.2 — Soil bearing pressure under beams

This step is a subroutine for Step 6, *Flow Chart 3.*

Allowable soil pressure: q_{allow} = 1500 psf

Loading Applied to the Soil:

Slab weight = 100 × 50 × 0.33 × 0.150 = 247.5 kips *(Slab is assumed to be 4 inches thick)*
Beam weight = 4 × 100 × 1.0 × 1.67 × 0.150 = 100.2 kips
Beam weight = 6 × 46 × 1.0 × 1.67 × 0.150 = 69.1 kips
Perimeter load (total) = 0.9 × 300 = 270.0 kips
Total live load = 100 × 50 × 125 = 625.0 kips
Total Loading 1311.8 kips

If we assume the total load is transmitted to the soil through contact at the beam bottoms, we can use the spacing shown in *Figure 62* to find the contact area. Assuming that the beams are 1 foot wide, the area is

(4 × 100 × 1) + (6 × 46 × 1) = 676 square feet

The soil bearing pressure is then:

w = 1311.8/676 = 1.94 kips per square foot
1.94 > 1.50 (the allowable); therefore bearing capacity is insufficient.

To gain sufficient bearing area to distribute the load on the under side of the grade beams alone, the beams could be widened. Divide the total load by the allowable soil pressure:

Required bearing area = 1311.8/1.5 = 874 square feet
Required beam width = 874/676 = 1.29 feet

This would make the beams 16 inches wide.

However, the authors choose to make the analysis using a 1-foot-wide beam, since it is reasonable to expect T-beam action. Also, increasing the beam width would place more load on the soil.

Commentary:
With a prestressed post-tensioned slab on ground, the enhanced T-beam action can be very helpful in distributing load. Observations indicate that using the T-beam section for load distribution is reasonable and effective. A distance of 8 times the slab thickness has been used successfully for computing the bearing area. For edge grade beams, 6 times the slab thickness on one side would comply with the ACI Building Code. Figures 67A *and* 67B *show this condition. Shear should be reviewed at the interface when this procedure is followed. If shear becomes critical, a thickened slab is a reasonable solution.*

Obviously, if the following analysis of a 12-inch-wide beam is successful, the 16-inch-wide wide beam would likewise be satisfactory.

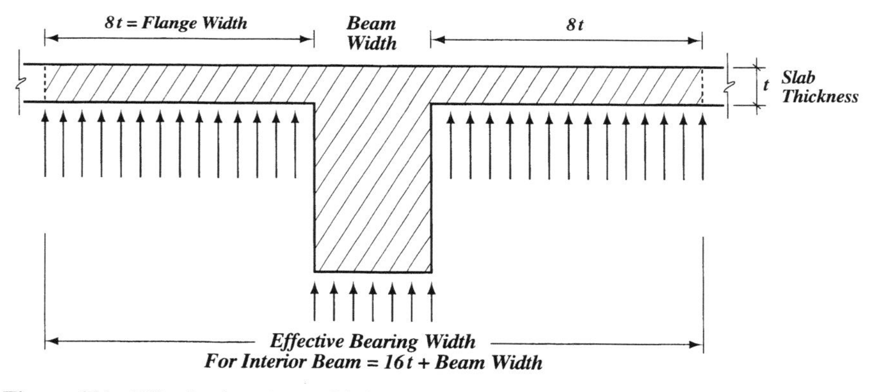

Figure 67A *Effective bearing width for interior grade beam, considering T-beam action.*

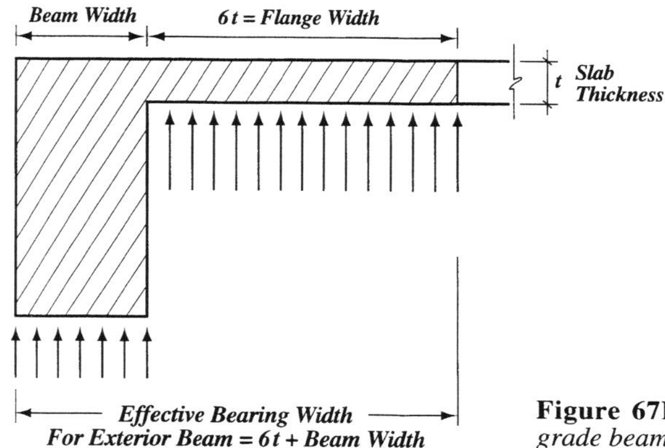

Figure 67B *Effective bearing width for exterior grade beam, considering T-beam action.*

8.4.5—Determine section properties for full slab width: Step 7, Flow Chart 3.

Dimensions already determined (Figure 62)	*Long Direction*	*Short Direction*
Beam depth, d	24 in.	24 in.
Individual beam width	12 in.	12 in.
Number of beams	4	6
Total beam width	48 in.	72 in.
Slab thickness	4 in.	4 in.

Commentary:
In calculating section properties, it can be seen that the transformed sections add significant stiffness to the system. Thus, the desirability of an uncracked section can be easily recognized. The greatest moments for compressible soils will be positive moments. This further enhances the desirability of using prestressing tendons as additional tendons can be draped if necessary in the bottom of grade beams to increase positive moment capacity.

Calculate section properties in the *long* direction:

Cross section	Area, in.^{2}	y, in.	Ay, in.^{3}
Slab = 46 × 12 × 4	2208	2	4,416
Beam = 12 × 24 × 4	1152	12	13,824
	3360		18,240

Distance from top of slab to neutral axis, $\bar{y}$ = 18,240/3360 = 5.43 inches

Moment of inertia (1/12 bh^3), beam and slab sections:

	Io	*A*	*d*	*Ad*^{2}	*TI*
Slab = 1/12 × 46 × 12 × 4^3	2944	2208	3.42	25,825	28,769
Beam = 1/12 × 4 × 12 × 24^3	55,296	1152	6.58	49,877	105,173
					133,942

S_T = 133,942/5.43 = 24,667 in.^{3}
S_B = 133,942/18.57 = 7213 in.^{3}

Calculate section properties in the *short* direction:

Cross section	Area, in.^{2}	y, in.	Ay, in.^{3}
Slab = 94 × 12 × 4	4512	2	9024
Beam = 12 × 24 × 6	1728	12	20,736
	6240		29,760

Distance from top of slab to neutral axis, $\bar{y}$ = 29,760 /6240 = 4.77 inches

Moment of inertia (1/12 bh^3), beam and slab sections:

	Io	*A*	*d*	*Ad*^{2}	*TI*
Slab = 1/12 × 94 × 12 × 4^3	6016	4512	2.77	34,370	40,386
Beam = 1/12 × 12 × 6 × 24^3	82,944	1728	7.24	90,577	173,521
					213,907

S_T = 213,907/4.77 = 44,844 in.^{3}
S_B = 213,907/19.23 = 11,123 in.^{3}

These same section properties are useful in any conventionally reinforced slab using this technique as it will be seen that stresses can often fall within rupture modulus limits, thus simplifying a conventional design.

Summary of Section Properties

	Long direction	*Short direction*
Cross sectional area, A, sq. in.	3360	6240
Centroid of slab strands, inches from top fiber	−2.00	−2.00
Centroid of beam strands, inches from top fiber	−21.00	−21.00
Depth to neutral axis, inches from top fiber	5.43	4.77
Section modulus, S_T, in.^{3}	24,667	44,844
Section modulus, S_B, in.^{3}	7,213	11,123
Allowable concrete tensile stress, $f_t = 6\sqrt{3000}$ = 329 psi	0.329 ksi	0.329 ksi
Allowable concrete compressive stress, f_c = 0.45 (3000) = 1350	1.350 ksi	1.350 ksi

8.4.6 — Calculate minimum number of tendons required, Step 8, Flow Chart 3

Commentary:
These figures are based on stress-relieved 270k strand. Low-relaxation strand will permit a higher value. As the difference in number of tendons required is relatively minor, it is prudent to consider stress-relieved strand in calculations and permit low-relaxation as a substitute rather than vice versa.

Stress permitted per tendon, $f_{ps} = 0.7 \times f_{pu} = 0.7 \times 270 = 189$ ksi
Stress in tendon after losses: $f_{ps} = 189 - 30 = 159$ ksi
Force P_r per tendon:
Area per 1/2-inch-diameter tendon = 0.153 square inches
$P_r = 0.153 \times 159 = 24.33$ kips per tendon

The value of 0.7 f_{pu} is assumed as a reasonable beginning approximation of f_{ps}, just as the losses of 30 ksi are an estimate. The number of tendons needed for this design procedure is generally low enough to encourage a conservative design approach.

Sufficient tendons must be installed to overcome slab-subgrade friction as well as to keep the minimum average prestress at 50 psi. Determine the number of tendons necessary for a minimum average prestress of 50 psi in the concrete.

$$\text{Number of tendons} = \frac{\text{concrete area} \times \text{average prestress}}{\text{force per tendon}}$$

The coefficient of friction used for polyethylene sheeting is 0.75 with a ribbed and stiffened section. This has been determined in the field to be not only realistic, but conservative. Care in subgrade preparation or a double layer of polyethylene can result in lowering this value.

$$N_t\,(long) = \frac{50 \text{ psi} \times 3360 \text{ sq.in.}}{24{,}330} = 6.90$$

$$N_t\,(short) = \frac{50 \text{ psi} \times 6240 \text{ sq.in.}}{24{,}330} = 12.82$$

Number of tendons to overcome slab-subgrade friction on polyethylene sheeting:

Weight of beams and slab = 416.8 kips (from Section 8.4.4.2)
$N_t = 0.5\,(\mu \times W_{slab})/24.330$
$= 0.5\,(0.75 \times 416.8)/24.33$
= 6.42 strands in each direction

Although 20 tendons are acceptable and meet all the criteria of the PTI Manual, the authors' experience is that tendon spacing greater than 5 feet on center is neither practical nor prudent. Also, it is desirable to have at least one tendon in each stiffening beam in addition to holding to a maximum spacing of 5 feet on center.

Total number of tendons needed is the sum of those required to maintain minimum prestress and those needed to overcome subgrade friction.

$N_t\,(long) = 6.90 + 6.42 = 13.32$ use 14 tendons
$N_t\,(short) = 12.82 + 6.42 = 19.24$ 20 tendons acceptable

Recheck minimum number of tendons in each direction, following the guidelines of the commentary:

$N_t\,(long) = 50$ ft/5 + number of beams
$= 10 + 4 = 14$ tendons — OK

$N_t\,(short) = 100$ ft/5 + number of beams
$= 20 + 6 = 26$ tendons > 20.

Therefore use 26 tendons in the short direction.

Design prestress forces

Since maximum moments occur near the slab perimeter, friction losses will be minimal at points of maximum moments. Therefore, assume total prestressing force effective for structural calculations:

Long direction: $P_r = 14 \times 24.3 = 340.2$ kips
Short direction: $P_r = 26 \times 24.3 = 631.8$ kips

If positive moment capacities are marginal in the compressible clay design, attention should be given to losses, as this moment does occur in the center of the slab.

8.4.7 — Check design moments against allowable moments for edge lift condition, Step 9, Flow Chart 3

To determine the design moment resulting from constructing over compressible soil, we must first determine moment and differential deflection for the no-swell condition for use in calculating the design moments.

Moment for no-swell condition, edge lift, *long* direction

$$M_{ns\ell} = \frac{d^{1.35}\ S^{0.36}}{80 \times L^{0.12} \times P^{0.10}}$$

$$M_{ns\ell} = \frac{24^{1.35} \times 16.67^{0.36}}{80 \times 100^{0.12} \times 900^{0.10}}$$

$$M_{ns\ell} = \frac{73 \times 2.75}{80 \times 1.74 \times 1.97}$$

$$M_{ns\ell} = \frac{200.75}{274.22} = 0.732 \text{ foot-kips per foot}$$

The perimeter load does not include the weight of the grade beam because this is a static condition not being acted upon by the soil. There is no soil function acting on the perimeter as found in a purely plastic soil condition.

Differential deflection for no-swell condition, *long* direction, edge lift

$$\Delta_{ns\ell} = \frac{L^{1.28}\ S^{0.80}}{133 \times d^{0.28} \times P^{0.62}}$$

$$\Delta_{ns\ell} = \frac{100^{1.28} \times 16.67^{0.80}}{133 \times 24^{0.28} \times 900^{0.62}}$$

$$\Delta_{ns\ell} = \frac{363 \times 9.49}{133 \times 2.54 \times 67.86}$$

$$\Delta_{ns\ell} = \frac{3444}{22{,}924} = 0.15 \text{ inches}$$

Design moment — Construction over compressible soil, edge lift condition
Long direction:

$$M_{cs\ell} = \left[\frac{\delta}{\Delta_{ns}}\right]^{0.50} \times M_{ns\ell}$$

$$M_{cs\ell} = \left[\frac{1.25}{0.15}\right]^{0.50} \times 0.732$$

$$M_{cs\ell} = 8.33^{0.50} \times 0.732$$

$$M_{cs\ell} = 2.89 \times 0.732 = 2.113 \text{ foot-kips per foot}$$

Short direction:

$$M_{css} = \left[\frac{970 - d}{880}\right] \times M_{cs\ell}$$

$$M_{css} = \frac{970 - 24}{880} \times 2.113$$

$$M_{css} = 1.075 \times 2.113 = 2.27 \text{ foot-kips per foot}$$

The design moments calculated above must now be compared with the allowable moments.

Commentary:
The quantity $P_r e$ *accounts for the moment associated with the eccentricity of the post-tensioning tendons. It has been standard practice to drape beam tendons to within 3 inches of the bottom of stiffening beams in order to maximize positive moment capacity. This is because negative moment capacity is almost always more than ample due to the significantly higher section modulus. With this consideration, this eccentricity factor can therefore be either positive or negative as in* Figure 68.

Allowable service moments, *long direction*, tension in bottom fiber, edge lift condition

$$(12 \times 50)\ {}_pM_t = S_B\,(P_r/A + f_t) - P_r e$$

Note that the long direction has a 50-foot-wide cross section; hence the design moment is multiplied by 50 feet. It is also multiplied by 12 inches per foot to make units of the left side of the equation compatible with the right.

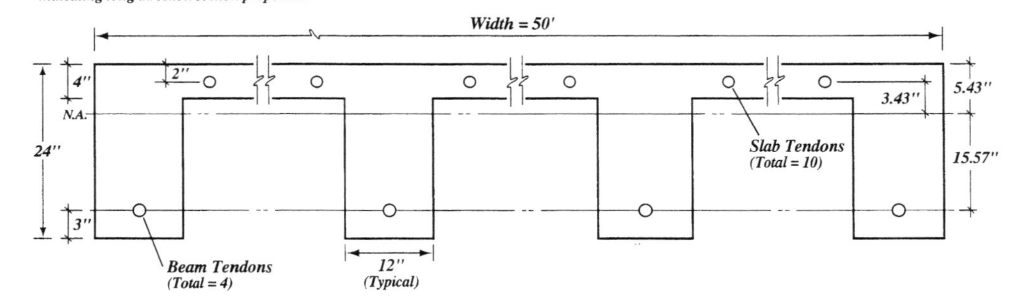

Figure 68 *Cross section of slab taken through the short direction, indicating long direction section properties. Location of neutral axis and eccentricities of beam and slab tendons are shown.*

$$\begin{aligned} P_r e &= \big[[N_t\,(\text{top}) \times 3.43] - [N_t\,(\text{beam}) \times 15.57]\big] \times 24.3 \\ &= [10 \times 3.43 - 4 \times 15.57] \times 24.3 \\ &= [3.43 - 62.28] \times 24.3 \\ &= -27.98 \times 24.3 \\ P_r e &= -679.9 \text{ inch-kips} \end{aligned}$$

Now substitute this value for $P_r e$ in the equation above for tension in bottom fiber.

$$\begin{aligned} (12 \times 50)\ {}_pM_t &= 7213\ [(340.2/3360) + 0.329] - (-679.9) \\ 600\ {}_pM_t &= 7213\ [0.101 + 0.329] + 679.9 \\ 600\ {}_pM_t &= 7213\ (0.430) + 679.9 \\ {}_pM_t &= (3102 + 679.9)/600 \\ &= 6.30 \text{ foot-kips per foot} \\ &\quad 6.30 > 2.21 \text{ — OK} \end{aligned}$$

The design moment is less than the allowable moment in the long direction

Allowable service moments, *long* direction, compression in top fiber, edge lift condition

$$\begin{aligned} (12 \times 50)\ {}_pM_c &= S_t\,[f_c - (P_r/A)] - P_r e \\ 600\ {}_pM_c &= 24{,}667\ [1.350 - 340.2/3360] - P_r e \\ 600\ {}_pM_c &= 24{,}667\ [1.350 - 0.101] - (-679.9) \\ {}_pM_c &= [30{,}809 + 679.9]/600 \\ &= 52.48 \text{ foot-kips per foot} \\ &\quad 52.48 > 2.21 \text{ — OK} \end{aligned}$$

Allowable service moment, *short* direction, tension in bottom fiber, edge lift condition

Again the quantity $P_r e$ must be calculated to account for the prestressing moment associated with the eccentricity of the post-tensioning tendons. *Figure 69* shows this cross section. Note that the neutral axis is slightly different for the two cross sections. Since the calculation of $P_r e$ depends on the location of the neutral axis, there are two sets of values *(Figures 68* and *69).*

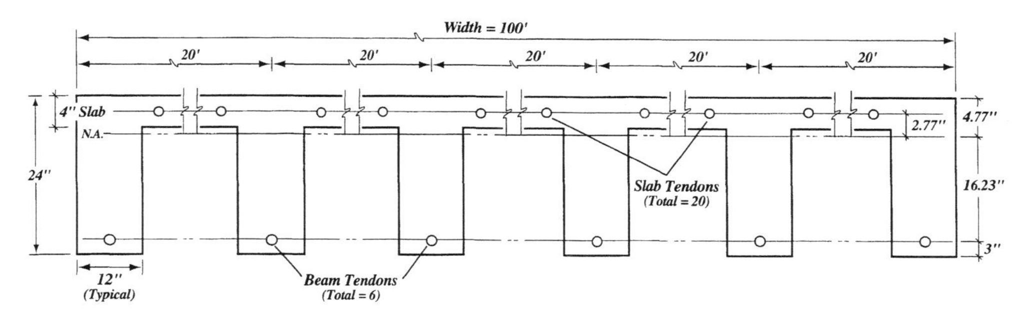

Figure 69 *Cross section of slab taken through the long direction, indicating short direction section properties. Neutral axis and eccentricities of slab and beam tendons are shown.*

$$(12 \times 100)\ {}_pM_t = S_B\,(P_r/A + f_t) - P_r e$$

$$\begin{aligned} P_r e &= \big[[N_t\,(\text{top}) \times 2.77] - [N_t\,(\text{beam}) \times 16.23]\big] \times 24.3 \\ &= [20 \times 2.77 - 6 \times 16.23] \times 24.3 \\ &= [55.4 - 97.38] \times 24.3 \\ &= -41.98 \times 24.3 \\ P_r e &= -1020.1 \text{ inch-kips} \end{aligned}$$

$$\begin{aligned} (12 \times 100)\ {}_pM_t &= 11{,}123\,[631.8/6240 + 0.329] - (-1020.1) \\ {}_pM_t &= [(11{,}123 \times 0.430) + 1020.1]/1200 \\ {}_pM_t &= 5803/1200 \\ &= 4.84 \text{ foot-kips per foot} \\ &\quad 4.84 > 2.27 \text{ — OK} \end{aligned}$$

Allowable service moments, *short* direction, compression in top fiber, edge lift condition

$$\begin{aligned} (12 \times 100)\ {}_pM_c &= 44{,}844\,[1.350 - 631.8/6240] - (-1020.1) \\ {}_pM_c &= [(44{,}844 \times 1.249) + 1020.1]/1200 \\ &= 57{,}030/1200 = 47.52 \text{ foot-kips per foot} \end{aligned}$$

Since 47.52 is greater than 2.27, section is OK.

8.4.8—Deflection calculations, edge lift condition

For Step 10, *Flow Chart 3*, it is necessary to calculate and compare the expected deflections with the allowable deflection for both long and short directions.

Proper deflection control is essential. It appears to be even more important when 6β governs the design instead of actual slab length, L. Since beam depth directly affects deflection, it is always advisable to be conservative in depth selection and unyielding in jobsite inspections.

Commentary:
Residential construction is particularly sensitive to deflection control because wall finishes such as stucco or masonry can be unmerciful in revealing cracks due to deflection in an otherwise sound design.

Allowable differential deflection, *long* direction, edge lift

$$\beta = \frac{1}{12}\sqrt[4]{\frac{E_c I}{E_s \times \delta/\Delta_{ns}}}$$

$$\beta = \frac{1}{12}\sqrt[4]{\frac{1{,}500{,}000 \times 133{,}942}{1000 \times 1.25/0.15}}$$

$$\beta = \frac{1}{12}\sqrt[4]{\frac{199.3 \times 10^9}{8.33 \times 10^3}}$$

$$\beta = \frac{1}{12}\sqrt[4]{24{,}119{,}208}$$

$$\beta = \frac{1}{12} \times 70.07 = 5.84 \text{ feet}$$

$6\beta = 35.03$ feet < 100 feet so 6β governs

$$\Delta_{allow} = [12 \times 35.03]/800 = 0.53 \text{ inches}$$

Expected differential deflection, *long* direction and *short* direction, edge lift

$$\Delta_{cs} = \delta \, expZ$$

where $Z = 1.78 - 0.103d - 1.65 \times 10^{-3}P + 3.95 \times 10^{-7}P^2$

$$Z = 1.78 - (0.103 \times 24) - 1.65 \times 10^{-3} \times 900 + 3.95 \times 10^{-7} \times 900^2$$

$$Z = 1.78 - 2.47 - 1.48 + 0.31$$

$$Z = -1.86$$

Commentary:
Note that exp Z refers to the natural antilog of a function Z; the base is e, *not 10.*

$$\Delta_{cs} = 1.25e^{-1.86} = 1.25 \times 1/6.42 = 0.19 \text{ inches}$$

The 0.19 inches expected deflection is less than the allowable 0.53 inches; section is OK for deflection.

Allowable differential deflection, *short* direction, edge lift

$$\beta = \frac{1}{12}\sqrt[4]{\frac{E_c I}{E_s \times \delta / \Delta_{ns}}}$$

$$\beta = \frac{1}{12}\sqrt[4]{\frac{1{,}500{,}000 \times 213{,}907}{1000 \times 1.25/0.15}}$$

$$\beta = \frac{1}{12}\sqrt[4]{\frac{320.86 \times 10^9}{8.33 \times 10^3}}$$

$$\beta = \frac{1}{12}\sqrt[4]{38{,}518{,}667}$$

$$\beta = \frac{1}{12} \times 78.78 = 6.56 \text{ feet}$$

$6\beta = 39.38$ feet < 50 feet so 6β governs

$$\Delta_{allow} = [12 \times 39.38]/800 = 0.59 \text{ inches}$$

Expected differential deflection, edge lift, *short* direction

$$\Delta_{cs} = 0.19 \text{ inches}$$

$$0.19 < 0.59 \text{ — OK}$$

Deflections are less than allowable in both long and short directions. Note that the expected differential deflection is the same for both the long and the short directions. This is because the expected differential deflection is a function of the expected differential settlement δ, which is the same for each direction.

8.4.9 — Shear calculations for edge lift condition

Step 11 of *Flow Chart 3* calls for comparison of the expected shear stress due to service loads with the allowable or permissible shear stress. According to PTI recommendations,

permissible shear $v_c = 1.5\sqrt{f_c'} = 1.5\sqrt{3000} = 82.2$ psi

The following steps determine design shear stress in both directions for comparison with the allowable 82.2 psi.

Design shear stress, *long* direction, edge lift condition

Design shear stress is calculated using expected service shear in the long direction. Note that both V_{ns} (no-swell condition) and V_{cs} (compressible soil) are determined, and the larger of the two values used to find design shear stress.

Expected service shear:

$$V_{ns\ell} = \frac{d^{0.90}\,[P \times S\,]^{0.30}}{550 \times L^{0.10}}$$

$$V_{ns\ell} = \frac{24^{0.90}\,[\,900 \times 16.67\,]^{0.30}}{550 \times 100^{0.10}}$$

$$V_{ns\ell} = \frac{17.46 \times 17.9}{550 \times 1.58}$$

$$V_{ns\ell} = \frac{312.53}{869} = 0.359 \text{ kips per foot}$$

$$V_{cs\ell} = \left[\frac{\delta}{\Delta_{ns\ell}}\right]^{0.30} \times V_{ns\ell}$$

$$V_{cs\ell} = \left[\frac{1.25}{0.15}\right]^{0.30} \times 0.359$$

$$V_{cs\ell} = 1.89 \times 0.359 = 0.678 \text{ kips per foot}$$

Design shear stress:

$$v = \frac{V \times W}{n\,db} = \frac{0.678 \times 1000 \times 50}{4 \times 12 \times 24}$$

$$v = \frac{33{,}900}{1152} = 29.42 \text{ psi}$$

This design stress of 29.42 psi is well below the allowable 82.2 psi, and so the slab is OK for shear in the long direction.

Design shear stress, *short* direction, edge lift condition

Expected service shear and design shear stress:

$$V_{css} = \left[\frac{116 - d}{94}\right] \times V_{cs\ell}$$

$$V_{css} = \left[\frac{116 - 24}{94}\right] \times 0.678$$

$$V_{css} = 0.97 \times 0.678 = 0.66 \text{ kips per foot}$$

$$v = \frac{V \times W}{n\,db} = \frac{0.66 \times 1000 \times 100}{6 \times 12 \times 24}$$

$$v = 66{,}000 / 1728 = 38.19 \text{ psi}$$

The expected design shear in the short direction 38.19 psi is also substantially below the allowable 82.2 psi, and thus section is OK for shear in both directions for the edge lift condition. This completes all of the checks for edge lift condition for the slab on compressible soil.

Commentary:

Using only the beams to resist shear is a conservative approach and in keeping with the PTI Manual. A strong case can be made for including a portion of the T-beam for shear resistance also.

Although permissible shear is 1.5 $\sqrt{f'_c}$ *according to PTI, a strong case has been made to permit a total allowable shear calculated at 1.5* $\sqrt{f'_c}$ + P/A. *Some available computer programs also use this enhanced allowable. This, naturally, only applies to prestressed slabs.*

8.4.10 — Center lift design

Although the predominant mode of movement in compressible soils is edge lift (also called "saucering"), center lift design must be considered because most compressible soils are highly susceptible to shrinkage due to a variation in moisture content. Long term conditions will more than likely be identical to those encountered in plastic soils.

Since soil conditions and preliminary design are the same as for the edge lift conditions just completed, we can begin at Step 9 of *Flow Chart 3*. Note: the edge beam weight (250 pounds per foot) is included in perimeter load P, making it 1150 pounds per foot wherever soil functions e_m or y_m also appear in the calculation.

Design moments for *long* direction, center lift condition, Step 9, *Flow Chart 3*

$$M_\ell = A_o\,[B \times e_m^{1.238} + C]$$

$$\text{where } A_o = \frac{1}{727}\,[L^{0.013}\,S^{0.306}\,d^{0.688}\,P^{0.534}\,y_m^{0.193}]$$

$$A_o = \frac{1}{727}\,[100^{0.013}\,16.67^{0.306}\,24^{0.688}\,1150^{0.534}\,0.48^{0.193}]$$

$$A_o = \frac{1}{727}\,[1.06 \times 2.36 \times 8.90 \times 43.09 \times 0.87]$$

$$A_o = \frac{833}{727} = 1.15$$

From the soils data, page 123, e_m, edge moisture variation distance is 4 ft. for the center lift condition. According to the PTI procedure, for $e_m \leq 5$, $B = 1.0$ and $C = 0$. Using these values, determine M_ℓ:

$$M_\ell = 1.15\,[1 \times 4^{1.238} + 0]$$

$$M_\ell = 1.15 \times 5.56 = 6.39 \text{ foot-kips per foot}$$

Design moments, *short* direction, center lift

$$M_s = \left[\frac{58 + e_m}{60}\right] \times M_\ell$$

$$M_s = \left[\frac{58 + 4}{60}\right] \times 6.39 = 6.60 \text{ foot-kips per foot}$$

Allowable moments, *long* direction, center lift

Allowable moments must be calculated and compared with design moments. First, calculate negative bending moments.

- Tension in top fiber. NOTE: From page 130, $P_r e = -679.9$ inch-kips

$$(12 \times 50)\;{}_nM_t = S_T\left[\frac{P_r}{A} + f_t\right] + P_r e$$

$$600\;{}_nM_t = 24{,}667\left[\frac{340.2}{3360} + 0.329\right] + (-679.9)$$

$${}_nM_t = \frac{[24{,}667 \times 0.430] - 679.9}{600} = 16.54 \text{ foot-kips per foot}$$

Since the allowable moment of 16.54 foot-kips per foot is greater than the design moment of 6.39, section is OK for tension in the top fiber.

• Compression in bottom fiber

$$(12 \times 50)\ {}_nM_c = S_B \left[f_c - \frac{P_r}{A} \right] + P_r e$$

$$600\ {}_nM_c = 7213 \left[1.350 - \frac{340.2}{3360} \right] + (-679.91)$$

$${}_nM_c = \frac{7213\,(1.249) - 679.91}{600} = 13.88 \text{ foot-kips per foot}$$

Since 13.88 is greater than 6.39 design moment, section is OK in long direction for compression in bottom fiber.

Allowable moments in *short* direction, center lift design

• Tension in top fiber   NOTE: From page 131, $P_r e = -1020.1$ inch-kips

$$(12 \times 100)\ {}_nM_t = S_T \left[\frac{P_r}{A} + f_t \right] + P_r e$$

$$1200\ {}_nM_t = 44{,}844 \left[\frac{631.8}{6240} + 0.329 \right] + (-1020.1)$$

$${}_nM_t = \frac{44{,}844\,(0.101 + 0.329) - 1020.1}{1200}$$

$${}_nM_t = \frac{19{,}283 - 1020.1}{1200}$$

$$= 15.22 \text{ foot-kips per foot}$$

Since the allowable 15.22 foot-kips per foot is greater than the design moment of 6.60, section checks OK.

• Compression in bottom fiber

$$(12 \times 100)\ {}_nM_c = S_B \left[f_c - \frac{P_r}{A} \right] + P_r e$$

$${}_nM_c = \frac{11{,}123\,(1.350 - 0.101) + (-1020.1)}{1200}$$

$${}_nM_c = \frac{12{,}872.53}{1200}$$

$${}_nM_c = 10.73 \text{ foot-kips per foot}$$

Since the allowable moment, 10.73 foot-kips per foot is greater than the design moment of 6.60, section is OK. Moment capacities exceed expected service moments for center lift in both long and short directions.

Center lift deflection calculations, *long* direction (Step 10, *Flow Chart 3*)

• Allowable differential deflection

$$\Delta_{allow} = 12(L \text{ or } 6\beta)/360$$

Use 6β, since it is less than L (= 100 ft). β, the relative stiffness length, is 5.84 feet as calculated on page 131 and 132.

$$\Delta_{allow} = 12(35.03)/360$$
$$= 1.17 \text{ inches}$$

• Expected differential deflection

$$\Delta = \frac{[y_m L]^{0.205} S^{1.059} P^{0.523} e_m^{1.296}}{380 \times d^{1.214}}$$

$$\Delta = \frac{[0.48 \times 100]^{0.205} 16.67^{1.059} 1150^{0.523} 4^{1.296}}{380 \times 24^{1.214}}$$

$$\Delta = \frac{2.21 \times 19.68 \times 39.88 \times 6.03}{380 \times 47.37}$$

$$\Delta = \frac{10{,}465}{18{,}004} = 0.58 \text{ inches}$$

The expected differential deflection, 0.58 inches is less than the allowable of 1.17 inches so section is OK for deflection in the long direction.

Center lift deflection in the *short* direction

• Allowable differential deflection

$$\Delta_{allow} = 12(L \text{ or } 6\beta)/360$$

Here 6β (= 39.38 feet as determined on page 132) is smaller than L of 50 feet so it is used in the calculation of the allowable:

$$\Delta_{allow} = 12(39.38)/360 = 1.31 \text{ inches}$$

• Expected differential deflection

Substitute in the same expression used above for expected deflection, noting that only the values of S and L are different:

$$\Delta = \frac{[0.48 \times 50]^{0.205} 20^{1.059} 1150^{0.523} 4^{1.296}}{380 \times 24^{1.214}}$$

$$\Delta = \frac{1.92 \times 23.87 \times 39.88 \times 6.03}{380 \times 47.37}$$

$$\Delta = \frac{11{,}021}{18{,}000} = 0.61 \text{ inches}$$

Since 0.61 inches is less than the allowable of 1.31 inches, section is OK for deflection. Deflections in both long and short directions are much less than the allowable deflection for center lift loading.

Shear calculations, center lift condition

This is Step 11 of *Flow Chart 3*. Again, according to the PTI Manual permissible shear stress

$$v_c = 1.5\sqrt{f_c'} = 1.5\sqrt{3000} = 82.2 \text{ psi}$$

We calculate design shear stress in both directions and compare with this value.

• Design shear in the *long* direction:

$$V_\ell = \frac{1}{1940}\left[L^{0.09} S^{0.71} d^{0.43} P^{0.44} y_m^{0.16} e_m^{0.93}\right]$$

$$V_\ell = \frac{1}{1940}\left[100^{0.09} 16.67^{0.71} 24^{0.43} 1150^{0.44} 0.48^{0.16} 4^{0.93}\right]$$

$$V_\ell = \frac{1}{1940}\left[1.51 \times 7.37 \times 3.29 \times 22.22 \times 0.89 \times 3.63\right]$$

$$V_\ell = \frac{3132}{1940} = 1.61 \text{ kips per foot.}$$

• Design shear stress

$$v = \frac{V \times W}{ndb} = \frac{1.61 \times 50 \times 1000}{4 \times 12 \times 24}$$

$$v = \frac{80{,}500}{1152} = 69.88 \text{ psi}$$

The stress of 69.88 is less than the allowable 82.2. Section is OK.

• Design shear in the *short* direction:

$$V_s = \frac{1}{1350} \left[L^{0.19} \, S^{0.45} \, d^{0.20} \, P^{0.54} \, y_m^{\,0.04} \, e_m^{\,0.97} \right]$$

$$V_s = \frac{1}{1350} \left[50^{0.19} \, 20^{0.45} \, 24^{0.20} \, 1150^{0.54} \, 0.48^{0.04} \, 4^{0.97} \right]$$

$$V_s = \frac{1}{1350} \left[2.10 \times 3.85 \times 1.89 \times 44.95 \times 0.97 \times 3.84 \right]$$

$$V_s = \frac{2558}{1350} = 1.89 \text{ kips per foot}$$

• Design shear stress

$$v = \frac{V \times W}{ndb} = \frac{1.89 \times 100 \times 1000}{6 \times 12 \times 24}$$

$$v = \frac{189{,}000}{1728} = 109.37 \text{ psi}$$

109.37 > 82.2 psi permissible shear stress;
this does not meet the PTI recommendation.

However, as explained earlier in these calculations, a strong case has been made for use of a shear value of

$$1.5 \sqrt{f_c'} + P/A$$

If this were applied here, the permissible shear would increase by P/A = 63,800/6240 or 101.3 to 183.52 psi, well above the calculated 109.37 psi shear stress.

A second consideration is that only the stem of the beam has been taken into consideration, and not the flange. By observation, consideration of the flange would provide a unit shear stress of 38.28 psi, a value well under the allowable.

A third possibility—if it is desired to follow the PTI recommendation rigorously—would be to alter the cross section to comply with the more rigid shear standards. This can be done by one of the following:

1. Increase beam width
2. Increase beam depth
3. Decrease beam spacing
4. Use web reinforcement over a distance of $1.5\,\beta$ at each end of the beam
5. A combination of 1, 2, and 3.

Increasing the short direction beam width to 14 inches would reduce the stress to 83.3 psi when calculated by the most conservative method shown above. This is close enough to the permissible of 82.2 psi to represent satisfactory compliance with the PTI recommendations.

However, the authors' choice is to keep the short direction beam width at 12 inches for simplicity of construction, relying on their experience that shear is rarely if ever critical for this type of construction.

This completes the problem for center lift. Edge lift and center lift design results are summarized in *Table 25* (Step 13, *Flow Chart 3*).

Edge lift design	***Design***	***Allowable***	**Center lift design**	***Design***	***Allowable***
Moment, ft.-kips per ft.			Moment, ft.-kips per ft.		
Long Direction			*Long Direction*		
Tensile	2.11	< 6.30	Tensile	6.39	16.54
Compressive	2.11	< 52.48	Compressive	6.39	13.88
Short direction			*Short direction*		
Tensile	2.27	< 4.84	Tensile	6.60	15.22
Compressive	2.27	< 47.52	Compressive	6.60	10.73
Differential deflection, in.			Differential deflection, in.		
Long Direction	0.19	< 0.53	*Long Direction*	0.58	1.17
Short direction	0.19	< 0.59	*Short direction*	0.61	1.31
Shear stress, psi			Shear stress, psi		
Long Direction	29.42	< 82.2	*Long Direction*	69.88	82.2
Short direction	38.19	< 82.2	*Short direction**	109.37	82.2
			** NOTE: See comments regarding shear in Sections 8.4.9 and 8.4.10 of this example.*		

Tendons and beam requirements *(see Figure 70)*

Long direction

Fourteen 1/2-inch-diameter 270k strands total. Ten tendons in the slab, 2 inches below the top, beginning 30 inches from each end, then 5 feet on center. One tendon in each beam, 3 inches from the bottom fiber. Four beams 12 inches wide, 24 inches deep, evenly spaced at 16 feet 8 inches on center.

Short direction

Twenty-six 1/2-inch-diameter 270k strands total. Twenty in the slab, beginning 30 inches from each end then 5 feet on center. One tendon in each of the 6 beams, which are 12 inches wide, 24 inches deep, and evenly spaced at 20 feet on center.

Table 25 *Design Summary for the Post-Tensioned Slab on Compressible Clay*

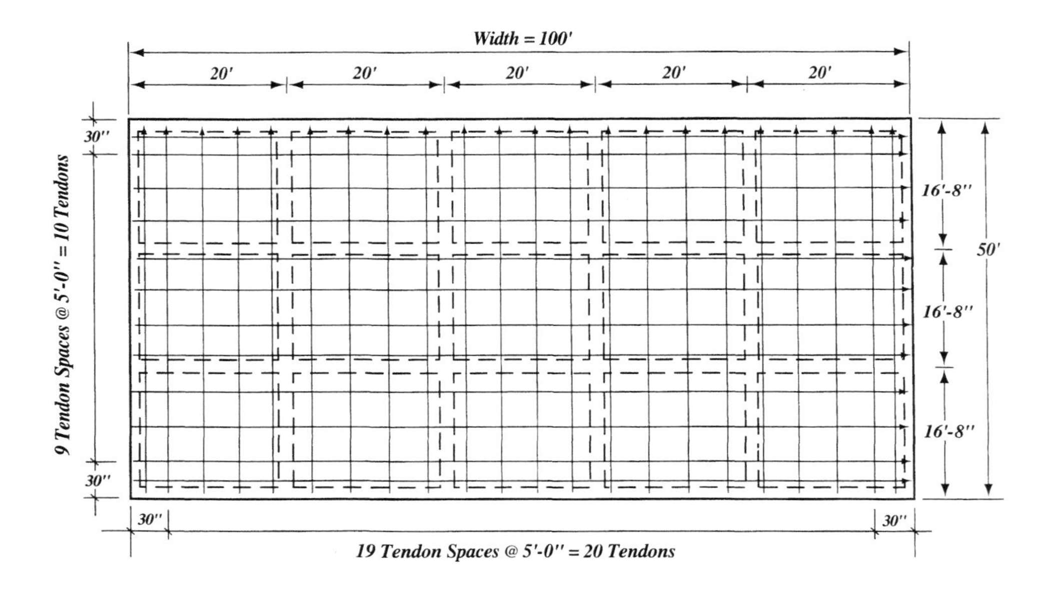

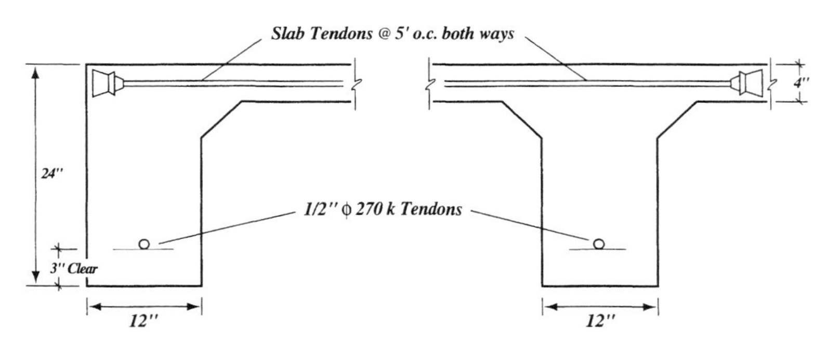

Figure 70 *Beam and tendon locations for the post-tensioned slab on compressible clay.*

8.5 — DESIGN EXAMPLE: Simple rectangle, conventionally reinforced slab on compressible clay with uniform and perimeter load, Step 12A, Flow Chart 3

8.5.1 — Introduction

The PTI design procedure shown in Section 8.4 is the state-of-the-art solution for slabs on compressible clays. The authors' choice for slabs on compressible clays is to use post-tensioning tendons for two distinct reasons:

1. The available design procedure is initially geared to post-tensioning, and
2. In many regions, post-tensioning is the most economical and most forgiving solution.

However, this is does not rule out the use of conventional reinforcement with this design procedure. Reinforcing bars can be readily adapted to the solution of the problem, as we will show here. Familiarity with the previous example of this chapter is necessary for the solution because the refinement necessary for a conventional reinforcement alternative is based on all previous work.

The primary difference between the two solutions is that the post-tensioned slab is analyzed as an uncracked section, and stiffness is based on the gross moment of inertia. With conventional reinforcement, a cracked section will have to be considered—at least for positive moment capacity.

In negative moment considerations, stresses will often be low enough for the gross uncracked moment of inertia to be considered.

8.5.2 — Modeling the problem, long direction

The easiest way to provide an equivalent conventionally reinforced solution is to provide a slab which has both an equivalent strength, and equally important, an equivalent stiffness. *Figure 71* shows a model of the post-tensioned section of the previous problem in the long direction.

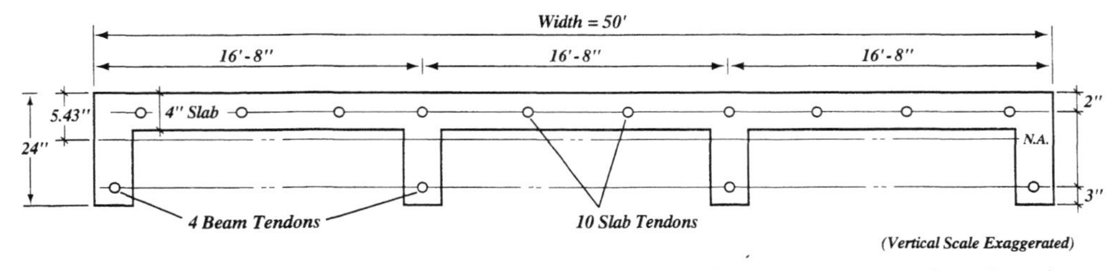

Figure 71 *Model of post-tensioned section obtained in solution of problem in Section 8.4. This section is cut across the short direction for use in analysis of the stiffened slab in the long direction.*

Cross section properties previously obtained are:

I = 133,942 in.^{4}
A = 3360 square inches
Depth to neutral axis (NA) = 5.43 inches from top
S_B = 7213 in.^{3}
S_T = 24,667 in.^{3}

As is the case with other slab solutions, the rupture modulus is often one of the primary considerations in design strength. This can also be considered in a conventional design when looking at the negative moment requirements. This is because of a top section modulus which is on the order of three times as large as the bottom section modulus.

For positive moment capacity, the slab is likely to crack when conventional reinforcement is implemented. The cross section of a conventionally reinforced slab will look something like the one shown in *Figure 72*.

Commentary:
Center lift produces a negative moment. Since the neutral axis of the slab is quite high, the top fiber stress rarely exceeds the concrete rupture modulus. In addition, the conventionally reinforced design almost invariably results in a greater beam depth, thereby providing a greater section modulus. However, the authors always recommend conventional deformed reinforcement or heavy flat-sheet welded wire fabric in the slab to insure its integrity for negative moments.

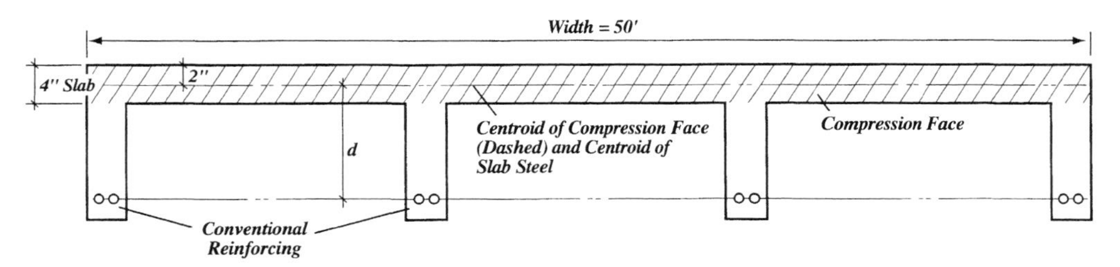

Figure 72 *Conventionally reinforced slab cross section (long direction analysis).*

For a simple, practical solution to the problem, assume the center of the 4-inch slab as the center of the compression face. Then use a modular ratio $N = 9$ in computing the effective transformed area for conventional reinforcing steel.

The authors have found that with this design method the critical design parameter appears to be deflection. This simplifies the conventional design procedure to determining a cracked section stiffness that satisfies the deflection requirements of the edge lift design condition.

8.5.3 — Values needed to solve the problem with conventional reinforcement

We need two values from the previous post-tensioned design to determine deflection requirements:

- Allowable differential deflection
- Expected differential deflection

Trial and error have indicated that a good place to start is with a cracked moment of inertia requirement 1.33 times the minimum now acceptable by direct proportion.

Most likely, the cracked moment of inertia for the conventional design will be less than the original uncracked post-tensioned section. The decreased moment of inertia, however, provides a lower allowable differential deflection. This creates a potential loop where a direct proportion determining the minimum acceptable moment of inertia is invalid.

Use the edge lift condition values from the post-tensioned slab design (summarized in *Table 25*):

Previous long direction allowable differential deflection = 0.53 inches
Previous long direction expected differential deflection = 0.19 inches

Therefore the apparent minimum acceptable moment of inertia is 0.19/0.53 or 0.35 times previous I of 133,942 in.^{4} :

$0.35 \times 133{,}942 = 46{,}879$ in.^{4}
Use $1.33 \times 46{,}879 = 62{,}349$ in.^{4}

A cracked section with approximately this moment of inertia in the long direction is the starting point for the reinforced concrete solution.

8.5.4 — The reinforced concrete solution for slab on compressible clay

Long direction section:

Choose a section for the long direction *(Figure 73)* that would approximate the desired moment of inertia from the previous example. Try three #8 bars in the bottom of each of the four beams.

Total steel area = 3 × 4 × 0.78 = 9.36 square inches.

Develop the idealized section shown in *Figure 74*. This assumes the 4-inch slab is the total compression flange and the tension flange is a transformed area equal to 9 × 9.36 = 84.24 square inches.

Commentary:
Selection of the 30-inch depth shown in Figure 73 *is based on the authors' experience. Several trial depths may be necessary; however, it is reasonable to select something greater than the post-tensioned depth, due to the lower stiffness found in the cracked section.*

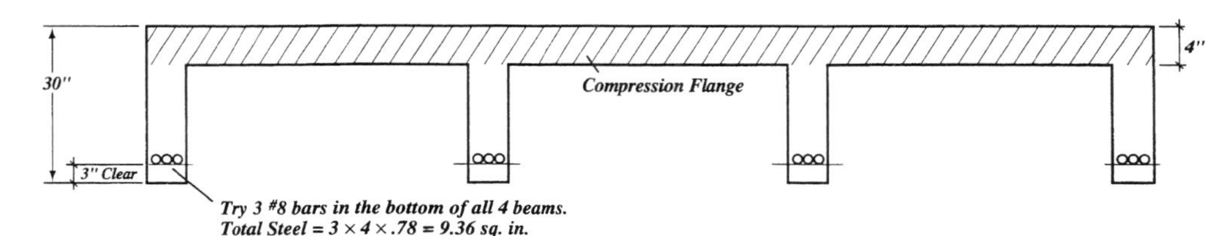

Figure 73 *Cross section selection approximating the desired moment of inertia.*

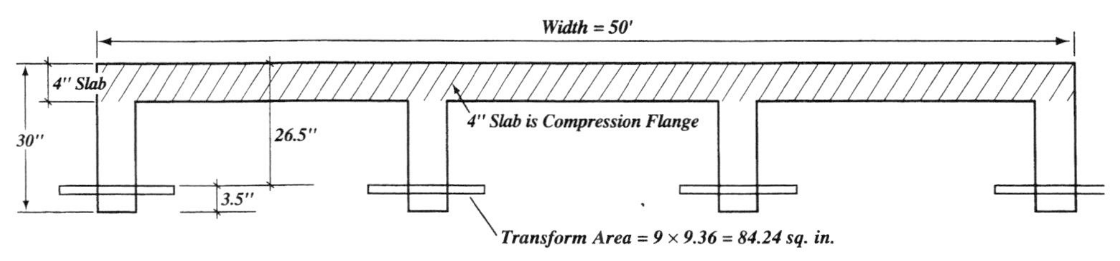

Figure 74 *Idealized section showing transformed steel area.*

Determine neutral axis and section properties:

	Area, in.²	y, in.	Ay, in.³
Concrete (4 × 50 × 12)	2400	2	4800
Steel *(transformed)* (9 × 9.36)	84.24	26.5	2232
TOTAL	2484.24		7032

$\bar{y}$ = 2.83 inches (distance from top fiber to neutral axis)

Determine moment of inertia:

	Io	*d*	*Ad²*	*TI*
Concrete (1/12 bh^3)	3200	0.83	1653	4853
Steel *(transformed)*	0	23.67	47,197	47,197
			TOTAL	52,050

Check chosen section for allowable deflection:

$$\beta = \frac{1}{12}\sqrt[4]{\frac{E_C I}{E_S}}$$

$$\beta = \frac{1}{12}\sqrt[4]{\frac{1{,}500{,}000 \times 52{,}050}{1000}} = \frac{1}{12} \times 94.00 = 7.83 \text{ feet}$$

$$6\beta = 6 \times 7.83 = 47.00 \text{ feet}$$

$$\Delta_{allow} = \frac{12 \times 47}{800} = 0.70 \text{ inches}$$

Determine expected differential deflection (*long* direction analysis):

Check the reinforced section obtained by proportioning against post-tensioned section for determining expected differential deflection.

I of post-tensioned section = 133,942 in.^{4}
I of cracked conventional section = 52,050 in.^{4}

Previous expected differential deflection (post-tensioned solution) = 0.19 inches
Expected deflection of conventionally reinforced section = 133,942 / 52,050 × 0.19 = 0.48 in.
Allowable = 0.70 inches > 0.48 inches — OK

Select four beams 12 inches wide and 30 inches deep, each reinforced with 3 #8 bars, 3 inches from the bottom.

For slab steel, it is recommended that #3 deformed bars be used at mid-depth of the slab for crack control. Authors suggest #3 bars 12 inches on center, both ways.

8.5.5 — Short direction section with conventional reinforcement

From the previous post-tensioned design example solution summarized in *Table 25*:

Previous allowable deflection, short direction = 0.59 inches
Previous expected deflection, short direction = 0.19 inches
Previous I, short direction 213,907 in.^{4}

To start, determine the approximate acceptable moment of inertia from the above values:

0.19/0.59 × 213,907 = 68,885 in.^{4}

Assume a slightly larger section as was done in analyzing the long direction:

68,885 × 1.33 = 91,617 in.^{4} (approximate value needed)

For simplicity of design use the same bars and beams determined for the long direction, that is:

12 × 30 inches, with three #8 bars per beam

As shown in *Figure 75*, there are six beams in the assumed section for the short direction.

Three #8 bars = 3 × 0.79 = 2.37 square inches of steel per beam
2.37 × 6 = 14.22 square inches for the entire section

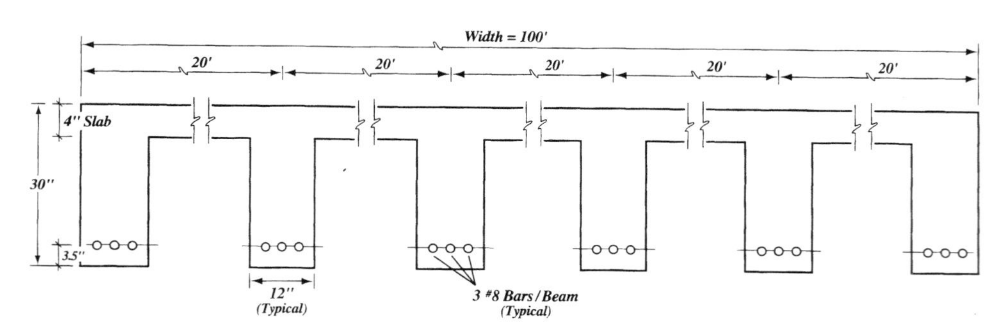

Figure 75 *Assumed section for short direction analysis. Cross section taken through the long direction.*

Determine the neutral axis and section properties of the assumed section:

	Area, in.^{2}	y, in.	Ay, in.^{3}
Concrete (4 × 100 × 12)	4800	2	9600
Steel (9 × 14.22)	128	26.5	3392
TOTAL	4928		12,992

$\bar{y}$ = 12,992 / 4928 = 2.63 inches

Determine moment of inertia:

	Io	*d*	Ad^2	*TI*
Concrete ($1/12 \times 100 \times 4^3$)	6400	0.63	1905	8305
Steel	0	23.87	72,920	72,920
			TOTAL	81,255

Determine allowable deflection of the chosen section (*short* direction):

$$\beta = \frac{1}{12}\sqrt[4]{\frac{E_c I}{E_s}}$$

$$\beta = \frac{1}{12}\sqrt[4]{\frac{1{,}500{,}000 \times 81{,}225}{1000}}$$

$\beta = 8.75$ feet

$6\beta = 52.2$ feet > 50 so use 50 feet

$$\Delta_{allow} = \frac{12 \times 50}{800} = 0.75 \text{ inches}$$

Determine the expected differential deflection:

Check the reinforced section obtained as a proportion of the post-tensioned design section to determine the expected differential deflection:

I of the post-tensioned section (short direction) = 213,907 in^4
I of the cracked conventional section = 81,225 in^4

Previous expected differential deflection (for the post-tensioned design): = 0.19 inches

Expected deflection of the conventionally reinforced section in the short direction = $213{,}907/81{,}225 \times 0.19 = 0.50$ inches

The allowable deflection computed above for the short direction is 0.75 inches.
$0.50 < 0.75$, so section is OK for deflection.

The final design with conventional reinforcement has the same beam layout as the post-tensioned design (see *Figure 76*). The beams are 12 x 30 inches, reinforced with three #8 bars at the bottom. Stirrups are optional for construction purposes only. The slab should be reinforced with #3 bars at 12 inches on center both ways.

Figure 76 is found on the following page.

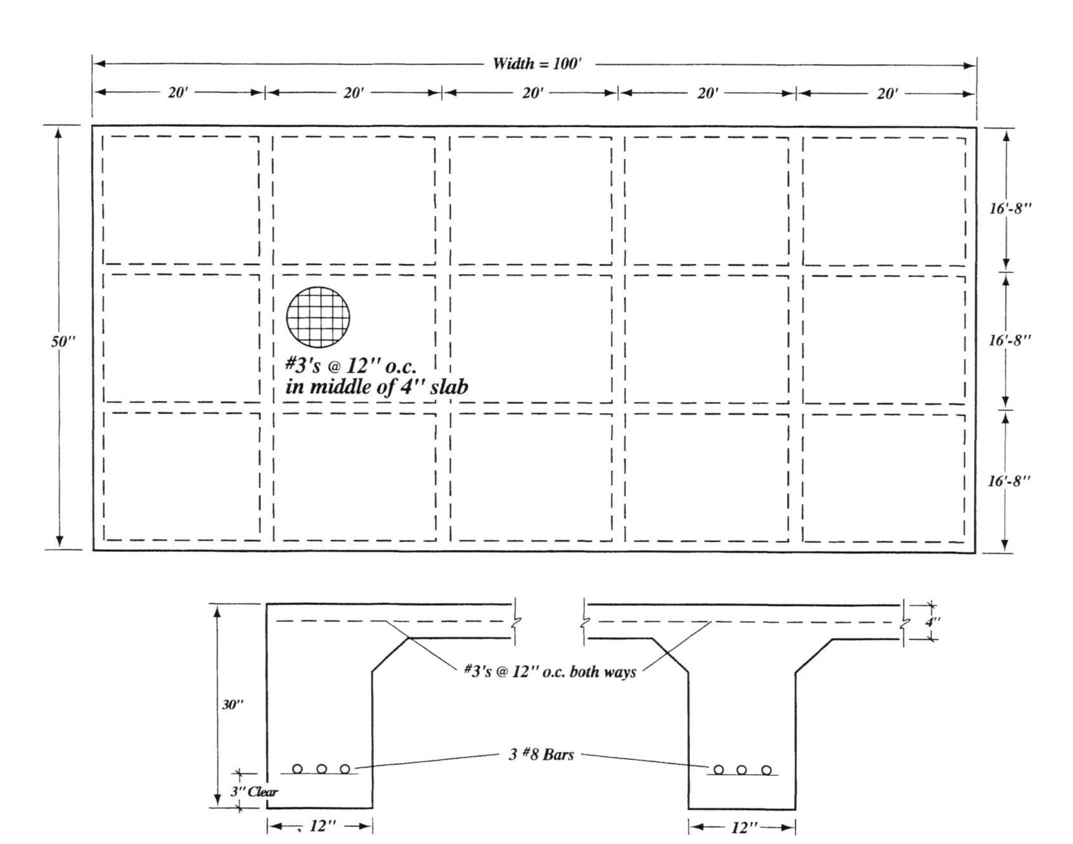

Figure 76 *Final design of conventionally reinforced slab on compressible clay.*

CHAPTER 9
THE HYBRID POST-TENSIONED SLAB: COMBINING STIFFENING ELEMENTS WITH REGIONS OF UNIFORM THICKNESS

9.1 — Introduction

In slab on grade design with uniform thickness, the design philosophy is that the slab places a load on the soil, and the soil provides support based on a uniform spring constant or modulus of subgrade reaction, known as *k*. In highly plastic soils, changes in soil moisture content can make the soil move, which imparts a nonuniform loading condition back on the slab. These soil moisture changes generally have an effect only on the perimeter of very large slabs. It is therefore reasonable in some instances that a hybrid type of slab combining stiffening elements where volume change exists (the slab's perimeter) and a uniformly thick slab where no change exists (the slab's interior) may be a desirable, economical solution. The designer must take great care in selecting this procedure as there are a number of factors that can render the design both ineffective and difficult.

First, the slab must be large enough to warrant consideration for design as a hybrid slab. It is reasonable to assume that a slab should be in excess of 100 feet long in any direction to be considered. Second, it is necessary that there be a sufficient geotechnical investigation to establish that the soil under the entire site is of uniform character; no discontinuities are to be permitted. Third, it is paramount that there be **no** water lines, drain lines, or any other sources of water present under the slab.

These constraints are needed because any elimination of stiffening elements in a plastic clay design exposes the slab to the possibility of severe bending with a change in moisture content. A broken water line under the slab is not simply a plumbing problem; it can cause severe damage to the slab due to soil volume change. With these caveats understood, a hybrid slab or "donut" design can offer many clients an economical solution in plastic soils.

This design procedure is essentially a combination of the design of slabs supported on plastic clays found in *Chapter 7* and a thickness design selection found in *Chapters 3*, *4*, and *6*. *Flow Charts 5* and *6* outline the necessary steps for the slab design. *Flow Chart 6* is a subroutine to Step 4 of *Flow Chart 5*, to be used when information from the soils specialist is incomplete. The design equations supporting *Flow Chart 5* are listed in *Appendix A.4*. These should aid both in following the example and in future design solutions.

9.2 — Plastic clay conditions

Plastic clays are fine grained materials that have a high potential to shrink and swell. The plasticity index, *PI*, of these materials is generally over 10%, and can easily approach 40% to 60%. Such conditions will very often indicate the presence of high-volume change-material. Adequate geotechnical information on the characteristics of these soils should be sought.

Commentary:
After the thickness is selected, apply the same equations as for other cases of slabs on plastic clay.

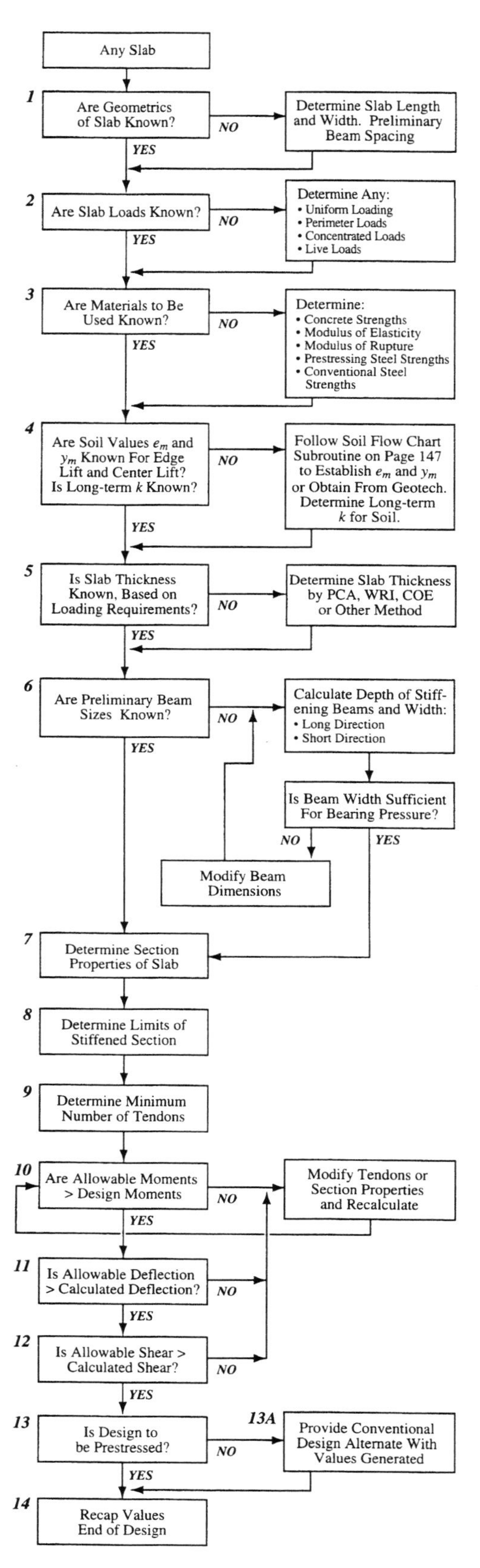

Flow Chart 5 *Procedure for design of hybrid slabs on plastic clays.*

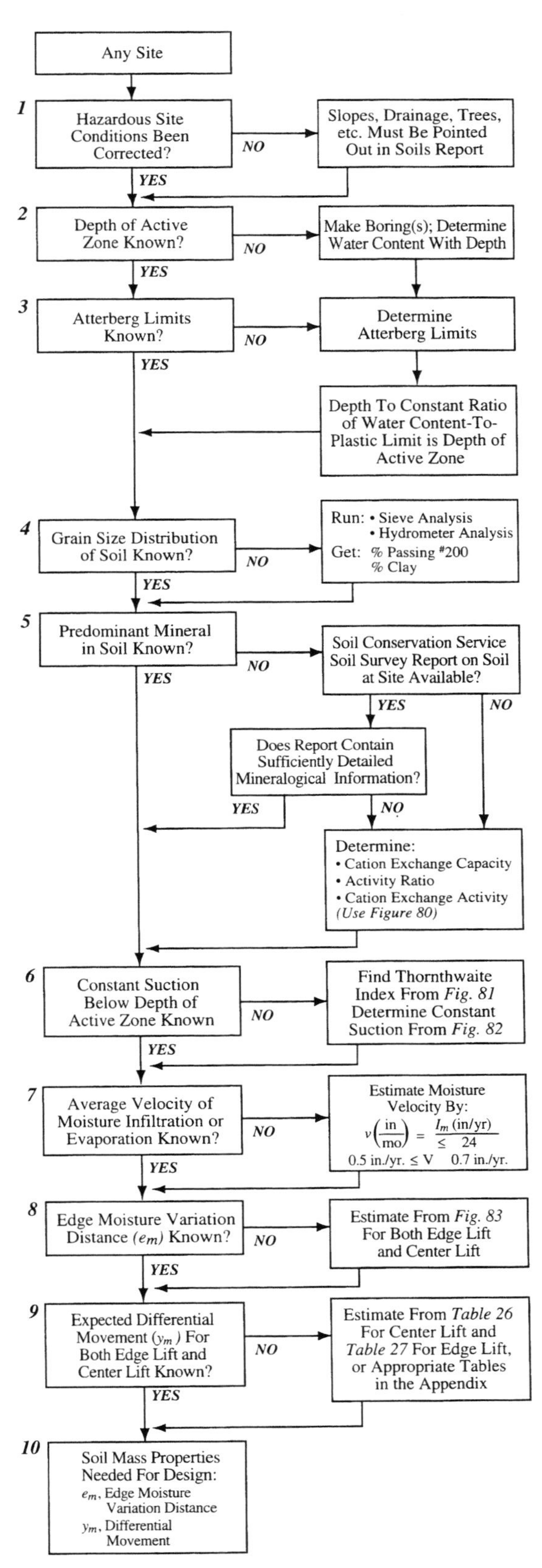

Flow Chart 6 *Subroutine for determination of soil properties, from the PTI Manual (Reference 10).*

For the uniform thickness portion of this design, a value for subgrade modulus k based on field measurement is also encouraged. It is important to realize that the long term k of a fine grained material is often lower than a short term k determination. This is due to the plastic nature of the soil. An effective k should be requested from the geotechnical engineer.

Commentary:
Always get site-specific information on long term and short term k *values, since plastic clays may vary appreciably with time.*

It is the authors' opinion that if a clay shows a potential for volume change in excess of 4 inches or more vertically, consideration should be given to another mode of addressing the design problem. This can be one of the following:

- A pier-supported slab built over collapsible voids
- Changing the nature of the soil through lime stabilization
- Removal of the problem soil

With potential differential vertical movement, y_m, less than 4 inches, a ribbed and stiffened slab on the perimeter and an interior slab isolated from moisture changes is a viable solution. *Flow Chart 5* provides an overview of this solution.

9.3 — The PTI method for slab design on plastic clay

The Post Tensioning Institute (PTI) sponsored research at Texas A&M University resulting in the publication of *Design and Construction of Post Tensioned Slabs-on-Ground* (*Reference 10*). This publication, which the authors refer to as the PTI Manual, presents a procedure for stiffening a slab with interior grade beams in order to control differential deflection and movement associated with shrink-swell potential. The PTI procedure is followed here for developing a sufficiently stiff perimeter system to meet these requirements. The interior portion of the slab that is satisfactorily isolated from changing moisture conditions is designed as a uniformly thick slab following a more conventional procedure such as PCA or WRI. The band width of the stiffened perimeter is a function of the required stiffness and soil properties. The relative stiffness length, referred to as β, is used to define the stiffened envelope.

In the design example the slab is analyzed as a single monolithic unit. Hence, the most common approach to the problem is to assume an uncracked post-tensioned slab analyzed as a large T beam. This is valid when the slab is post-tensioned in both directions. If a conventionally reinforced slab is desired, the reader is referred to the procedure found in Section 7.6.

9.3.1 — Analysis of slab loads

There are two basic considerations in this hybrid slab design procedure: the stiffened perimeter section and the uniform thickness slab in the interior. For the stiffened portion of the slab, all attributable live and dead loads are added, and the total considered to be applied uniformly over the stiffened area. This is a reasonable approach since there will be a grid of stiffening beams no more than 20 feet on center in both directions. The interior of the "donut" is analyzed as a simple uniformly thick slab, using one of the methods given in *Chapters 3* through *6*, depending on the loading types.

9.3.2 — Values needed to solve the problem

Since plastic clay is a fairly elastic material whose properties vary widely depending on moisture content and clay content, significantly more soil information is required to define the problem.

In dealing with this hybrid slab design, four primary areas must be considered:

- Slab geometry, including preliminary beam spacing
- Loads on the slab
- Materials to be used in this slab
- Soil conditions

These are shown as Items 1 through 4, *Flow Chart 5* (Page 146). If soil values e_m (edge moisture variation distance) and y_m (differential movement) are not known, then several steps in *Flow Chart 6* must also be followed. Information on the long term modulus of subgrade reaction k should also be obtained.

9.3.3 — Design objectives

The design in the following example checks the ability of the slab to resist the positive and negative bending moments based on the shrink-swell potential of the soil. The slab can take either a domed or dished shape (*Figure 77*) depending on the relative shrinkage or swelling across the edge moisture variation distance e_m.

The mathematical model used in the PTI design procedure resulted in the observation that the maximum moment and maximum shear for both edge and center lift did not occur at the actual soil-slab separation but at some distance further toward the interior. The distance is closely estimated as a value β, a length that depends on the relative stiffness of the soil and the stiffened slab. After the moment and shear values rapidly reach a maximum at a distance of β, the magnitude reduces towards the midpoint of the slab. The mathematical model indicates that there is little difference in the reduction in the midpoint region for slabs 48 feet or longer. This phenomenon can be seen in *Figure 78* which is taken from *Reference 10*. From these observations, it is reasoned that the width of the stiffened perimeter can be set at a value of 3β. β, the relative stiffness length (in feet), is defined as follows:

$$\beta = \frac{1}{12}\sqrt[4]{\frac{E_c I}{E_s}}$$

where:

E_c = creep modulus of elasticity of concrete, psi
E_s = modulus of elasticity of soil, psi
I = gross moment of inertia of section, inches^4

The hybrid slab design example presented here follows the flow chart sequence and references by number the applicable steps in the charts. For the convenience of the user, all PTI formulas required for this design case are presented in *Appendix A.4*. Charts used for the determination of e_m and y_m can also be found in the appendix.

When designing a hybrid slab it is best to obtain values of e_m, y_m, and k from a qualified geotechnical engineer who is not only responsible for establishing these values, but can verify the acceptability of this modified design procedure for the project being considered.

9.3.4 — Computer solutions

A number of vendors, as well as the Post Tensioning Institute, have developed computer programs for producing these PTI solutions. The flow chart for this chapter provides a good outline for developing a computer program for the plastic clay solution. The detailed problem that follows provides all formulas needed to develop either simplifying subroutines for hand-held calculators or a full-scale computer program. If this slab design procedure is used on a regular basis, purchase of one of the existing programs or the preparation of either a program or spread sheet is highly recommended.

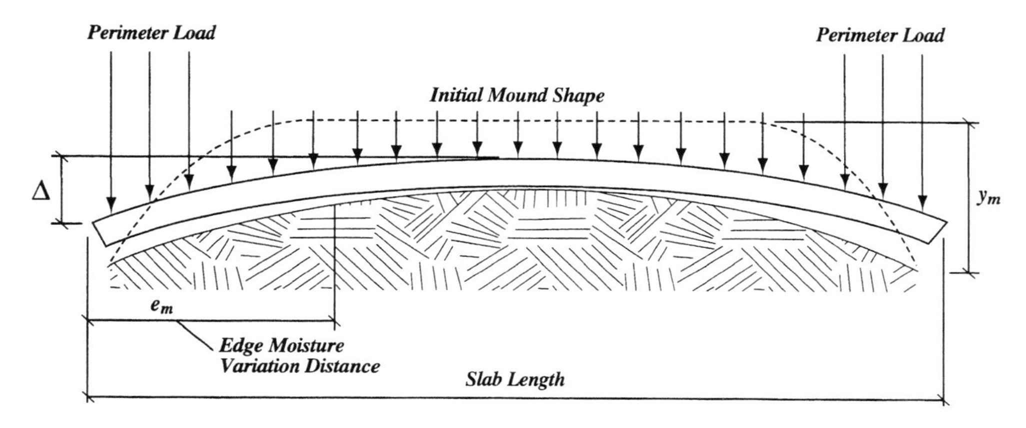

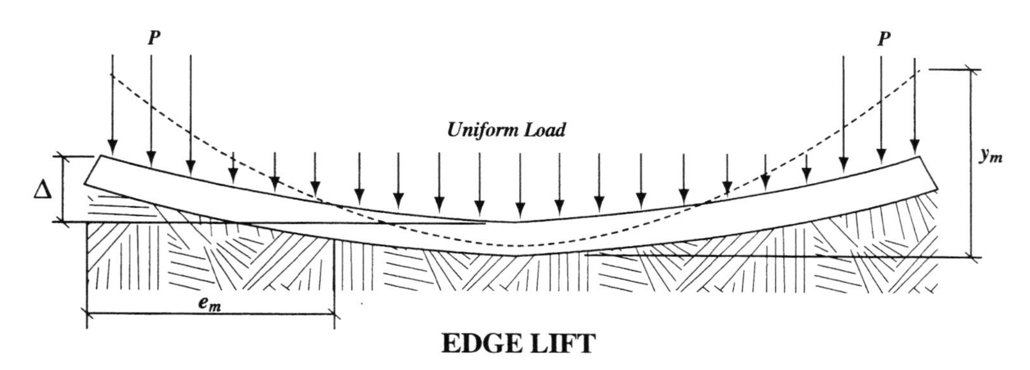

Figure 77 *Soil-structure interaction models, from the PTI Manual* (Reference 10). *The domed shape is called "center lift condition," and the dished shape is the "edge lift condition."*

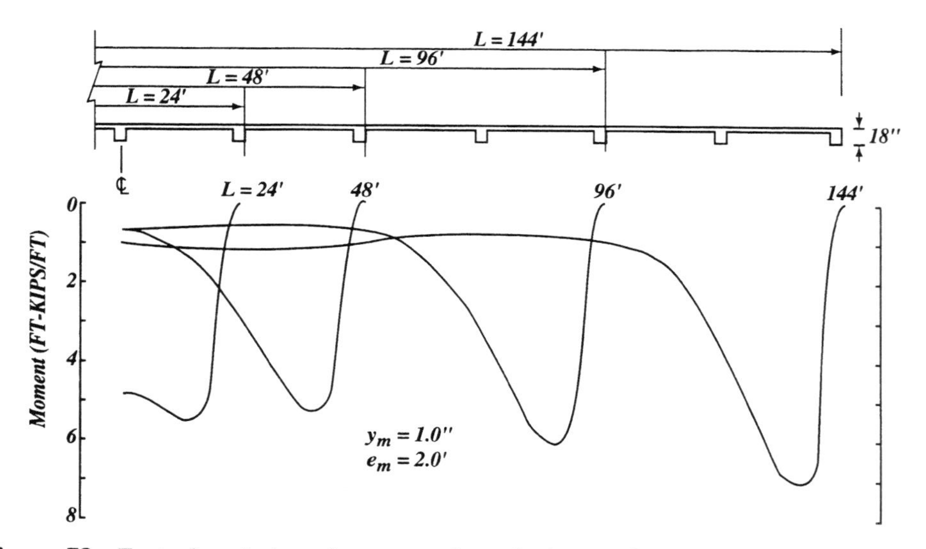

Figure 78 *Typical variation of moment along the longitudinal axis as slab length increases* (from Reference 10).

9.4 — DESIGN EXAMPLE: Post-tensioned slab combining stiffening elements with uniform thickness on plastic clay with uniform and perimeter loads

9.4.1 — Symbols and notation

A area of gross concrete cross section, square inches

A_o a co-dependent variable used in factoring center lift moment design, dependent on the physical properties of the slab, loading conditions, and soil properties

Ac activity ratio of clay

b width of an individual stiffening beam, inches

B a nondimensional constant used in factoring center lift design, dependent on soil properties

C a nondimensional constant used in factoring center lift design, dependent on loading condition and soil properties

$CEAc$ cation exchange activity of soil

CEC cation exchange capacity of soil = $PL^{1.17}$

d depth of stiffening beam measured from top surface of slab to bottom of beam, inches

e eccentricity of post-tensioning force, inches

e_m edge moisture variation distance, feet

E_c long-term or creep modulus of elasticity of concrete, psi

E_s modulus of elasticity of soil, psi

f_c allowable compressive stress in concrete, psi

f_c' 28-day compressive strength of concrete, psi

f_{ps} permissible stress in prestressing tendon, psi

f_{pu} ultimate stress in prestressing tendon, psi

f_t allowable tensile stress in concrete, psi

I gross moment of inertia, in.^{4}

I_m Thornthwaite index, moisture velocity in inches per year

k subgrade modulus, pci (pounds per cubic inch)

L total slab length in the direction being considered, feet

LL liquid limit

M_ℓ design moment in long direction, foot-kips per foot

M_s design moment in the short direction, foot-kips per foot

${}_nM_t$ ${}_nM_c$ negative and positive bending moments including

${}_pM_t$ ${}_pM_c$ tension or compression in the extreme fibers, foot-kips per foot

n number of beams in a cross section

N modular ratio (modulus of elasticity of steel to modulus of elasticity of concrete)

N_t number of tendons

pF constant soil suction value

P perimeter loading on the slab, pounds per foot

P_r prestressing force, kips

P_re moment due to post-tensioning eccentricity, inch-kips

P/A prestress force resulting from tendon load divided by gross concrete area, psi

PI plasticity index
PL plastic limit
q_{allow} allowable soil bearing pressure, psf
q_u unconfined compressive strength of soil, psf
S beam spacing, feet
S_B section modulus with respect to the bottom fiber, in.^{3}
S_T section modulus with respect to the top fiber, in.^{3}
t slab thickness, inches
v design shear stress, psi
v_c allowable concrete shear stress, psi
V design shear force, kips per foot
V_s expected shear force in short direction, kips per foot
V_ℓ expected shear force in long direction, kips per foot
w soil bearing pressure, kips per square foot
W slab width, feet
W_{slab} slab weight, pounds
$\bar{y}$ neutral axis location of stiffened cross section, inches
y_m maximum differential soil movement, inches; also referred to as differential swell
Z depth to constant suction, feet
β relative stiffness length, feet
Δ expected differential deflection under service load, inches
Δ_{allow} allowable differential deflection of slab, inches
μ coefficient of friction

Commentary:
You can run an analysis of the transformed section of the edge beam to show whether the concentrated load of the rigid frame can be satisfactorily distributed along its length from point to point. A uniform load is a reasonable assumption in many instances. If concentrated loads are questionable, they should be checked and accommodated with a modified section.

If a soil is extremely stable and does not necessarily require a plastic design analysis, a stiffened design may still be desirable. In such instances, stiffening beams may be placed to accommodate rigid frames, bearing partitions, or other elements which may be spaced more than 20 feet apart. Provided that a perimeter grade beam exists, stiffening elements may not be required on extremely stable soils.

9.4.2 — The problem and initial assumptions; materials data

A single story rigid metal frame building in Jackson, Mississippi (*Figure 79*), has a perimeter wall loading of 2000 pounds per lineal foot. The concentrated load from the rigid frame has been included in this perimeter load value. Uniform floor loading on the slab is 125 pounds per square foot.

The slab measures 150x150 feet, and the assumed center-to-center spacing of stiffening beams is 15 feet in both directions. The rigid frames are 30 feet on center. Relative stiffness length β is assumed to be 10 feet, and the width of the stiffened perimeter is set at 30 feet, equal to 3β as discussed in Section 9.3.3. (Note: These are values required by Steps 1 and 2 of *Flow Chart 5*.)

Racks are placed on the interior of the slab with post loads on a 4x8-foot grid. Maximum post load is 5 kips. Base plates are 4x4 inches.

The beam spacing should not exceed 20 feet as a matter of standard practice. The rigid frames are 30 feet on center, which dictates an intermediate beam. Thus it is reasonable to select a beam spacing of 15 feet along the perimeter. Since the building is square, the same spacing is selected for both directions.

The properties of materials to be used (Step 3 in *Flow Chart 5*) are as follows:

Concrete compressive strength $f_c' = 3000$ psi
Concrete creep modulus of elasticity $E_c = 1{,}500{,}000$ psi
Concrete tensile strength $f_t = 6\sqrt{f_c'} = 329$ psi (PTI recommendation)
Concrete modulus of rupture $MOR = 493$ psi (for use with uniform slab thickness in Section 9.4.4
Prestressing steel: 270 k, 1/2-inch diameter, 7-wire strand

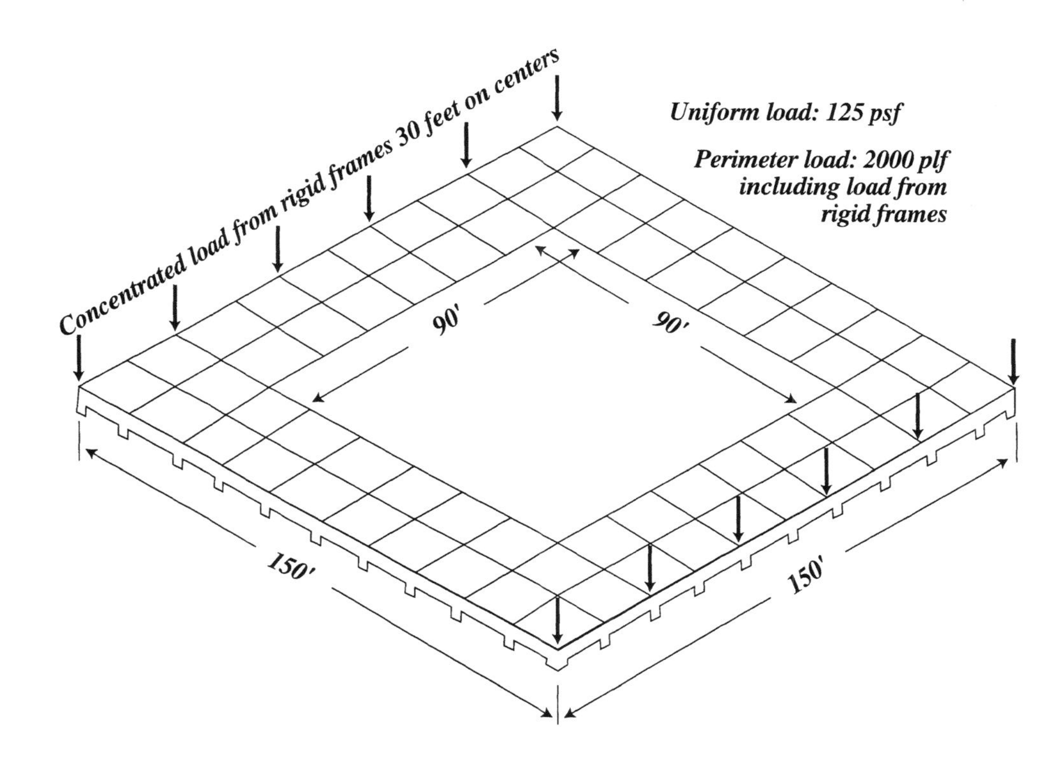

Figure 79 *Hybrid slab has stiffening beams at 15 feet both ways in a 30-foot-wide perimeter band; uniform thickness slab in center of donut.*

9.4.3 — Soils investigation

The design procedure requires the determination of the amount of climate-controlled differential movement of expansive soils, Step 4, *Flow Chart 5*. We have the following soils information:

Atterberg limits:

Plastic limit $PL = 30$

Liquid limit $LL = 70$

Plasticity Index $PI = 40$

Clay content = 60%

Allowable soil pressure, $q_{allow} = 3000$ psf

Soil modulus of elasticity, $E_s = 1000$ psi

Depth to constant suction, $Z = 5$ feet

Long term $k = 50$ pci

Location: Jackson, Mississippi

This information, as it stands, is insufficient to complete Step 4, *Flow Chart 5*. There is sufficient raw data to determine e_m and y_m for both center lift and edge lift. In order to secure these values, we must go to the soils subroutine in *Flow Chart 6*. Assuming there are no hazardous site conditions, there is enough information to satisfy Steps 2, 3, and 4 in *Flow Chart 6*. The reader can then proceed with additional steps outlined in *Flow Chart 5*.

It is strongly recommended that the consent of the geotechnical engineer be secured in the geotechnical report for the use of a hybrid or donut design such as this. With this consideration, it would appear reasonable to also require the geotechnical investigation to provide values of e_m and y_m for the project. Regardless, determinations of e_m and y_m should proceed as follows.

Commentary:
With fine grained soils, it is reasonable to assume that the long term value of k *may be different from the surface value of* k*. This is because long term* k *is affected by underlying soils which may have* in situ k *values lower than the surface materials.*

9.4.3.1 — Determine the predominant clay mineral in the soil

This is Step 5, *Flow Chart 6*. To make the determination of predominant clay mineral, we use *Figure 80*, but first we need to know the cation exchange activity *CEAc* and the activity ratio *Ac* of the clay. Both can be calculated using the known values of plastic limit *PL* and percent of clay. The cation exchange capacity *CEC* is taken as $PL^{1.17}$ based on information from *Reference 10*.

$$CEAc \text{ (cation exchange activity)} = \frac{\text{cation exchange capacity}}{\text{percent clay}}$$

$$= \frac{PL^{1.17}}{60}$$

$$= \frac{30^{1.17}}{60} = \frac{53.48}{60}$$

$$CEAc = 0.89$$

Find the clay activity ratio *Ac*, using the known value of the plasticity index, *PI*.

$Ac = PI/\text{percent clay}$

$Ac = 40/60$

$Ac = 0.67$

With these two values, we can enter the clay classification chart (*Figure 80*) to determine the predominant clay mineral in the soil. Enter the chart from the bottom, drawing a vertical line through $Ac = 0.67$. Draw a horizontal line through $CEAc = 0.89$. The two lines intersect in the area labeled "montmorillonite," indicating that montmorillonite is the principal clay mineral.

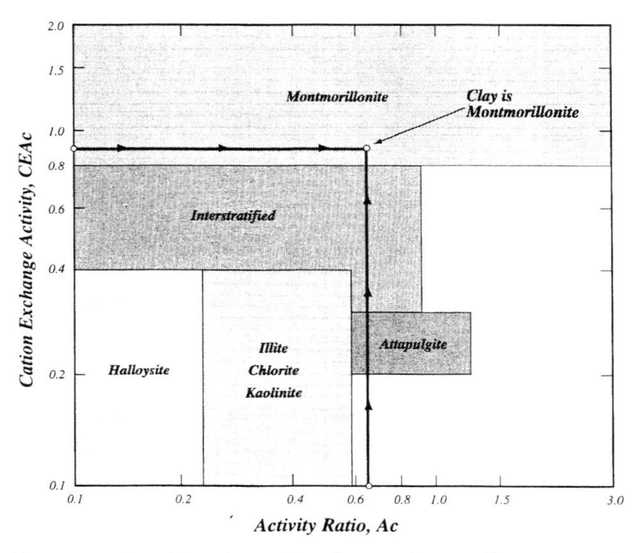

Figure 80 *Clay type classification related to cation exchange activity and clay activity ratio, from the PTI Manual* (Reference 10).

9.4.3.2 — Find the constant suction value for the soil

This is Step 6, *Flow Chart 6*. The constant suction value is needed for Step 9 below. First is is necessary to select a Thornthwaite moisture index from the map of *Figure 81*. By observation, the moisture index, $I_M = +40$.

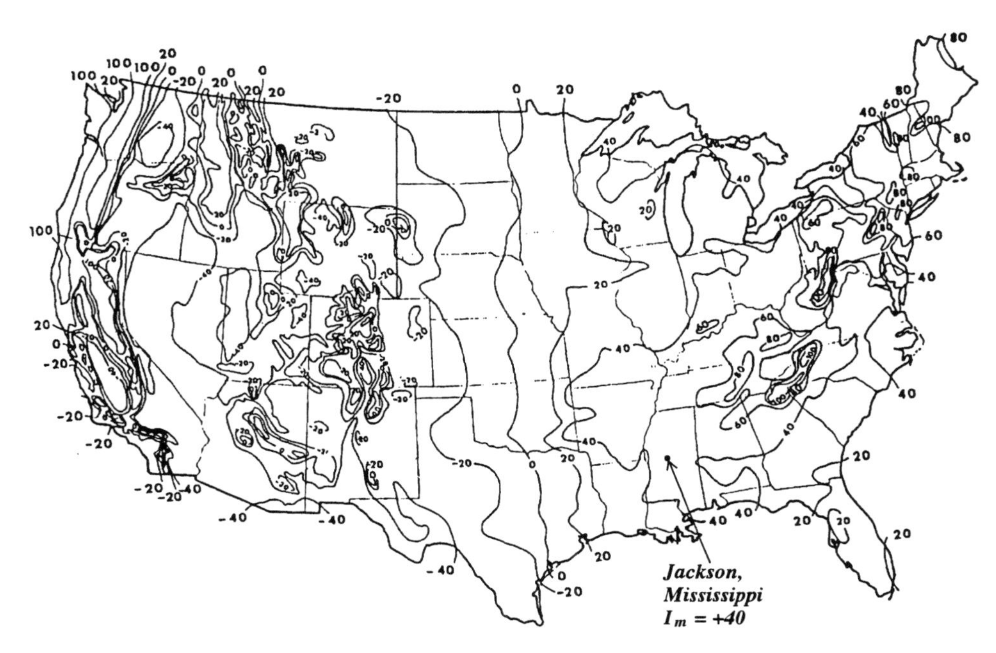

Figure 81 *Thornthwaite moisture index distribution in the United States, from the PTI Manual* (Reference 10).

Enter the chart of *Figure 82* at the bottom with the Thornthwaite index value of +40. Draw a vertical line to the intersection with the curve, then move left to read the soil suction value *pF* as 3.2.

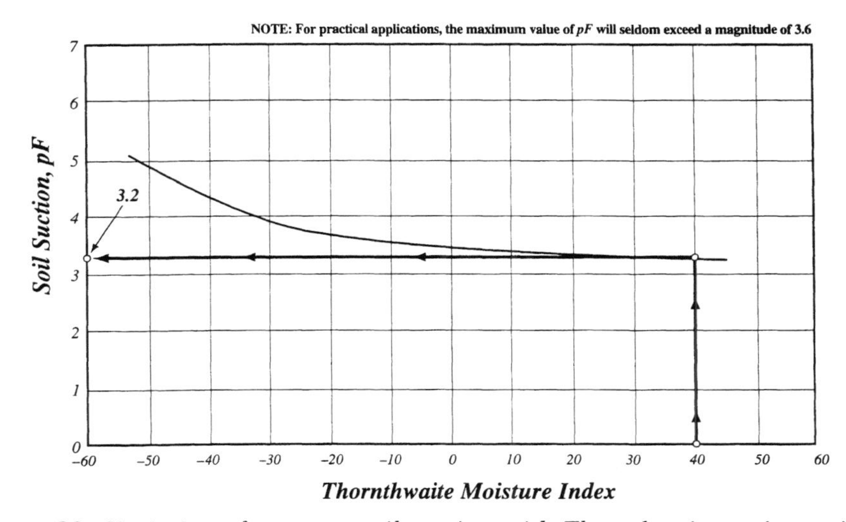

Figure 82 *Variation of constant soil suction with Thornthwaite moisture index, from* Reference 10.

9.4.3.3 — Determine the average moisture movement velocity

This is Step 7, *Flow Chart 6*. The estimated velocity of moisture flow is calculated using the Thornthwaite moisture index I_M of +40 obtained in the previous step.

$$\text{Moisture velocity} = 0.5 \times I_M/12$$
$$= 0.5 \times 40/12 = 1.66 \text{ inches per month}$$

However, according to the PTI procedure (*Reference 10*), the maximum moisture velocity shall be 0.7 inches per month. Therefore, use 0.7 inches per month in this problem.

9.4.3.4 — Find the edge moisture variation distance

With the data given and developed in the previous steps, we can now go to Step 8, *Flow Chart 6*. With the Thornthwaite moisture index of +40 we enter the chart of *Figure 83* at the bottom. Note that the edge lift and center lift bands appear to be asymptotic on the far right of the chart. Since the chart does not go to a value of +40, and the values become asymptotic, simply enter the chart on the far right and draw a vertical line to the middle of the center lift band, then proceed horizontally to the left to read:

e_m edge moisture variation distance = 3.8 feet (center lift)

Then continue the vertical line to the middle of the band for edge lift condition and again proceed horizontally to the left, reading:

e_m edge moisture variation distance = 5.2 feet (edge lift)

Commentary:
*The example uses a value midway across the band (*Figure 83*), but if the designer is extremely familiar with the local conditions, the chart bands allow some leeway for interpretation suited to site conditions. For example, if a particular vicinity is known to be a cause for concern, be more conservative by moving to higher values in either of the bands. To be less conservative, move to values closer to the bottom of the band.*

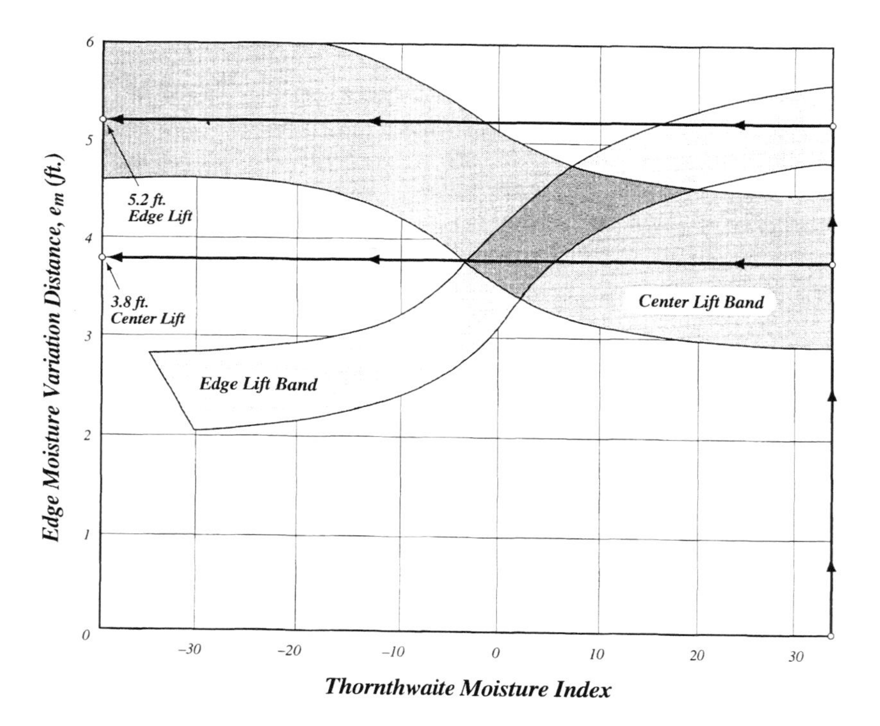

Figure 83 *Approximate relationship between Thornthwaite index and edge moisture variation distance, from* Reference 10. *Note that extremely active clays may generate larger values of edge moisture variation than reflected by the above curves and related tables. Therefore these curves should be used only in conjunction with a site-specific soils investigation by knowledgeable geotechnical engineers.*

9.4.3.5 — Determine expected differential swell for edge lift and center lift conditions

This is *Step 9* of *Flow Chart 6*. In the previous steps we have determined all of the values needed to use the charts of *Tables 26* and *27* to find y_m, the maximum differential soil movement, for both edge lift and center lift conditions. Note that this is referred to as *differential swell* in the tables. Additional tables for other soil conditions are found in *Appendix A.4*.

Commentary:

As mentioned previously, it would be important for the slab designer to have e_m *and* y_m *values supplied by an expert soils authority. Without this kind of site-specific information, calculations must be simplified in the conservative direction. Always assume that the clay mineral is montmorillonite. Use a conservative value of depth Z to constant suction. If the clay percentage varies, using the higher percentage is more conservative.*

Center lift:

Since the soil is 60% montmorillonite clay, the center lift condition of *Table 26* is used.

Work from the left at a depth to constant suction $Z = 5$ feet, selecting the constant soil suction $pF = 3.2$, and the moisture velocity of 0.7 inches per month. Since the edge moisture variation distance e_m is 3.8 feet, move to the right to columns for 3-foot and 4-foot penetration distances. Interpolating between values of 0.158 and 0.217, we find $y_m = 0.205$.

Percent Clay	Depth of Constant Suction	Constant Suction	Velocity of Moisture Flow (inches	DIFFERENTIAL SWELL (INCHES)							
				EDGE DISTANCE PENETRATION (FT)							
(%)	(FT)	(pF)	/month)	1 FT	2 FT	3 FT	4 FT	5 FT	6 FT	7 FT	8 FT
60	3	3.2	0.1	0.003	0.006	0.009	0.013	0.016	0.019	0.022	0.026
			0.3	0.009	0.019	0.028	0.038	0.048	0.059	0.069	0.080
			0.5	0.015	0.031	0.048	0.065	0.082	0.101	0.120	0.140
			0.7	0.022	0.044	0.068	0.092	0.119	0.147	0.176	0.209
		3.4	0.1	0.006	0.013	0.019	0.026	0.033	0.040	0.047	0.054
			0.3	0.019	0.039	0.060	0.082	0.105	0.130	0.157	0.186
			0.5	0.031	0.065	0.102	0.143	0.189	0.240	0.300	0.372
			0.7	0.044	0.093	0.147	0.211	0.286	0.380	0.503	0.683
		3.6	0.1	0.015	0.031	0.048	0.065	0.083	0.102	0.122	0.144
			0.3	0.044	0.095	0.152	0.219	0.300	0.403	0.543	0.758
			0.5	0.073	0.161	0.272	0.420	0.645	1.056	2.037	4.865
			0.7	0.101	0.229	0.407	0.689	1.246	2.689	6.912	-
	5	3.2	0.1	0.008	0.015	0.023	0.030	0.038	0.045	0.053	0.060
			0.3	0.022	0.044	0.067	0.090	0.113	0.138	0.162	0.187
			0.5	0.037	0.074	0.113	0.153	0.194	0.237	0.282	0.329
			0.7	0.050	0.103	0.158	0.217	0.278	0.343	0.412	0.487
		3.4	0.1	0.015	0.030	0.043	0.061	0.077	0.093	0.109	0.126
			0.3	0.045	0.091	0.140	0.191	0.245	0.302	0.363	0.426
			0.5	0.073	0.152	0.237	0.331	0.435	0.551	0.686	0.846
			0.7	0.102	0.214	0.341	0.485	0.655	0.865	1.140	1.541
		3.6	0.1	0.035	0.071	0.109	0.149	0.190	0.233	0.279	0.326
			0.3	0.103	0.219	0.349	0.499	0.678	0.901	1.200	1.652
			0.5	0.169	0.370	0.618	0.945	1.425	2.280	4.284	9.923
			0.7	0.234	0.526	0.922	1.528	2.686	5.574	-	-
	7	3.2	0.1	0.013	0.027	0.041	0.055	0.069	0.083	0.097	0.111
			0.3	0.041	0.082	0.125	0.168	0.212	0.256	0.302	0.349
			0.5	0.068	0.137	0.209	0.283	0.360	0.441	0.524	0.612
			0.7	0.093	0.191	0.294	0.402	0.516	0.637	0.767	0.907
		3.4	0.1	0.027	0.055	0.083	0.112	0.142	0.171	0.201	0.232
			0.3	0.082	0.167	0.256	0.351	0.449	0.555	0.666	0.786
			0.5	0.135	0.280	0.438	0.609	0.799	1.013	1.260	1.553
			0.7	0.188	0.395	0.627	0.892	1.204	1.587	2.092	2.840
		3.56	0.1	0.053	0.107	0.163	0.221	0.281	0.342	0.407	0.474
			0.3	0.156	0.326	0.512	0.720	0.957	1.232	1.566	1.994
			0.5	0.256	0.549	0.895	1.320	1.879	2.702	4.182	8.216
			0.7	0.354	0.779	1.317	2.059	3.247	5.677	-	-

Table 26 *Center lift condition, differential swell at the perimeter of a slab in predominantly montmorillonite clay soil, 60 percent clay (from* Reference 10*).*

Edge lift:

This time we use the edge lift table for 60% montmorillonite clay (*Table 27*).

As before, work from the left at a depth to constant suction Z = 5 feet, selecting the constant soil suction pF = 3.2, and the moisture velocity of 0.7 inches per month. Since the edge moisture variation distance e_m is 5.2 feet for this edge lift condition, move to the right to columns for 5-foot and 6-foot penetration distances. Interpolating between values of 0.202 and 0.240, we find y_m = 0.210.

Percent Clay (%)	Depth to Constant Suction (FT)	Constant Suction (pF)	Velocity of Moisture Flow (inches/month)	DIFFERENTIAL SWELL (INCHES) EDGE DISTANCE PENETRATION (FT) 1 FT	2 FT	3 FT	4 FT	5 FT	6 FT	7 FT	8 FT
60	3	3.2	0.1	0.003	0.006	0.008	0.011	0.014	0.016	0.019	0.022
			0.3	0.008	0.016	0.025	0.033	0.040	0.048	0.056	0.064
			0.5	0.014	0.027	0.041	0.054	0.066	0.079	0.091	0.104
			0.7	0.019	0.038	0.056	0.074	0.091	0.109	0.125	0.142
		3.4	0.1	0.006	0.012	0.017	0.023	0.029	0.035	0.040	0.046
			0.3	0.018	0.035	0.051	0.068	0.084	0.099	0.115	0.130
			0.5	0.029	0.057	0.084	0.110	0.135	0.160	0.183	0.206
			0.7	0.041	0.079	0.116	0.151	0.184	0.216	0.247	0.277
		3.6	0.1	0.014	0.029	0.043	0.056	0.070	0.083	0.096	0.109
			0.3	0.043	0.083	0.122	0.159	0.194	0.228	0.260	0.291
			0.5	0.071	0.136	0.195	0.251	0.303	0.352	0.399	0.433
			0.7	0.098	0.185	0.264	0.336	0.402	0.463	0.521	0.575
		3.8	0.1	0.035	0.069	0.102	0.133	0.163	0.191	0.219	0.246
			0.3	0.104	0.195	0.277	0.352	0.421	0.484	0.544	0.599
			0.5	0.169	0.309	0.428	0.533	0.627	0.712	0.790	0.863
			0.7	0.233	0.413	0.562	0.690	0.802	0.903	0.994	1.077
	5	3.2	0.1	0.006	0.012	0.018	0.024	0.030	0.036	0.042	0.048
			0.3	0.018	0.036	0.054	0.071	0.089	0.106	0.123	0.140
			0.5	0.030	0.060	0.090	0.118	0.146	0.174	0.202	0.229
			0.7	0.042	0.083	0.124	0.163	0.202	0.240	0.278	0.314
		3.4	0.1	0.013	0.026	0.039	0.051	0.064	0.076	0.089	0.101
			0.3	0.039	0.077	0.114	0.151	0.187	0.222	0.256	0.291
			0.5	0.065	0.127	0.188	0.246	0.303	0.359	0.413	0.465
			0.7	0.090	0.177	0.259	0.339	0.415	0.488	0.559	0.628
		3.6	0.1	0.032	0.064	0.095	0.126	0.156	0.186	0.215	0.244
			0.3	0.096	0.188	0.276	0.360	0.440	0.518	0.593	0.665
			0.5	0.160	0.308	0.446	0.575	0.697	0.812	0.921	1.025
			0.7	0.224	0.425	0.607	0.775	0.931	1.077	1.214	1.343
		3.8	0.1	0.080	0.156	0.230	0.301	0.369	0.436	0.500	0.562
			0.3	0.238	0.450	0.642	0.817	0.980	1.132	1.274	1.407
			0.5	0.395	0.724	1.009	1.261	1.488	1.695	1.886	2.063
			0.7	0.551	0.984	1.345	1.657	1.933	2.181	2.407	2.614
	7	3.2	0.1	0.010	0.021	0.031	0.042	0.052	0.062	0.072	0.083
			0.3	0.031	0.062	0.093	0.124	0.154	0.184	0.214	0.243
			0.5	0.052	0.104	0.155	0.205	0.254	0.303	0.351	0.398
			0.7	0.073	0.145	0.215	0.284	0.352	0.419	0.484	0.548
		3.4	0.1	0.023	0.045	0.067	0.090	0.112	0.134	0.155	0.177
			0.3	0.068	0.135	0.200	0.264	0.328	0.390	0.451	0.511
			0.5	0.113	0.223	0.330	0.434	0.535	0.633	0.729	0.823
			0.7	0.159	0.311	0.458	0.598	0.734	0.865	0.992	1.115
		3.6	0.1	0.057	0.113	0.168	0.222	0.275	0.328	0.380	0.431
			0.3	0.171	0.334	0.490	0.640	0.785	0.924	1.058	1.188
			0.5	0.286	0.551	0.799	1.031	1.251	1.459	1.658	1.847
			0.7	0.402	0.764	1.095	1.400	1.684	1.950	2.200	2.437
		3.8	0.1	-	-	-	-	-	-	-	-

Table 27 *Edge lift condition, differential swell at the perimeter of a slab in predominantly montmorillonite clay soil, 60 percent clay (from* Reference 10).

9.4.4 — Check slab thickness based on loading requirements

Having completed the soil properties subroutines of *Flow Chart 6*, we are now ready to proceed to Step 5 of *Flow Chart 5*. Use the design procedure for slabs with post loads as presented in *Chapter 4*. Data necessary for the solution (as given in Sections 9.4.2 and 9.4.3) can be summarized as follows:

From materials, site, and designer:

Concrete compressive strength: $f_c' = 3000$ psi

Modulus of rupture: $MOR = 9\sqrt{f_c'} = 493$ psi NOTE: this value is appropriate for use with this PCA design procedure, but should not be used in other calculations.

Average precompression in slab from post-tensioning tendons = 75 psi

Subgrade modulus: $k = 50$ pci

Safety factor: SF = 1.7

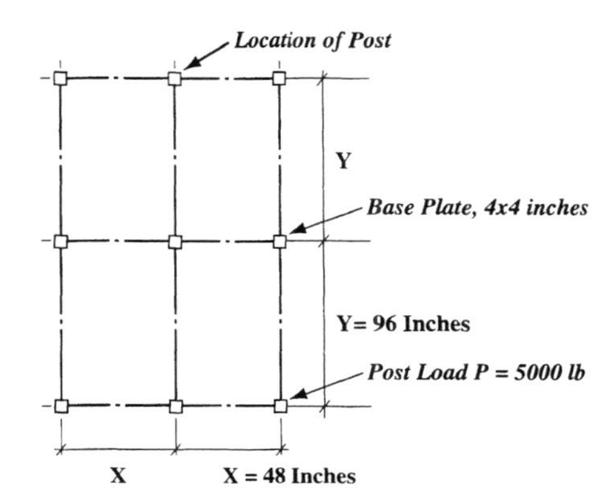

Figure 84 *Layout of rack storage posts on hybrid slab on grade.*

From loading specifications:

Short post spacing: $X = 48$ inches (4 feet)

Long post spacing: $Y = 96$ inches (8 feet)

Base plate area: 16 square inches (4x4 inches)

Post load: $P = 5000$ pounds or 5 kips

The appropriate PCA design chart for a subgrade modulus k of 50 pci is reproduced on the following page as *Figure 85*. Determine the required slab thickness from the chart in the following manner.

• First calculate the allowable stress per 1000 pounds (1 kip) of post load. The allowable stress is

$$\frac{\text{modulus of rupture} + \text{precompression from tendons}}{\text{factor of safety}}$$

or

$$(493 + 75)/1.7 = 334 \text{ psi}$$

Since each post load is 5 kips, the allowable stress per 1000 pounds of post load is 334/5 = 66.8 psi per kip.

Commentary:
In many post-tensioned slabs, 4 inches is considered acceptable. The authors believe it is reasonable in most commercial and industrial applications to maintain a minimum thickness of 5 inches. In hybrid (donut) designs such as this one, it is also prudent not to extrapolate beyond chart recommendations.

• Draw a line diagonally down from the left side of the chart to represent the load of 66.8 psi per kip. Also on the left side of the chart, draw a vertical line representing the effective contact area of 16 square inches, extending it to a point of intersection with the diagonal.

• From the intersection point draw a horizontal line to the right, intersecting the curve for Y = 96 inches. You must go to the Y-line first (see *Chapter 4*).

• Then draw a line to the curve representing X = 48 inches, as close as graphically possible. In this case, the solution is beyond the range of the chart, and a minimum slab thickness of 5 inches is recommended. As explained in *Chapter 4*, where the solution is in a range covered by the chart, another horizontal line to the right is required to read off the slab thickness.

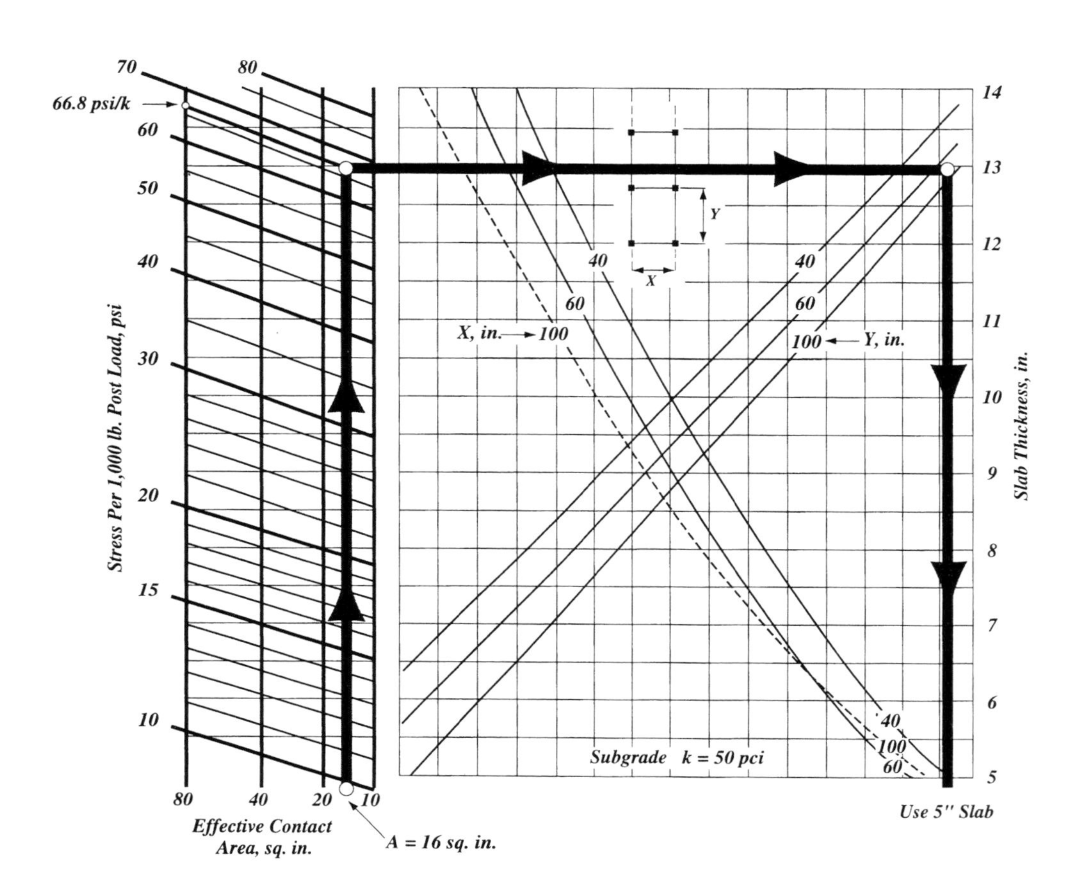

Figure 85 *PCA chart for slab thickness selection when using post loading with subgrade* k = *50 pci.*

9.4.5 — Check preliminary beam sizes, Step 6 of Flow Chart 5

With the soil values for edge moisture variation distance e_m and maximum differential soil movement y_m determined in Section 9.4.3.4, we can now proceed to determine beam sizes and tendon requirements. Only the edge lift condition is checked when selecting cross section and tendons, since it is the more critical condition. However, both edge lift and center lift conditions will be checked later for deflection, bending, and shear.

9.4.5.1 — Preliminary determination, stiffening beam depth d, edge lift condition

$$d = x^{1.176}$$

where

$$x = \frac{L^{0.35} \times S^{0.88} \times e_m^{0.74} \times y_m^{0.76}}{12 \times \Delta_{allow} \times P^{0.01}}$$

Long direction (same as short direction):

Beam length $L = 150$ feet
Beam spacing $S = 15$ feet
Perimeter load $P = 2000$ pounds per foot

Assume β, the relative stiffness length according to PTI procedure, is 10 feet. The maximum distance over which differential deflection will occur is L or 6β, whichever is smaller. $6\beta = 60$ feet $< L = 150$ feet, so 6β governs.

Since this slab is square the short and long deflection calculations are the same. However, with a rectangular slab when 6β controls in both directions, only long direction calculations are needed since the short direction is simply a function of the long one. Hybrid slabs usually are 100 feet or longer, and 6β generally governs.

Commentary:
Using L/1700 for allowable deflection is an empirical approach which provides a reasonable starting point for beam selection. Later, allowable deflection is calculated using L/800 for edge lift and L/360 for center lift, as recommended by PTI.

Use the assumed value for 6β and calculate the allowable deflection:

$$\Delta_{allow} = (12 \times 60)/1700 = 0.424 \text{ inches}$$

This value of Δ is substituted in the equation above, along with the given values and soil properties to find x:

$$x = \frac{150^{0.35} \times 15^{0.88} \times 5.2^{0.74} \times 0.210^{0.76}}{12 \times \Delta_{allow} \times 2000^{0.01}}$$

$$= \frac{5.77 \times 10.84 \times 3.38 \times 0.31}{12 \times 0.424 \times 1.08}$$

$$x = \frac{65.48}{5.49} = 11.93$$

$$d = x^{1.176} = 11.93^{1.176} = 18.45 \text{ inches}$$

Assume a beam depth of 18 inches. This assumption is reasonable for both directions since each direction is governed by 6β.

9.4.5.2 — Soil pressure under the beams (a subroutine for Step 6, Flow Chart 5)

The allowable soil pressure $q_{allow} = 3000$ psf is given in Section 9.4.3. The load on the ground consists of the weight of the slab and beams plus applied uniform load and the perimeter load.

For simplicity in determining the applied loading, take a bay section as indicated in *Figure 86*. The slab is 5 inches or 0.42 foot thick. Assuming the beams are 1 foot wide, reduce the slab area to allow for the beams when calculating the weight of slab as applied load. Total length of beams in the bay is $(15 \times 3) + 30 = 75$ feet.

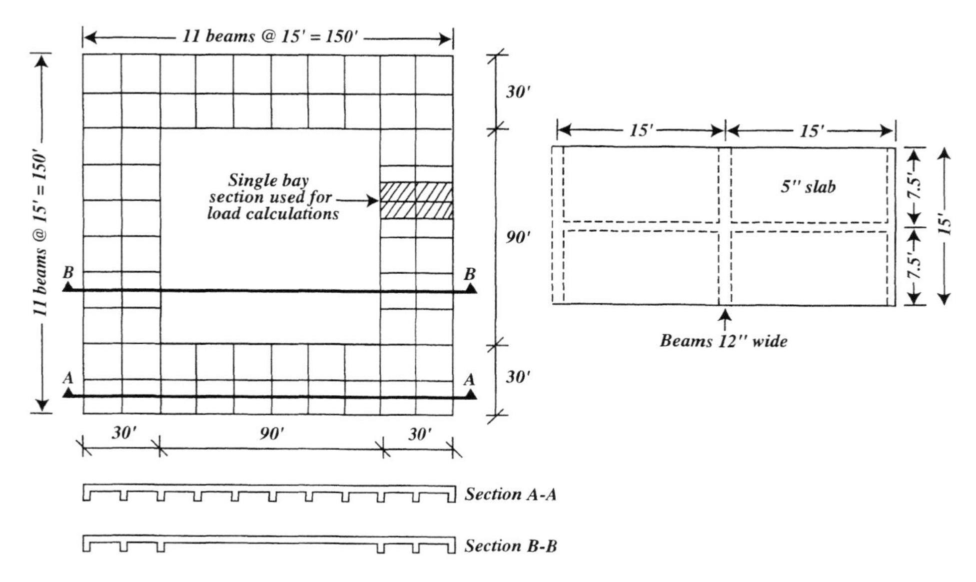

Figure 86 *Plan and cross section of hybrid slab floor, with detail of bay as used for load calculations.*

Applied loadings:

Slab weight = [(15 × 30) − (75 × 1)] × 0.42 × 0.150 = 23.63 kips
Long direction beams = 15 × 3 × 1.5 × 1 × 0.150 = 10.13 kips
Short direction beam = 30 × 1 × 1.5 × 1 × 0.150 = 6.75 kips
Perimeter load = 15 × 2.0 = 30.0 kips
Uniform live load = 15 × 30 × 0.125 = 56.25 kips
Total load applied to soil = 126.75 kips

We assume that in the volume-sensitive zone all of the load is transmitted through contact of beam bottoms. Therefore calculate the contact area of the beams based on the assumed spacing shown in *Figure 86*. The beam bottoms are 1 foot wide.

(15 × 3) + 30 = 75 feet of beam length × 1 foot of beam width
= 75 sq. ft. of beam bottom contact area.

The soil bearing pressuring is then

w = 126.75/75 = 1.69 kips per square foot

1.69 < the allowable 3.0; therefore bearing pressure is OK.

Alternative procedure, based on T-beam action

The above procedure, considering all of the load applied through the bottom of the rectangular beams is conservative and simple to follow. However, if the simplified method shows soil bearing pressures that are borderline, or questionable, the enhanced T-beam action of the post-tensioned slab on the ground can be very helpful in distributing load. Observations indicate that using the T-beam section for load distribution is reasonable and effective.

A distance of eight times the slab thickness has been used successfully for computing the bearing area. For edge grade beams, six times the slab thickness on one side of the beam would comply with the ACI Building Code. *Figure 87* shows these larger bearing widths. Shear should be reviewed at the interface when this procedure is followed. Should shear become critical, a thickened slab is a reasonable solution.

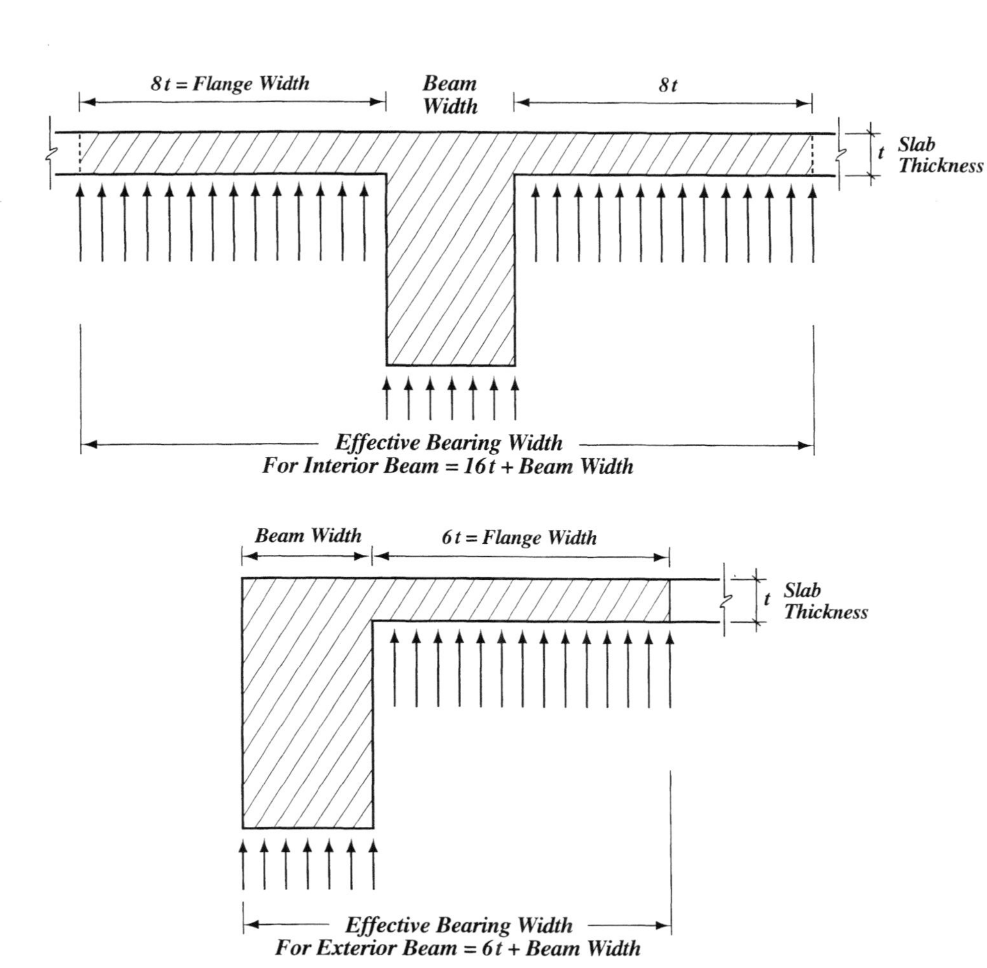

Figure 87 *Effective bearing width for interior and exterior grade beam considering T-beam action.*

9.4.6 — Determine section properties for full slab width, Step 7, Flow Chart 5

DIMENSIONS ALREADY DETERMINED	LONG DIRECTION (SAME AS SHORT)
Beam depth d	18 in.
Individual beam width	12 in.
Total number of beams	11
Total beam width	132 in.
Slab thickness	5 in.

CALCULATE SECTION PROPERTIES

Cross Section	Area, in.^{2}	y, in	Ay, in.^{3}
Slab = 139 × 12 × 5	8340	2.5	20,850
Beam = 11 × 18 × 12	2376	9	21,384
	10,716		42,234

Distance from top of slab to neutral axis $\bar{y}$ = 42,234/10,716 = 3.94 inches

Moment of inertia ($1/12bh^3$), beam and slab sections:

	I_O	A	d	Ad^2	TI
Slab = $1/12 \times 139 \times 12 \times 5^3$	17,375	8340	1.44	17,293.8	34,668.8
Beam = $1/12 \times 11 \times 12 \times 18^3$	64,152	2376	5.06	60,834.2	124,986.2
					159,655.0

$S_T = 159{,}655/3.94 = 40{,}521$ in^3
$S_B = 159{,}655/14.06 = 11{,}355$ in^3

9.4.7 — Determine adequacy of stiffened section in relation to the β distance chosen

Commentary:
Since only the perimeter is being stiffened, it becomes necessary to check the stiffness of the slab relative to the length β assumed. Too great a stiffness of the perimeter can result in possibly a greater transition distance.

β was assumed to be 10 feet. We must check using the PTI formula to determine if the actual β is 10 feet or less. If β is more than 10 feet, either the transition distance must be modified from 30 feet upwards, or a less stiff section must be chosen.

Determine actual β:

$$\beta = \frac{1}{12}\sqrt[4]{\frac{E_c I}{E_s}}$$

$$\beta = \frac{1}{12}\sqrt[4]{\frac{1{,}500{,}000 \times 159{,}655}{1000}}$$

$$\beta = \frac{1}{12}\,(124.40)$$

$$\beta = 10.36 \text{ feet}$$

The value of 3β is $10.36 \times 3 = 31$ feet.

Since the 30-foot width of stiffened perimeter is measured to the center line of the interior grade beam, the stiffened section is actually 30 feet 6 inches wide. The authors consider this adequate and we may proceed since we are within 6 inches of the desired dimension (a deviation of less than 2%). Any value within 5% would be acceptable.

If the difference were greater than 5%, there would be two options:

Decrease the slab thickness slightly
Increase the width of stiffened area to 3β

Either option is acceptable.

In the present example, if one wishes to be precisely accurate, the grade beam pattern can be shifted slightly to establish a 3β width of stiffened perimeter, as shown in *Figure 88.*

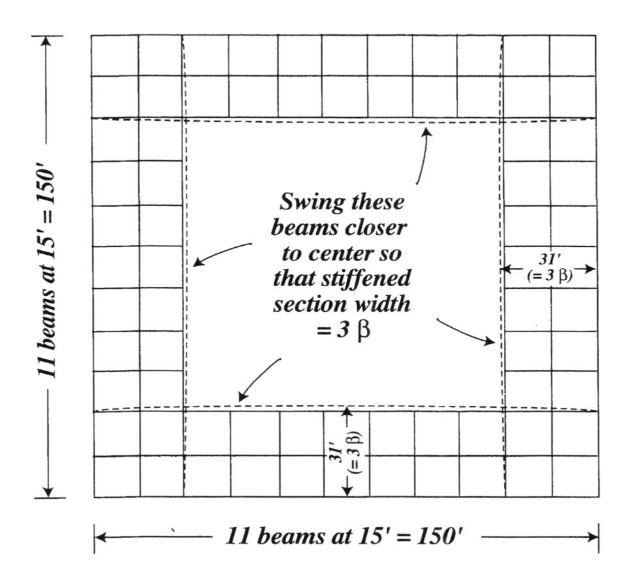

Figure 88 *Modified grade beam location so that stiffened perimeter area is 3β wide.*

9.4.8 — Calculate minimum number of tendons required, Step 9, Chart 5

Stress permitted per tendon: $f_{ps} = 0.7 \times f_{pu} = 0.7 \times 270 = 189$ ksi
Stress in tendon after losses: $f_{ps} = 189 - 30 = 159$ ksi
Force P_r per tendon:
Area per 1/2-inch diameter tendon = 0.153 square inches
$P_r = 0.153 \times 159 = 24.33$ kips per tendon

Sufficient tendons must be installed to overcome subgrade drag as well as to keep an average prestress in the slab of 75 psi.

Determine the subgrade drag (friction) force to be overcome at the center of the slab, using a one-bay section (*Figure 89*) for the analysis, then converting values to a per-foot-of-slab-length basis. Note that subgrade drag theory (Section 1.5.1.3) assumes the slab will shrink equally from its free ends on both sides of the center, so only half of the slab width is used in the analysis.

Commentary:
50 psi is permissible as a residual prestressing force according to the PTI manual, but the authors prefer to maintain a minimum force of 75 psi.

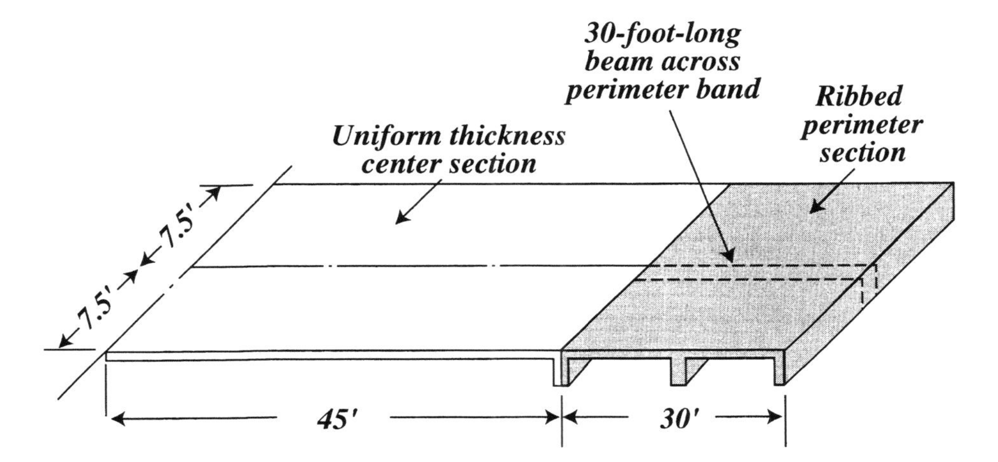

Figure 89 *Half of a typical one-bay section used in subgrade friction determination (not to scale).*

Ribbed section:

Weight = 40.5 kips per 15x30-foot section (calculated in Section 9.4.5.2)
Weight per foot = 40.5 / 15 = 2.70 kips

Uniform thickness section:

Weight per foot of slab = width × thickness × weight
= 45 × 5/12 × 0.15 = 2.812 kips/foot

Use of friction coefficient μ of 0.5 for the uniform thickness section and 0.75 for the ribbed section, assuming that the slab is place on a single layer of polyethylene sheeting.

Total friction per foot = (0.75 × 2.7) + ((0.5 × 2.812)
= 2.02 + 1.41 = 3.43 kips per foot

To this we must add the force per foot necessary to maintain the desire average prestress of 75 psi:

5 inches × 12 inches × 75 psi = 4500 pounds per foot
= 4.5 kips per foot

Commentary:
The coefficients of friction used are reasonable and may be slightly conservative. With a double layer of plastic sheeting, the values used may possibly be cut in half.

The total force required per foot to maintain the desired 75 psi prestress and to overcome subgrade drag is:

3.43 + 4.5 = 7.93 kips per foot

Determine total number of tendons in each direction to maintain this force:

(7.93 × 150) / 24.33 kips per tendon = 48.89 tendons
Use 49 tendons.

Since there are 11 stiffening beams including the perimeter, place one tendon in each beam and equally space the other 38 tendons approximately 4 feet on center across the slab. It is customary to drape beam tendons within 3 inches of the beam bottom. Slab tendons will be placed at the center of the 5-inch slab.

Design prestress forces:

Since maximum moments occur near the slab perimeter, friction losses will be minimal at points of maximum moments. It is therefore reasonable to assume the total prestressing force as effective for structural calculations.

For both long and "short" directions: P_r = 49 × 24.3 = 1190.7 kips

Separate short direction calculations will not be necessary for this example because the slab is square. However, when both directions of a large rectangular slab are calculated based on 6β, the complementary short direction calculations are not relevant. This is because this slab is too long for one side to affect the other.

9.4.9 — Check design moments against allowable moments for edge lift condition, Step 10, Flow Chart 5

Design moment "long" direction, edge lift condition

$$M_\ell = \frac{S^{0.10}\,(d\,e_m)^{0.78}\,y_m^{(0.66)}}{7.2 \times L^{0.0065}\,P^{0.04}}$$

$$M_\ell = \frac{15^{0.10} \times (18 \times 5.2)^{0.78} \times 0.21^{0.66}}{7.2 \times 150^{0.0065} \times 2000^{0.04}}$$

$$M_\ell = \frac{1.31 \times 34.48 \times 0.35}{7.2 \times 1.033 \times 1.35}$$

$$M_\ell = \frac{15.8}{10.04} = 1.57 \text{ foot-kips per foot}$$

Allowable service moments, "long" direction, tension in bottom fiber, edge lift condition

The allowable positive bending moment is basically equal to the section modulus times the allowable stress, with an adjustment made for the moment due to eccentricity of the post-tensioning tendons, $P_r e$. The allowable stress in the concrete is increased by the amount of the applied post-tensioning stress. The equation is:

$$(12 \times 150)\ _pM_t = S_B\ (P_r/A + f_t) - P_r e$$

Note that the long direction has 150-foot-wide cross section; hence the design (allowable) moment is multiplied by 150 feet. It is also multiplied by 12 inches per foot to make units on the left side of the equation compatible with those on the right. Use the dimensions from *Figure 90* to calculate $P_r e$ in inch kips.

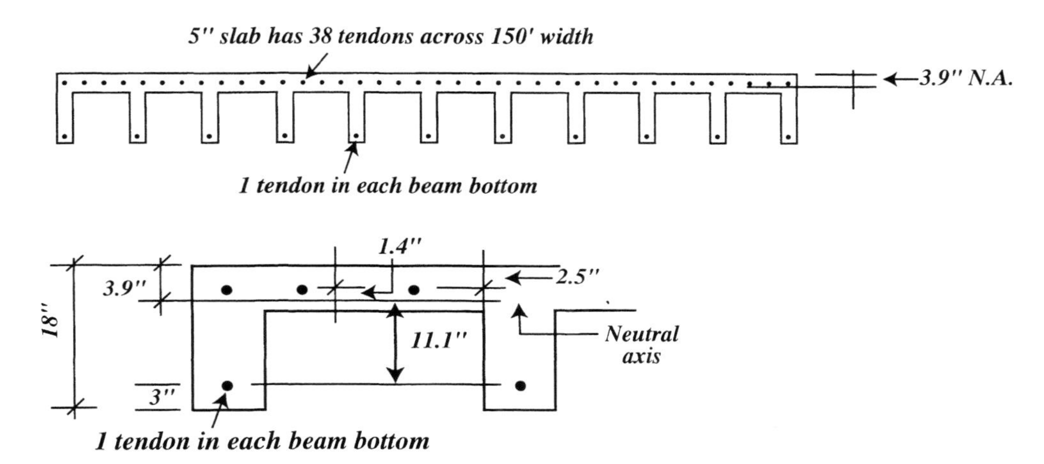

Figure 90 *Cross section of slab shows long direction section properties, with neutral axis location and beam and slab tendon eccentricities.*

$$\begin{aligned}
P_r e &= [(N_t(top) \times 1.4) - (N_t\,(beam) \times 11.1)] \times 24.3 \\
&= [(38 \times 1.4) - (11 \times 11.1)] \times 24.3 \\
&= (52.3 - 122.1) \times 24.3 \\
&= -68.9 \times 24.3 \\
P_r e &= -1674.27 \text{ inch-kips}
\end{aligned}$$

Now substitute this value of $P_r e$ in the equation given above for tension in the bottom fiber to find the allowable moment. Use section properties as calculated in Section 9.4.6. The prestressing force P_r was determined in Section 9.4.8, and f_t the allowable tensile stress in the concrete was given as 329 psi in Section 9.4.2.

Commentary:
The f_t value here agrees with the PTI recommendation of $f_t = 6\sqrt{f_c'}$. In the determination of uniform slab thickness in Section 9.4.4, a different value of allowable tension in the concrete was used for consistency with the PCA design method being applied there.

$$\begin{aligned}
(12 \times 150)\ _pM_t &= 11{,}355\ (1192/10{,}716 + 0.329) - (-1674) \\
1800\ _pM_t &= 11{,}355\ (0.111 + 0.329) - (-1674) \\
1800\ _pM_t &= 11{,}355\ (0.44) + 1674 \\
1800\ _pM_t &= 4996 + 1674 = 6670 \\
_pM_t &= 3.71 \text{ foot-kips per foot}
\end{aligned}$$

The allowable 3.71 > 1.57 design value — OK

Commentary:
The allowable concrete compressive stress f_c is taken as 1350 psi, based on the PTI recommendation to use $0.45\sqrt{f'_c}$.

Allowable service moments, "long direction," compression in top fiber, edge lift condition

Again multiplying by the 150-foot width of cross section and the factor 12 for compatibility with units on the right side of the equation

$$(12 \times 50)\ {}_pM_c = S_T\,(f_c - P_r/A) - P_r e$$

$$1800\ {}_pM_c = 40{,}521\ (1.350 - 1192/10{,}716) - (-1674)$$

$$= 40{,}521\ (1.350 - 0.111) - (-1674)$$

$$= 40{,}521\ (1.239) + 1674$$

$$1800\ {}_pM_c = 50{,}205 + 1674$$

$${}_pM_c = 28.82 \text{ foot-kips per foot}$$

This allowable of 28.82 exceeds the design value of 1.57 foot-kips per foot — OK.

9.4.10 — Deflection calculations, edge lift condition Step 11, Flow Chart 5

Allowable differential deflection, "long" direction, edge lift

$$\beta = \frac{1}{12}\sqrt[4]{\frac{E_c I}{E_s}}$$

From Section 9.4.7, $\beta = 10.36$ feet, $6\beta = 62.16$ feet

$$\Delta_{allow} = (12 \times 62.16)/800 = 0.93 \text{ inches}$$

Expected differential deflection, "long" direction, edge lift

$$\Delta = \frac{L^{0.35}\,S^{0.88}\,e_m^{\ 0.74}\,y_m^{\ 0.76}}{15.90\,d^{0.85}\,P^{0.01}}$$

$$\Delta = \frac{150^{0.35} \times 15^{0.88} \times 5.2^{0.74} \times 0.21^{0.76}}{15.90 \times 18^{0.85} \times 2000^{0.01}}$$

$$\Delta = \frac{5.77 \times 10.83 \times 3.38 \times 0.305}{15.90 \times 11.66 \times 1.07}$$

$$\Delta = \frac{64.42}{198.37} = 0.324 \text{ inches}$$

The expected differential deflection of 0.32 inches is less than the allowable deflection of 0.93 inches. Cross section is OK for deflection.

9.4.11 — Shear calculations for edge lift condition, Step 12, Flow Chart 5

Expected shear force, "long" direction, edge lift condition

$$V_\ell = \frac{L^{0.07}\,d^{0.40}\,P^{0.03}\,e_m^{\ 0.16}\,y_m^{\ 0.67}}{3 \times S^{0.015}}$$

$$V_\ell = \frac{150^{0.07} \times 18^{0.40} \times 2000^{0.03} \times 5.2^{0.16} \times 0.21^{0.67}}{3 \times 15^{0.015}}$$

$$V_\ell = \frac{1.42 \times 3.17 \times 1.26 \times 1.30 \times 0.35}{3 \times 1.04}$$

$$V_\ell = \frac{2.58}{3.12} = 0.83 \text{ kips per foot}$$

Total design shear stress *v* "long" direction, edge lift

$$v = \frac{V_\ell \times W}{n\,d\,b}$$

Note: Only beams are considered to resist shear stress.

$$v = \frac{0.83 \times 1000 \times 150}{11 \times 18 \times 12}$$

$$v = \frac{124{,}500}{2376} = 52.4 \text{ psi}$$

Allowable shear stress:

The second edition of the PTI Design Manual, in preparation at this writing, permits a more liberal shear stress than the first edition. The formula in the second edition is

$v_c = 1.7\sqrt{f_c'} + 0.20\,P_r/A$

Allowable shear stress v_c

$v_c = 1.7\sqrt{3000} + 0.20\,(111)$

$v_c = 93.11 + 22.2 = 115.13$ psi

52 psi is less than this 115 psi allowable, so shear is OK.

Commentary:
The previous formula simply permitted $1.5\sqrt{f_c'}$. As will be seen in this example, the new formula sanctions certain elements that would previously been considered overstressed.

9.4.12 – Center lift design

A check of bending, deflection, and shear must be made for the center lift condition just as has been done for the edge lift condition completed above. Since soil conditions and preliminary design are the same, we can begin at Step 10 of *Flow Chart 5.*

Design moments for "long" direction, center lift design (Step 10, Flow Chart 5)

$M_\ell = A_o\,[B \times e_m^{1.238} + C]$

where

$$A_o = \frac{1}{727}\,[L^{0.013}\,S^{0.306}\,d^{0.688}\,P^{0.534}\,y_m^{0.193}]$$

$$A_o = \frac{1}{727}\,[150^{0.013} \times 15^{0.306} \times 18^{0.688} \times 2000^{0.534} \times 0.205^{0.193}]$$

$$A_o = \frac{1}{727}\,[1.067 \text{ x } 2.29 \text{ x } 7.305 \text{ x } 57.90 \text{ x } 0.73]$$

$$A_o = \frac{754.43}{727} = 1.04$$

From the soils data, edge moisture variation distance e_m = 3.8 feet for the center lift condition (Section 9.4.3.4). According to the PTI procedure, for $e_m \leq 5$, B = 1.0 and C = 0. Using these values, determine M_ℓ

$M_\ell = 1.04\,[1 \times 3.8^{1.238} + 0]$

$M_\ell = 1.04 \times 5.22$

$M_\ell = 5.43$ foot-kips per foot

Allowable moments, "long" direction, center lift

Allowable moments must be calculated and compared with design moments. First, calculate negative bending moments.

• Tension in top fiber Note: from Section 9.4.9,

$P_r e = -1674.27$ inch-kips

$$(12 \times 150)\ {}_nM_t = S_T \left[\frac{P_r}{A} + f_t\right] + P_r e$$

$$1800\ {}_nM_t = 40{,}521 \left[\frac{1192}{10{,}716} + 0.329\right] + (-1674.27)$$

$$1800\ {}_nM_t = 40{,}521\ (0.44) - 1674$$

$$1800\ {}_nM_t = 17{,}829 - 1674 = 16{,}155$$

$${}_nM_t = \frac{16{,}155}{1800} = 8.97 \text{ foot-kips per foot}$$

Since the allowable moment of 8.97 foot-kips per foot is greater than the design moment of 5.37, the section is OK for tension in the top fiber.

Commentary:
Beam tendons are normally located 3 inches from the bottom. If the allowable compression in the bottom fiber is critical, beam tendons can be varied in elevation. Location of beam tendons from 3 to 12 inches from the bottom is not uncommon. Location closer to the neutral axis will reduce compression in the bottom fiber.

• Compression in the bottom fiber

$$(12 \times 150)\ {}_nM_c = S_B \left[f_c - \frac{P_r}{A}\right] + P_r e$$

$$1800\ {}_nM_c = 11{,}355 \left[1.350 - \frac{1192}{10{,}716}\right] + (-1674)$$

$$1800\ {}_nM_c = 11.355\ [1.239] - 1674$$

$${}_nM_c = 12{,}394/1800$$

$${}_nM_c = 6.88 \text{ foot-kips per foot}$$

Since the allowable 6.88 is greater than the 5.43 design moment, the section is OK for compression in the bottom fiber for the long direction.

When lengths are calculated in both directions using 6β, differential deflection tends to be sinusoidal. For this reason, "short" direction calculations are not necessary since the opposite direction does not affect the stiffness.

Center lift deflection calculations, "long" direction, Step 11, Flow Chart 5

• Allowable differential deflection for center lift, per PTI recommendations.

$$\Delta_{allow} = 12(6\beta)/360$$

$$\Delta_{allow} = 12(6 \times 10.36)/360$$

$$\Delta_{allow} = 745.92/360 = 2.07 \text{ inches}$$

•Expected differential deflection

$$\Delta = \frac{[y_m L]^{0.205}\ S^{1.059}\ P^{0.523}\ e_m{}^{1.296}}{380 \times d^{1.214}}$$

$$\Delta = \frac{[0.205 \times 150]^{0.205} \times 15^{1.059} \times 2000^{0.523} \times 3.8^{1.296}}{380 \times 18^{1.214}}$$

$$\Delta = \frac{2.01 \times 17.59 \times 53.26 \times 5.64}{380 \times 33.41}$$

$$\Delta = \frac{10{,}620}{12{,}695} = 0.83 \text{ inches}$$

The expected differential deflection, 0.83 inches, is less than the allowable of 2.07 inches, so the section is OK for deflection.

Shear calculations, long direction, center lift condition Step 12, Flow Chart 5

• Design shear force and stress

$$V_\ell = \frac{1}{1940}\left[L^{0.09}\, S^{0.71}\, d^{0.43}\, P^{0.44}\, y_m^{0.16}\, e_m^{0.93}\right]$$

$$V_\ell = \frac{1}{1940}\left[150^{0.09} \times 15^{0.71} \times 18^{0.43} \times 2000^{0.44} \times 0.205^{0.16} \times 3.8^{0.93}\right]$$

$$V_\ell = \frac{1}{1940}\left[1.56 \times 6.83 \times 3.46 \times 28.34 \times 0.776 \times 3.46\right]$$

$$V_\ell = \frac{1}{1940}(2805)$$

$$V_\ell = 1.44 \text{ kips per foot}$$

$$v = \frac{1.44 \times 150 \times 1000}{11 \times 12 \times 18} = \frac{216{,}881}{2376} = 91.3 \text{ psi}$$

The design shear stress v of 91 psi is less than the allowable of 115 psi, which is determined from the revised PTI recommendation of $v_c = 1.7\sqrt{f_c'} + 0.2\, P_r/A$. The section is OK for shear.

This completes the check of moment, shear, and deflection for center lift. Edge lift and center lift design results are summarized in Table 28. Since the slab is square, calculations were based on "long" direction only.

Edge lift design	***Design***	***Allowable***	**Center lift design**	***Design***	***Allowable***
Moment, ft.-kips per ft.			Moment, ft.-kips per ft.		
Tensile	1.57	3.71	Tensile	5.43	8.97
Compressive	1.57	28.82	Compressive	5.43	6.88
Differential deflection, in.	0.32	0.93	Differential deflection, in.	0.83	2.07
Shear stress, psi	52	115	Shear stress, psi	91	115
Edge moisture variation distance e_m, ft.	5.2		Edge moisture variation distance e_m, ft.	3.8	
Maximum vertical movement y_m, in.	0.210		Maximum vertical movement y_m, in.	0.205	

Table 28 *Summary of design checks for hybrid slab*

Commentary:
Note that shear would have been slightly overstressed with the formula in the first edition of the PTI manual since the allowable would have been only 82 psi. Regardless as previously explained, the authors do not consider shear a major concern.

CHAPTER 10
SUPPORTING DESIGN INFORMATION

10.1 — Dowels for floor slabs on grade

10.1.1 — Purpose and function of dowels

Dowels are frequently used in slabs on grade, primarily as load-transfer devices. As such, dowels maintain nearly the same slab strength at the joint as the slab possesses internally. A secondary function of the dowel is to maintain vertical alignment between adjacent slabs. In this function, the dowel prevents faulting and resists curling at the joint. Dowels can prove economical where there is poor subgrade support or heavy and frequent vehicle traffic.

For the dowels to be effective, they must be properly supported during construction. The authors recommend the so-called dowel basket for positive support and alignment. Correct alignment is essential to keep the dowels parallel to the direction of expected motion. Dowels must also be lubricated on at least one complete end so as to prevent bonding with the concrete.

For grade slabs, the decision to use dowels depends on the type of joint, the motion expected, and the traffic intensity. Dowels are frequently used in traffic corridors for all types of joints except complete isolation joints. They are commonly used at construction joints and frequently at contraction joints where traffic warrants.

10.1.2 — Types of dowels

The most commonly used dowel is a solid, smooth steel rod supported by a dowel cage or basket. All dowels must be straight, smooth, and free of burrs at the ends. Where the dowel is greater than 1 inch in diameter, an extra-strength (thick-walled) pipe may be substituted, according to Army and Air Force technical manuals *(Reference 17)*. These manuals also consider the doweled butt joint as the preferred construction joint for purposes of load transfer and slab alignment.

Frequently, the construction pouring sequences will cause a newly placed slab to be placed against an existing slab. The new slab will frequently shrink, or expand and then shrink, laterally along the common joint as well as directly away from that joint. A square dowel with a sponge-like material attached to each side surface of the dowel will allow this lateral motion while maintaining vertical shear transfer (see *Figure 91*). The flexible material allows this motion without restraint and the tendency to locally crack the newer slab is eliminated.

The proper location of dowels is at mid-depth of the slab, with a vertical tolerance of one-half the dowel diameter (above or below mid-depth). One-half of the dowel must be coated to prevent bonding with the concrete. Where the dowel is at an expansion joint, the dowel must have a cap at one end to permit unrestrained motion and placement exactly at the mid-depth is much preferred.

Commentary:
Dowels for slabs must be detailed and placed so that movement of the second slab placed is not locked up or restrained in a direction at right angles to the dowel itself. Severe slab cracking can occur if the slab shrinks excessively in this right-angle direction and is restrained by the dowels.

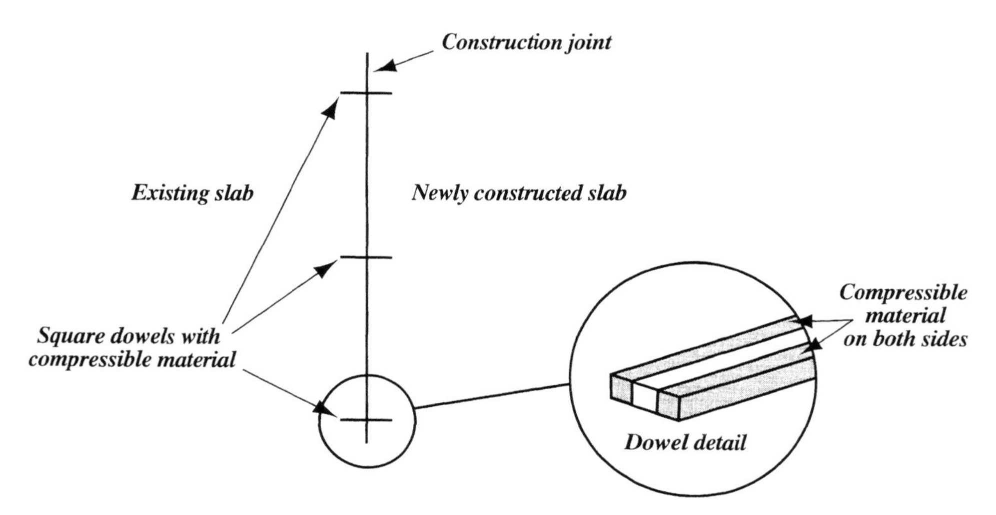

Figure 91 *Square dowels with compressible material on both sides permit lateral movement while maintaining vertical shear transfer.*

10.1.3 — Size and spacing of dowels

Three commonly used sources give recommendations of dowel bar sizes, lengths, minimum embedment (on one side of the joint) and spacings laterally along the joint. These sources are the ACI Committee 302 guide *(Reference 26)*, the ACI Committee 330 report *(Reference 27)*, and technical manuals of the Army and Air Force *(Reference 17)*. While the three recommendations differ somewhat, the authors believe they all fall within a reasonable range. *Table 29* summarizes the recommended values.

Slab Thickness	Dowel Diameter ACI 330	Dowel Diameter ACI 302	Dowel Diameter TMs	Minimum Embedment ACI 330	Dowel Length ACI 330	Dowel Length ACI 302	Dowel Length TMs	Spacing ACI 330 & ACI 302
5	5/8	3/4	3/4	5	12	16	15	12
6	3/4	3/4	3/4	6	14	16	15	12
7	7/8	1	3/4	6	14	18	15	12
8	1	1	1	6	14	18	16	12
9	1 1/8	1 1/4	1	7	16	18	16	12
10	1 1/4	1 1/4	1	7 1/2	18	18	16	12
11	1 3/8	1 1/4	1	8	18	18	16	12
12	1 1/2	—	1 1/4	9	20	18	18	12

NOTE: Only ACI 330 and 302 recommend spacings of 12 inches. The Army/Air Force technical manual (TM) recommends a variable spacing from 9 inches to 24 inches depending upon the type of joint. Only ACI 330 recommends a minimum embedment.

Table 29 *Comparison of recommended dimensions and spacing for dowels (all values in inches).*

10.2 — Joints in slabs on grade

10.2.1 — Purpose of joints

The basic purpose of joints in concrete slabs on grade is to accommodate slab motion and thereby prevent or control cracking of the concrete. Joints are also necessary to allow the placement process to occur within the practical limits of column and joint spacings, screed dimensions, flatness criteria, etc. Cracks are the net result of several effects including drying shrinkage of the concrete and stresses due to applied external loading.

Crack control is the basic purpose of joints, and inherent in this purpose is that the joint must provide load transfer from slab unit to slab unit. Proper load transfer is particularly important for vehicle loadings to prevent excess maintenance as well as to prevent cracking due to the imposed loading *(Reference 6)*. Load transfer, also called shear transfer, is commonly provided by aggregate interlock, dowels, or keys depending upon the joint construction.

Commentary:
The commonly accepted purpose of joints is to prevent or control cracking in the slab. Other factors equally important in crack control, but frequently not given adequate attention, are curing, concrete mix proportions (particularly cement content), and timing the cutting of sawed joints.

10.2.2 — Types and functions of joints

Joint types are classified in two basic ways. The first classification generally describes the function of the joint:

- Isolation
- Construction
- Contraction *(control)*
- Expansion

The second classification generally describes the construction of the joint:

- Dowel
- Tied
- Key *(tongue and groove)*
- Tooled
- Insert *(strips)*
- Butt-type joints
- Saw-cut, early: before concrete has fully hardened *(wet-cut)*
- Saw-cut, conventional: after initial hardening of the concrete *(dry-cut)*

Isolation joints—Drying shrinkage in concrete is normal, even in shrinkage-compensating concrete. Joint spacing is a key item in control of cracks due to drying shrinkage, but it is actually the restraint to shrinkage that causes the tension and thus the crack. Any construction feature that adds undue restraint will tend to cause more cracking. This includes standpipes, interior footings for special uses, dock leveling frames, sump pits, wall footings at grade level, and the like. All of these should be isolated from the so-called floating slab.

Isolation joints are intended to completely separate the concrete slab from an adjacent slab or from any of these restraints. They transfer nothing vertically nor horizontally. In a broader sense, they are a type of control joint. Depending on the width of the joint, which is the gap between slabs, isolation joints may also act as expansion joints where needed.

Construction joints are always needed in slabs. They may be at the outer edge of the slab or they may be throughout the interior of the slab, as is the case for large floor areas. These are basically stopping places during construction and frequently provide the support for the moving screed. The construction joint can be doweled, keyed, or tied, or it can be a butt-type (isolation) construction joint.

Contraction joints are essential to slabs on grade for prevention of cracking due to drying shrinkage of the concrete. They are frequently called control joints. They may be produced by hand tooling, by placement of a flexible (sometimes removable) strip, or by saw-cutting the hardened concrete slab.

The saw-cut may be made after the concrete has hardened and has started to gain strength, or it may be made before the concrete has hardened too much. All contraction joints are intended to relieve the tensile stress produced by restraint to slab motion. Their spacing is extremely important to this function.

The saw-cut (conventional) on the hardened concrete must not be made too late. Timing of the saw-cut varies, but four to 12 hours is common. The saw-cut (early) on the relatively fresh concrete is made immediately following the final finishing process, usually in less than one hour.

Expansion joints, while common in outdoor slabs, are not so common in interior slabs. They allow the concrete slab to expand relative to the adjacent slab or building component. Expansion joints may be doweled (dowels with bond breaks), but cannot be tied or keyed. The usual reason for requiring an expansion joint is temperature. Expansion joints are also necessary when an expansive-cement concrete is used, since that concrete expands significantly before it shrinks *(Reference 13)*.

10.2.3— Shear transfer

The designer should remember that most thickness determination procedures relate to an interior-of-the-slab loading condition. Therefore shear transfer requires additional consideration in slabs where design load traffic across joints is likely to occur, or where post loads are at or near joints.

As vehicle loads move across joints, their load effect is transferred from one slab section to another. Shear transfer is one phrase used to describe the ability of the slab, subgrade, and any load-transfer devices to support the load and permit it to be shared by two adjacent slabs. Stationary loads placed very close to joints create a similar situation.

The designer should consider shear transfer devices wherever loads are at joints or are likely to move across joints, particularly in defined traffic lanes. Both dowels and shear plates can be used. A keyed (tongue and groove) joint also transfers shear, but its effectiveness is markedly reduced as drying shrinkage progresses. Dowel selection information is given in Section 10.1.3. Shear plates are horizontally continuous and require separate analysis beyond the scope of this book.

Where loads and resulting shear forces are large, and the designer considers a shear-check computation desirable, the effective width of the concrete cross section used in the check should not exceed three times the slab thickness in any direction. Where this is done, the authors recommend that dowel spacings also do not exceed three times the slab thickness.

Commentary:
Close joint spacings will help reduce drying shrinkage cracks, but will have absolutely no effect on cracking due to superimposed loading. The only exception to this is where more joints may result in a greater likelihood that concentrated loadings such as posts may occur at the edge of a joint and therefore have less support than was anticipated.

*The authors consider this guide (*Table 30*) to be both conservative and relatively safe for most applications. However, it does not account for column spacings, plant layout such as aisle locations, or the joint lengths created as a result of spacings and locations. All must be considered. Note that this guide does not apply to post-tensioned or shrinkage-compensating concrete slabs.*

10.2.4— Joint spacings

The selection of joint spacing is one of the most critical decisions in the determination of the cost and behavior of a slab on grade. A very close joint spacing will virtually eliminate all drying shrinkage cracks in the concrete. Too close a spacing, however, may not prove economical due to the cost of the joint itself, its treatment (cleaning, curing, and sealing), and subsequent maintenance.

The most specific set of recommendations for joint spacings is given by the Portland Cement Association *(Reference 6)*. PCA cites slab thickness, shrinkage potential, base surface friction, curing efficiency and reinforcement as the factors upon which spacing is based. The PCA guide recommends that the joint spacing in feet be 2 to 3 times the slab thickness in inches for plain concrete *(Table 30)*.

Slab thickness, inches	***Slump 4-6 inches*** MCA < 3/4 in.	MCA > 3/4 in.	***Slump < 4 in.***
5	10	13	15
6	12	15	18
7	14	18	21
8	16	20	24
9	18	23	27
10	20	25	30

Table 30 *PCA recommendation for joint spacing (feet) in plain concrete slabs, from* Reference 6.

Joint spacing also depends on the design approach and the type of construction selected for the particular job. Further, the center-line-to-center-line spacings of columns, and other structural features, may be controlling distances in choosing joint spacings. For example, the column spacing may be 48 feet, but the slab contraction joint spacing may appear to be best set at 20 feet. If the construction joints are placed at the column center lines, then the contraction joints will not be possible at a 20 foot spacing. They will most certainly be placed at either 24 feet, which may be too much, or at 16 feet, which may create too many joints.

There are two possible solutions to this dilemma. One choice could be the 16-foot-spacing using plain concrete. The other decision could be the 24-foot spacing with reinforcement (bars, mesh, or steel fibers). There is no single correct answer. Either could work with good job control.

Sources other than PCA do not give specific recommendations for joint spacings. The Post-Tensioning Institute *(Reference 10)* states that the slab length is controlled by functional requirements or by the owner's requests. *Reference 10* refers to slab lengths (joint spacings) of 200 to 250 feet. No specific guidelines related to slab thickness are presented. ACI Committee 223 *(Reference 13)* gives no guide based on slab thickness for shrinkage-compensating concrete. ACI 223 does refer to placement areas up to as large as 16,000 square feet (if square, this would give a joint spacing of 126 feet) and mentions that spacings of 150 to 200 feet have been used. References on steel fiber reinforced concrete slabs have no recommendations on joint spacings except for one manufacturer *(Reference 28)* which states that joints are spaced up to just over 48 feet.

Commentary:
Slab units created by joint layouts should be as square as possible. Aspect ratios of 1.5 to 1 should be a maximum and an aspect ratio of 1.25 to 1 is preferred. L-shaped and T-shaped slab panels should be avoided, as should all re-entrant corners.

10.2.5—Joints as stress raisers

Although absolutely necessary in floor construction, joints interrupt the continuity of the floor and cause an increase in bending stresses in the floor when wheel forces or other loadings move across the joint. These stresses are commonly referred to as *edge stresses.* At a free edge, such as an isolation joint, there is no shear transfer and the edge stresses may be significantly greater (50% or more) than those at the interior of the floor (Section 3.2.3). Where the joint is doweled, the bending stresses are also greater than those at the slab's interior, but by a much smaller percentage. As a result of edge stresses, either the safety factor is effectively reduced or the floor designer increases the normal safety factor when determining thickness.

10.2.6—Other factors affecting joints

10.2.6.1 —Bond-breaking inter-layers

A bond-breaking layer is commonly used to reduce restraint. For this purpose, it is usually placed immediately beneath the concrete slab. This bond-breaking layer is usually essential in post-tensioned slabs and in other slabs where joint spacings are quite large. Although the material may be like that of a vapor barrier used to protect the slab from water or water vapor, the function is quite different. A vapor barrier need not be placed immediately beneath the concrete as the bond-breaker is, but rather may be beneath a base course *(Reference 23).*

10.2.6.2 —Shrinkage-compensating concrete

This type of concrete actually expands before it shrinks; however, its total shrinkage is similar in magnitude to that of conventional portland cement concrete *(Reference 13).* The important point here is that expansion joints are required as construction joints, even for completely interior slabs.

10.2.6.3 —Continuous steel

Distributed steel (reinforcing bars or welded wire fabric sheets), where used, must be continuous throughout the individual slab panels from joint to joint. This distributed steel should not be completely continuous through any joints intended to act as contraction, isolation, or expansion joints. No steel whatsoever should be continuous through isolation or expansion joints *(Reference 6).* Some small percentage of the distributed steel may be continuous through contraction joints; however, it must not be enough to inhibit the function of the joint.

10.2.6.4 — Steel fibers

The use of steel fibers creates a resistance to shrinkage cracking throughout the slab volume. It allows an increase in joint spacing somewhat beyond that of the completely unreinforced slab. The increase in joint spacing depends on the percentage of steel fibers used and on the expectations for drying shrinkage cracking *(Reference 24)*.

10.2.6.5 — Post-tensioned slabs

Post-tensioning creates a net pre-compression within the concrete slab. The location of the panel's edge joints may have a wide spacing, as shown in the examples of this book. The joint itself, however, must have sufficient space (a width of frequently 3 feet) so as to allow the post-tensioning process to take place. Basically, this creates two joints (one open joint) along a construction line within the slab's interior.

10.3— Control of shrinkage

Simply stated, concrete shrinkage is the change in volume that occurs when concrete loses water. This water loss is the result of both physical and chemical actions, and it occurs before, during, and after the hardening of concrete.

For most concretes, shrinkage after two years is between 500 and 700 millionths. The commonly accepted values are:

- Low shrinkage, 5×10^{-4} inches per inch
- Medium shrinkage, 6×10^{-4} inches per inch
- High shrinkage, 7×10^{-4} inches per inch

Poor practices that can cause increased shrinkage	***Equivalent increase in shrinkage (%)***	***Cumulative effect***
Temperature of concrete at discharge allowed to reach 80°F (27°C), whereas with reasonable precautions, temperature of 60°F (16°C) could have been maintained	8	1.00 × 1.08 = 1.08
Used 6 to 7 in. (150 to 180 mm) slump where 3 to 4 in. (75 to 100 mm) slump could have been used	10	1.08 × 1.10 = 1.19
Excessive haul in transit mixer, too long a waiting period at job site, or too many revolutions at mixing speed	10	1.19 × 1.10 = 1.31
Use of ¾ in. (19mm) maximum size aggregate under conditions where 1½ in. (38mm) could have been used	25	1.31 × 1.25 = 1.64
Use of cement having relatively high shrinkage characteristics	25	1.64 × 1.25 = 2.05
Excessive "dirt" in aggregate due to insufficient washing or contamination during handling	25	2.05 × 1.25 = 2.56
Use of aggregates of poor inherent quality with respect to shrinkage	50	2.56 × 1.50 = 3.84
Use of admixture that produces high shrinkage	30	3.84 × 1.30 = 5.00
Total increase	Summation = 183%	Cumulative = 400%

Table 31 *Cumulative effect of adverse factors on shrinkage, from* Reference 8.

Between 20% and 50% of shrinkage occurs in the first few days during the final set and hardening. Long-term studies indicate that approximately 75% of total shrinkage is complete in the first four years and 90% between six and eight years.

ACI 360 points out that drying shrinkage is the result of the cumulative effect of a number of factors shown in *Table 31*.

The first three factors on this table can be referred to as job control factors, and their cumulative effect on shrinkage as a departure from the best job practice is only 31%. Material selection, the last five factors, offers the most effective way to "fine tune" the control of shrinkage.

10.4— Construction tolerances for slabs on grade

There are three tolerance criteria that are part of the slab on grade design problem. They are the tolerance for flatness or smoothness of the top surface of the base or subgrade material; the tolerance for thickness of the slab as built; and the tolerance for flatness and levelness of the top surface of the slab. Each of the three has an effect upon the other. Particularly, the flatness and levelness of the base and the flatness and levelness of the top of the concrete directly affect the thickness of that slab.

10.4.1 —Tolerance for base construction

In the construction of a slab's subgrade, the upper surface of that subgrade, also called the top of the base, is rarely, if ever, perfectly flat. Normal construction procedures will produce a top surface with some measurable variation from flat. This part of floor construction is done in two parts. Rough grading is first, followed by fine grading which should be done immediately before concrete placement. Rough grading is almost never done by the floor contractor, although the floor contractor occasionally does the fine grading. Construction tolerances of both are critical to performance and cost.

Commentary:
Compliance to the specified tolerance should be confirmed by rod and level survey using 20 foot intervals (proposed recommendation, Reference 26*). Measurement by an independent party should be considered.*

Class	*Type of traffic and floor use*
1	Single course, foot traffic, exposed surface. Offices, commercial, residential, decorative.
2	Single course, foot traffic, covered surface. Offices, residential, commercial, institutional.
3	Two course, exposed or covered surface, foot traffic. Commercial, non-industrial.
4	Single course, exposed or covered surface, foot or light vehicular traffic. Institutional and commercial.
5	Single course, exposed surface, industrial vehicular traffic (pneumatic wheels).
6	Single course, exposed surface, industrial vehicular traffic (hard wheels).
7	Two course, exposed surface, heavy-duty industrial vehicular traffic (hard wheels). Bonded two course floors, heavy traffic and impact.
8	Two course (similar to Class 4), exposed or covered surface, vehicular traffic (hard wheels). Unbonded toppings on old or new floors.
9	Single course or topping, exposed surface, critical surface tolerance special materials handling. Narrow-aisle, high-bay warehouses, television studios.

Table 32 *Brief description of floor classes from ACI 302 (*Reference 26*).*

*Using these tolerances (page 180) will require more concrete yield for the floor. The extra amount will be very close to one-half of the negative tolerance. Contractors with refined grading equipment will see only a slight change in concrete yield required, while contractors with less refinement will see a greater yield requirement. Tolerances should be specified in the contract documents. Note that the ACI 302 tolerances differ from those recommended in the ACI 330 guide for parking lots (*Reference 27*), where the stated tolerance is +1/4 inch, −1/2 inch.*

Two tolerances are appropriate. ACI Committee 302 is currently (1995) considering recommending one tolerance for rough grading, which applies to all classes of floors (*Reference 26*), and two tolerances for fine grading according to the class of floor.

The recommended rough grading tolerance is +0 inches, − 1½ inches, where + means the surface is too high and − means it is too low. No portion of the top of the subgrade should be above its planned top elevation.

The recommended fine grading tolerance is +0 inch, − 1 inch for Floor Classes 1 through 3, and +0 inch, − 3/4 inch for Floor Classes 4 through 9. (See *Reference 26* for a complete description of floor classes.) As in rough grading, no portion of the fine grading material is above its top planned elevation. This material is placed on the rough-graded subgrade and should be granular, compactible, and trimmable.

The floor designer cannot assume that a perfectly flat base surface will be constructed. If a tolerance is appropriate for a particular job, then it should be specified. If a specified tolerance is not appropriate or is not desired, the designer must make an allowance for a variation in the slab's constructed thickness.

10.4.2 — Tolerance for slab thickness

The tolerance for the finished thickness of the concrete slab may be written into the specifications or placed on the drawings. The thickness tolerance can also be covered by incorporating by reference the ACI 117 tolerances standard (see *Reference 29*). ACI 117 (1990 Standard) sets the tolerance for slabs 12 inches or less in thickness at no more than +3/8 inch (thicker) and no more than – 1/4 inch (thinner) from the design thickness. For slabs thicker than 12 inches (up to 36 inches), the tolerances are +1/2 inch and – 3/8 inch.

As normally written in various documents, including those of ACI, this (ACI 117) tolerance on thickness is independent of any other tolerances as it is usually stated (when in fact it is stated) in specifications. In reality, the slab thickness is extremely dependent upon the other tolerances discussed in this section. The actual thickness of the concrete slab is absolutely controlled by the constructed top surface of the base and upon the constructed top surface of the slab. The authors believe that a better slab would result if tolerances were selected for the individual job and so specified.

Commentary:

It is the authors' opinion that slab thickness tolerances in ACI 117 (1990) may not have been intended to apply to slabs on grade. Nonetheless, in the absence of an explicit statement, most floor designers do apply ACI 117 for slabs on grade.

In general, the authors do not recommend design thicknesses less than 5 inches for plain concrete and 4 inches for post-tensioned slabs. Such slabs could well be constructed as thin as 4 inches and 3.5 inches, respectively. Thinner slabs on grade might be considered with exceptional quality control procedures and with careful design checks.

10.4.3 — Slab surface tolerances

The top surface of the floor slab is subject to variations inherent to the finishing process. The surface will possess a condition of levelness, which may be described as the slope of the overall plane of the floor unit. ACI 117 *(Reference 29)* gives a tolerance of ± 3/4 inch for level alignment. The surface will also possess a condition of flatness, which may be described as smoothness or, conversely, bumpiness or waviness.

The tolerances which control flatness are frequently specified in job documents. For slabs on grade, the important ACI publications are ACI 302 (*Reference 26*) and ACI 117 (*Reference 29*), both of which deal with the F-number system (F_F and F_L). Waviness indices ($WI_{2\text{-}10}$ and $SWI_{2\text{-}10}$) are discussed in *Reference 30.*

Surface tolerances can be specified by one of these criteria:

- F_F/F_L flatness and levelness numbers (ASTM E 1155)
- $WI_{2\text{-}10}$ and $SWI_{2\text{-}10}$ waviness indices (ASTM E 1486)
- 10-foot straightedge (This has no ASTM standard. Note that a computer-generated automation of the straightedge technique has been developed and may soon be available.)

Descriptive wordings (See *Table 33*) can also be used in the specifications as long as the specifier knows that selected words have a numerical equivalent.

Unless levelness is badly amiss, the condition of flatness is almost always more critical to the use and appearance of the floor. Floors with lift truck traffic need good rideability. Where considered important, flatness should be specified in preconstruction documents.

The 10-foot straightedge technique is strongly discouraged since ACI 117 states that "less definitive results are obtained with this method." (Section 7.15.1.2, Reference 29*).*

The surface tolerance configuration has three distinct parts. The first is what is specified in the design phase of the project. The second is the construction process in which the surface condition is produced. Helpful information on this part is found in ACI 117 and ACI 302, particularly in *Table 7.15.3* on page 29 of the latter document *(Reference 26)*. The third part is the measurement of the surface condition after floor construction. This is frequently required by specification and is the verification process. Information on this part is also found in the ACI 117 and ACI 302 documents.

Table 33 shows the general classifications now accepted by the design/construction profession.

Floor quality classification	***Specified Numbers*** F_F	F_L	***Minimum Numbers*** F_F	F_L	***Waviness Index*** $WI_{2\text{-}10}$ ***(in.)****
Conventional					
Bull-floated	15	13	13	10	0.330
Straightedged	20	15	15	10	0.220
Flat	30	20	15	10	0.140
Very flat	50	30	25	15	0.100
Superflat	100+	50+	—	—	0.080

** Local test line. See ASTM E 1486 for discussion of* $SWI_{2\text{-}10}$ *surface waviness index.*

Table 33 *Suggested F numbers (*from Reference 26*) and waviness index values (*from Reference 30*) for floors of the indicated classification.*

In the commentary for the ACI 117, Section 4.5.6, the following rough correlation is given between the F-number and the older straightedge methods:

F-number F_F	***Gap under an unleveled 10-foot straightedge***
12	1/2 inch
20	5/16 inch
25	1/4 inch
32	3/16 inch
50	1/8 inch

However, the commentary cautions that there is no direct equivalent between these two tolerance systems.

Commentary:
To determine what flatness and levelness to specify when the floor is designed, the use of the floor must be reviewed. From ACI 302.1R-89: "The selection of the proper F_F/F_L *tolerances for a project is best made by measurement of a similar satisfactory floor." This is true for any of the tolerance systems.*

In summary, from the design point of view, when the specifications and slab requirements are set forth, tolerances are critical. The floor designer should consider the three tolerances discussed here and decide which ones are most critical to the expected performance of the floor. A decision on appropriate tolerance magnitudes should be made and indicated in the contract documents. Avoid specifying tolerances that contradict one another.

Finding the appropriate specification for floor flatness in relation to its proposed use is not a clear-cut process. Measuring an existing satisfactory floor to find out what degree of flatness has worked well is the best option at this time (1995). Nonetheless, Tipping (*Reference 31*) gives some general guidelines which may help. These are summarized briefly in *Table 34*.

Typical Use	*F-Numbers*	*Description*
Carpeted areas of office buildings	F_F 25/F_L 20	Conventional
Public areas to receive thin-set coverings (ceramic or vinyl tile)	F_F 36/F_L 20	Flat
Specific use floors sensitive to waviness and levelness such as a warehouse with air pallets	F_F 45/F_L 30	Very flat

Table 34 *Flatness related to use from* Reference 31.

10.5—Effect of changes in the variables: What if?

Commentary:
Do not waste time refining a value whose effect on design thickness is less than 5 percent or 6 percent. Even this may be too little; that is, a 10 percent change may be considered too academic to be practical in this complex system. Further, do not create an overly expensive design by using excessive values on a number of variables; that is, do not compound safety factors by selecting a combination of overly safe values.

Each design procedure requires the selection of numerical values for the variables that control the design. Often, the value selected must be estimated or taken from general literature. Frequently we wonder what can be saved if we change one of the variables such as slab thickness or concrete strength.

Some changes have a drastic effect on the design results, while others have only a nominal effect. The common variables to be considered are:

- Subgrade modulus, k
- Concrete strength, f_c'
- Slab thickness, t_s
- Thickness of base placed on the subgrade, t_b
- Safety factor (SF) used with the modulus of rupture

The following shows how to estimate the percentage change in slab thickness or in the strength (load-carrying capacity) of the floor slab due to a change of each of the five listed variables. Two brief examples are included for each variable.

It is not correct to do the actual thickness selection (final design) using these relative percentage effects. The analysis required for thickness selection does not follow a simple mathematical root process as used here. Accuracy of the estimated effects of changing each of the five inputs varies. Section 10.5.6 explains the reasons for these variations. However, simplified estimates of slab thickness changes due to changes in the input variables can be very helpful in a study of floor construction costs.

When considering the effect of changing input variables, remember that with normal floor construction practices that meet standard tolerances, the as-built slab may differ up to 1 inch or more from the design thickness.

10.5.1—Effect of changing modulus of subgrade reaction, k

A change in the value of subgrade modulus k must be substantial in order to have a significant effect on slab thickness. However, the effect does not vary as a constant root (such as a square root or a fourth root). The percentage effect is variable as other problem values change. See Section 10.5.6.

To make a conservative estimate of the percentage change in floor thickness resulting from a change in k values, the 8th root of the ratio of the two k values is suggested for use as follows:

Increasing k: $k_1 = 200$ pci and $k_2 = 300$ pci

$$\sqrt[8]{200/300} = 0.951$$

Thus, if the k changes from 200 to 300, the slab may be approximately 5 percent thinner.

Decreasing k: $k_1 = 400$ pci and $k_2 = 50$ pci

$$\sqrt[8]{400/50} = 1.297$$

Thus, if k changes from 400 to 50, the slab should be approximately 30 percent thicker.

10.5.2 — Effect of changes in f_c'

In slab design, the modulus of rupture varies as the square root of the concrete's compressive strength. A change in the compressive strength has only a slight effect in the slab's load-carrying capacity (for the same slab thickness) or in the required design thickness. When the slab thickness is unchanged, the load-carrying capacity varies approximately as the square root of compressive strength. When the change is carried through the solution for required thickness, thickness varies approximately as the fourth root of relative compressive strength values. The following illustrates this effect.

Increasing compressive strength:

$f_c'(1) = 3000$ psi and $f_c'(2) = 5000$ psi

$$\sqrt[2]{5000/3000} = 1.291$$

Thus, for a slab whose thickness remains the same, the 67 percent increase in strength then increases the load-carrying capacity by 29 percent.

When this compressive strength change is carried through the solution for required thickness, we have:

$$\sqrt[4]{5000/3000} = 1.136 \text{ and } 1/1.136 = 0.880$$

Thus, the required thickness is reduced by approximately 12 percent.

Reducing the compressive strength:

$f_c'(1) = 4000$ psi and $f_c'(2) = 3500$ psi

$$\sqrt[2]{3500/4000} = 0.935$$

Thus, for a slab whose thickness remains the same, the 13 percent decrease in compressive strength then decreases the load-carrying capacity by approximately 7 percent.

When this compressive strength change is carried through the solution for required thickness, we have:

$$\sqrt[4]{3500/4000} = 0.967 \text{ and } 1/0.967 = 1.034$$

Thus, the required thickness is increased by approximately 3 percent.

Commentary:

The theory which controls these effects is based on the bending strength of the unreinforced concrete slab. The appropriate equation is:

$$\frac{\text{Moment in slab}}{\text{Section modulus}} \leq \frac{\text{Modulus of rupture}}{\text{Safety factor}}$$

which, in symbols, is:

$$\frac{M_{app}}{bt^2/6} \leq \frac{9\sqrt{f_c'}}{SF}$$

10.5.3 — Effect of changes in slab thickness, t_s

A change in the actual thickness of the slab has the most dramatic effect on the strength or load-carrying capacity of the floor. The relative strength varies nearly as the square of the thickness values. Assuming all other variables as constant (as is done throughout Section 10.5), the following shows how to predict the percentage change in strength.

Reducing the slab thickness:

$$t_s(1) = 7 \text{ inches and } t_s(2) = 5.5 \text{ inches}$$

$$(5.5/7)^2 = 0.617$$

Thus, if the thickness changes from 7 to 5.5 inches, the strength is reduced by approximately 38 percent.

Commentary:
An increase in slab thickness is the most effective way of increasing load-carrying capacity.

Increasing the slab thickness:

$$t_s(1) = 6 \text{ inches and } t_s(2) = 8 \text{ inches}$$

$$(8/6)^2 = 1.778$$

Thus, if the thickness changes from 6 to 8 inches, the strength is increased by approximately 78 percent.

10.5.4 — Effect of additional base thickness t_b on top of subbase

Adding graded, granular, and compacted base material will add a small amount of support stiffness; that is, it will increase k somewhat. *Figure 92* (from *Reference 14*) is a graph that can be used to obtain results like those in the examples, showing the effective changes in k due to a base placed on the existing subbase or subgrade.

Adding 6 inches of base:

k_1 = 100 pci on a subgrade beneath a 6-inch slab.

Add a 6-inch thickness t_b of granular, compacted base. Enter *Figure 92* at the bottom with the 6 inch thickness. Go up to the line for k = 100 pci, then horizontally left to read the resultant k_2 on top of the added base = 130 pci.

Remember that only a nominal change in k *results from adding the base material, and as shown in Section 10.5.1, it takes a substantial change in* k *to make a significant difference in the slab thickness. Thus, the overall effect on strength is quite nominal. On the other hand, the strength and stability of a good base surface will prove extremely beneficial in providing a proper working surface for construction operations.*

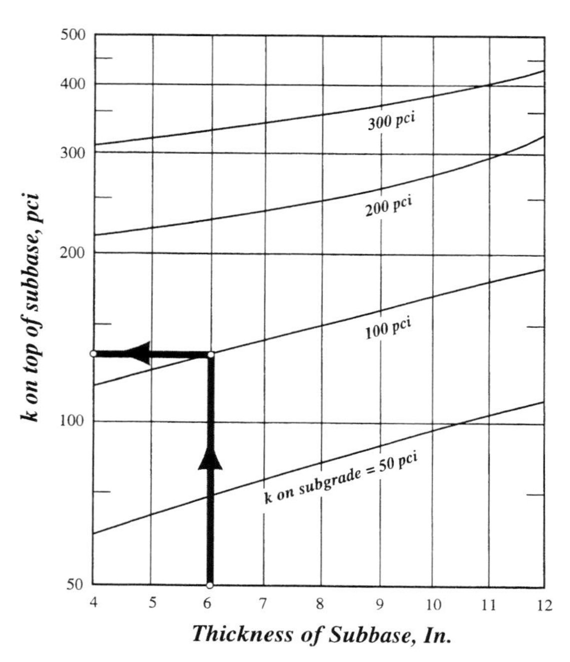

Figure 92 *Effect of granular subbase thickness on* k *value, from* Reference 14.

Adding 8 inches of base:

k_1 = 200 pci on a subgrade beneath an 8-inch slab.

Add an 8-inch thickness t_b of granular, compacted base. Enter *Figure 92* at the bottom with the 8-inch thickness value. Go up to the line for k = 200 pci, then horizontally to the left side, reading the resultant k_2 = 245 pci.

By the same process, it can be seen that a base of only 4 inches would have raised k in the second example to only 215 pci, which would permit less than 2 percent reduction of slab thickness (see Section 10.5.1).

Commentary:
It is frequently asked: How thick a base should be placed between subbase (subgrade) and slab? The authors believe that the thickness of the base should be nearly equal to the thickness of the concrete slab. This holds unless there is an extremely weak or unstable soil below. In such a case, there may have to be a thicker base and perhaps even a subbase between concrete slab and soil. Consultation with a geotechnical specialist familiar with the area is recommended in such cases.

10.5.5 — Effect of changes in safety factor, SF

A change in safety factor has only a slight effect on the change in slab thickness, or on the change in load-carrying capacity. When the slab thickness remains the same, the safety factor has a linear effect on load-carrying capacity, assuming all other variables constant. When the change in safety factor is carried through the solution for thickness, the effect is then according to the square root of that change. The following examples show how to calculate these effects.

Reducing the safety factor:

$SF_1 = 2$ and $SF_2 = 1.7$

$(2/1.7) = 1.176$

Thus, for a slab whose thickness remains the same, this reduction in safety factor increases the design value of load-carrying capacity by 17 percent.

When this change in safety factor is carried through the solution for required thickness, we have:

$\sqrt[2]{2/1.7} = 1.084$

Thus, the slab may be 8 percent thinner for the original load-carrying requirement.

Remember that the safety factor does not actually measure safety. It represents the effect of fatigue due to the repetition of a given stress level.

Increasing the safety factor:

$SF_1 = 1.4$ and $SF_2 = 2$

$(2/1.4) = 1.428$

Thus, for a slab whose thickness remains the same, the increase in safety factor will decrease the design value of load-carrying capacity by 43 percent.

When this change in safety factor is carried through the solution for required thickness, we have:

$\sqrt[2]{2/1.4} = 1.195$

Thus, the slab must be 20 percent thicker for the original load-carrying requirement.

10.5.6— Comments on theory controlling the effects

Floor design theory is based on the analysis of a two-way concrete slab acting as a plate supported continuously by a soil-support system. The problem is complex. Closed-form-equation solutions are not possible. Problems are solved by setting up a so-called model (a set of simultaneous equations) which is then solved by computer.

Floor design charts are based on computer-solved models. The terms within each computer model contain several variables. Changes in results that are generated by altering an input value are not linear nor are they the same for all variables. For example, the analysis usually involves the relative stiffness of the concrete slab and the subgrade material

The analysis of the effects of changing variables has been refined in this second edition. More refined studies were made possible due to availability of new software.

interacting with one another. When a single input value is changed, it alters more than one term in the mathematical model. This makes percentage changes based on equations or graphs variable.

Percentage changes in strength or thickness, when based on computer-generated graphs, depend on which computer model (method of analysis) was used to create the graph. These results can vary significantly. For these reasons, the effects are described as "nearly," "approximately," and "conservative estimate."

With respect to the previous examples, the effects of changing the safety factor are precise, as stated in Section 10.5.5. The effects of changing the concrete strength or slab thickness are estimated quite closely in Sections 10.5.2 and 10.5.3. The effects of a change in subgrade modulus k are quite variable, according to the other input values and the design method selected.

10.6— Freezer floors

10.6.1 — General

In numerous industrial plants the concrete slab on grade is used as the floor in an abnormally cold section of the plant. It may be an area somewhat cooler than normal plant temperature, or it can be an area where the temperature is well below 32° (Fahrenheit), such as in a freezer. Design of these floors for service at cold temperatures is basically the same as for floors at more usual temperatures. However, there are substantial differences in floor construction and base support. *Figure 93* shows a commonly encountered construction system.

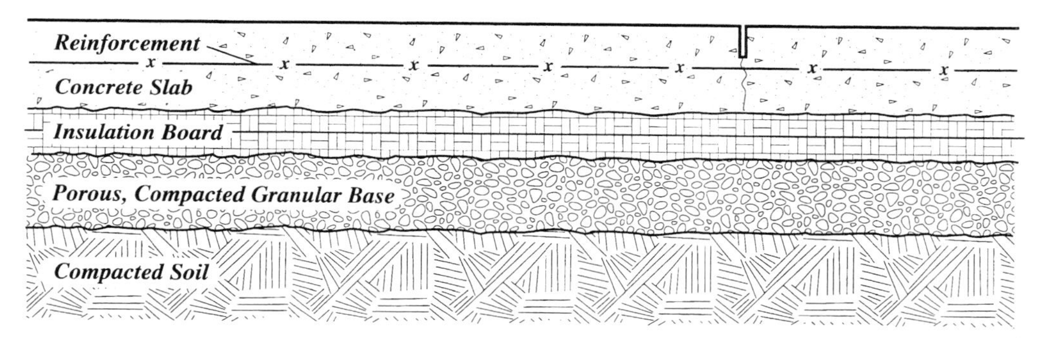

Figure 93 *Typical construction of freezer floor on insulation layer placed on top of compacted granular base material.*

10.6.2 — Thickness of freezer floors

Selection of the appropriate thickness of the floor slab for a cold area is the same as for any other floor. It is based on loading, subgrade modulus, base friction, and the concrete modulus of rupture. Design charts in this book are appropriate for thickness design of freezer floors or cold floors.

Commentary:
Where the cold area is at freezing temperature or below, the floor cannot be constructed without insulation boards beneath it. If insulation is omitted, the moisture in the granular base material will freeze, expand, and heave the floor, cracking it badly.

10.6.3 — Base support for freezer floors

It is essential that insulation be provided immediately beneath the concrete floor in cold areas. This is especially critical for freezer floors. Insulation boards 2 inches thick are commonly used, installed to provide a total thickness of board insulation from 4 to 8 inches. These boards are the base support for the floor, and they have a subgrade modulus (k in pci), which should be obtained from the manufacturer. The k value is usually taken as 100 to 150 pci.

One advantage of the insulation boards as a base support is that their surface is quite smooth. This provides a low value of the coefficient of friction between the slab and its base support. Tensile stresses due to drag restraint are reduced and the designer may consider a wider than normal joint spacing.

10.6.4 — Subgrade for freezer floors

The granular subgrade material provides the support surface for the insulation boards. It should be compacted material with sufficient porosity to prevent the wicking of water from deeper in the subgrade and soil. Since it supports the insulation material, it is part of the subgrade that provides the resultant subgrade modulus. The authors believe that the modulus of the compacted subgrade, k, should be higher than the manufacturer's k value for the insulation boards.

10.6.5 — Joints in the freezer floor

The freezer floor joints will commonly open wider than if the same floor construction were used at a higher temperature. There are two reasons for this. The temperature differential will cause the slab to shorten, depending on the magnitude of the temperature difference, an amount generally the same as that caused by drying shrinkage. Further, due to the smooth insulation surface just below the concrete, friction is reduced and the slab will move more readily. Both of these movements can be calculated with reasonable accuracy.

If it is planned to fill the joint, it should not be done until the joint has opened to near its maximum gap. This prevents the joint opening further by an amount that cannot be accommodated by the filler at the lower temperature.

Freezer floors are normal floors built at normal temperatures, but functioning at abnormally low temperatures. The designer may well consider shrinkage-compensating concrete, reinforced concrete, or prestressed concrete slabs. All of these can allow wider joint spacings as well as aid in effective control of concrete cracking.

10.6.6 — Curing needs

The concrete in freezer floors is placed at a high temperature relative to its later use. This concrete must be allowed to cure thoroughly before the area's temperature is lowered. Sixty or more days of curing at temperatures above 40° F are strongly suggested. If too much water remains in the concrete when the temperature is lowered, the surface will be destroyed due to freezing and scaling.

10.7— Radius of relative stiffness

10.7.1 — Introduction

Throughout the design procedures used with slabs on grade, there is frequent reference (for example, *References 1* and *14*) to the radius of relative stiffness, ℓ, a term expressed in inches. The radius of relative stiffness is quite useful in design because it relates the stiffness of the concrete slab to the stiffness of the subgrade.

Commentary:
The physical meaning of the radius, ℓ, for a single concentrated load is that the bending moment is maximum and positive (tension on the bottom) directly under the load. Along a straight line, the moment remains positive and decreases to zero at a distance of 1ℓ from the load. It then becomes negative and is a maximum at 2ℓ from the load. This maximum negative moment (tension on top) is significantly less than the maximum positive moment. The moment approaches zero at 3ℓ from the load. (Reference 9)

10.7.2 — How ℓ is determined

The radius of relative stiffness is defined as the fourth root of the results found by dividing the concrete plate stiffness by the subgrade modulus k which is discussed in *Chapters 1* and 2 and Section 10.7.3. The expression for the plate stiffness is:

$$\frac{E h^3}{12(1-\mu^2)}$$

where E is the modulus of elasticity of the concrete (usually assumed to be 4000 ksi), h is the slab thickness, and μ is Poisson's ratio for the concrete (usually assumed to be 0.15). The radius of relative stiffness, ℓ in inches, is then:

$$\ell = \sqrt[4]{\frac{E h^3}{12(1-\mu^2)k}}$$

Table 35 presents a range of calculated values for ℓ, based on k values from 25 to 500 pci and slab thicknesses from 4 to 14 inches.

10.7.3 — Significance of ℓ

From a practical point of view, the significance of the radius of relative stiffness is most pronounced when considering the effect of a set of loadings on the slab. Any load that is more than 3ℓ away from a given location has virtually no influence on slab stresses at that location. On the other hand, loads within a distance of 1ℓ from the location may have a significant influence on those slab stresses and must be included in the analysis. Where the slab has been considered continuous, any joint within 1.5ℓ will interrupt that continuity.

Values of the radius of relative stiffness can be used to indicate which loads must be considered in a particular design. They can also be used to select critical locations for joints.

k pci	*25*	*50*	*75*	*100*	*150*	*200*	*250*	*300*	*350*	*400*	*500*
t inches											
4	30.6	25.7	23.2	21.6	19.5	18.2	17.2	16.4	15.8	15.3	14.5
4.5	33.4	28.1	25.4	23.6	21.3	19.8	18.8	17.9	17.3	16.7	15.8
5	36.1	30.4	27.5	25.5	23.1	21.5	20.3	19.4	18.7	18.1	17.1
5.5	38.8	32.6	29.5	27.5	24.8	23.1	21.8	20.9	20.1	19.4	18.4
6	41.4	34.8	31.5	29.3	26.5	24.6	23.3	22.3	21.4	20.7	19.6
6.5	44	37	33.4	31.1	28.1	26.2	24.7	23.6	22.7	22	20.8
7	46.5	39.1	35.3	32.9	29.7	27.7	26.2	25	24	23.2	22
7.5	49	41.2	37.2	34.6	31.3	29.1	27.5	26.3	25.3	24.5	23.2
8	51.4	43.2	39.1	36.4	32.9	30.6	28.9	27.6	26.6	25.7	24.3
8.5	53.8	45.2	40.9	38	34.4	32	30.3	28.9	27.8	26.9	25.4
9	56.2	47.2	42.7	39.7	35.9	33.4	31.6	30.2	29	28.1	26.6
9.5	58.5	49.2	44.4	41.4	37.4	34.8	32.9	31.4	30.2	29.2	27.7
10	60.8	51.1	46.2	43	38.8	36.1	34.2	32.7	31.4	30.4	28.7
10.5	63	53	47.9	44.6	40.3	37.5	35.5	33.9	32.6	31.5	29.8
11	65.3	54.9	49.6	46.2	41.7	38.8	36.7	35.1	33.8	32.6	30.9
11.5	67.5	56.8	51.3	47.7	43.1	40.1	38	36.3	34.9	33.7	31.9
12	69.7	58.6	52.9	49.3	44.5	41.4	39.2	37.4	36	34.8	33
12.5	71.8	60.4	54.6	50.8	45.9	42.7	40.4	38.6	37.1	35.9	34
13	74	62.2	56.2	52.3	47.3	44	41.6	39.8	38.3	37	35
13.5	76.11	64	57.8	53.8	48.6	45.3	42.8	40.9	39.4	38.1	36
14	78.2	65.8	59.4	55.3	50	46.5	44	42	40.4	39.1	37

Table 35 *Radius of relative stiffness ℓ, in inches, for slabs on grade.*

10.8— Computer solutions

10.8.1 — Practical applications

Computer solutions can be considered in two categories. The first is the programming of equations and mathematical models already in use to create design graphs. This is most commonly done for personal computers (PCs). Individuals have elected to program certain equations, such as Westergaard's equations *(Reference 32)*, for their convenience. The second is a computer solution based on finite element analysis or some other mathematically more exact solution. This may be done using either a PC or a main-frame computer. Certainly, any graphical or table-based solution should be capable of verification by a more sophisticated computer solution. Agreement may not be exact, but should be within reasonable tolerances.

The floor designer may find computer solutions, particularly using PCs, desirable. Arithmetic accuracy is guaranteed, as long as the input is correctly keyed into the program. Alternate solutions are easily found since most programs allow rapid changing of the several variables influencing the design. As noted throughout this book, all graphs and tables have limitations by their very nature.

The use of finite element solutions for most slab problems is not common. Such computer programs provide arithmetic accuracy and variable input; however, they are usually somewhat demanding to use. They may well be useful for difficult or unusual loading situations. They may also be valuable for verification of more conventional methods or for forensic work.

Commentary:
The Portland Cement Association AIRPORT program which runs on IBM-compatible PCs is based on the program originally used to generate the PCA charts used in this book. It is convenient to use when the values needed for solution of a problem are outside the limits of the charts, such as a k value above 200 pci or a wheel contact area below 25 square inches.

PCA's MATS program (described and discussed in Section 2.2.6) is not as convenient to use but is far more powerful than AIRPORT. It allows analysis of slabs with variable thickness, variable support modulus, and combination loadings. Results output by MATS are moments, not thicknesses, and their exact values are controlled by the grid dimensions (mesh size) set by the analyst, as well as over how many grid intersections (nodes) the load is distributed.

10.8.2 — Software for slab on grade design

Software is available from a number of sources; however, the most commonly used and readily obtained software packages are marketed by the Portland Cement Association. They are the programs known as MATS and AIRPORT. The AIRPORT program (latest version 1987) is for a PC (IBM or an IBM true-compatible). While labeled as for airport and industrial pavement thicknesses, it is readily used for interior industrial floors as well *(Reference 9)*. The MATS program (Version 5.01, 1994) is described as for the "analysis and design of foundation mats and combined footings" *(Reference 18)*. It is a sophisticated microcomputer program, using the finite element process with specific elements and soil modeling.

10.8.3 — Significance of computer solutions

The more common slab on grade design problems can be adequately handled by design graphs, as is done in this book. More mathematically exact solutions are rarely required.

Except for use in intense analyses, forensic investigations, or extremely unusual design problems, the accuracy of the computer solution is beyond the accuracy of the input variables. This is especially true for finite element investigations. On the other hand, many designers find the arithmetic accuracy and the speed of multiple solutions quite convenient.

The overall accuracy of the slab on grade design and construction problem does not warrant extreme mathematical rigor. Design theory and thickness determination can be arithmetically exact; however, construction tolerances reduce the significance of this implied accuracy.

10.9— Understanding soil properties and classification

Chapter 1 outlines the knowledge of basic soil properties needed for slab design, pointing out the desirability of working with a soils specialist in the early planning of any slab on grade. This section gives more detailed information on the relevant soil properties, explaining how they are determined, and what influence they have on slab design.

10.9.1 — Soil properties

Soils first must be classified by a qualified person, usually the soils engineer, in order to define their physical properties. Different types of classification systems are used; besides those used for engineering purposes, there are classifications for agronomists and geologists. The common classifications encountered in slab on grade design delineate soils according to grain size, moisture content, and Atterberg limits.

10.9.1.1 — Grain size of soil particles

The U. S. Army Corps of Engineers *(Reference 2)* in its Unified Soil Classification defines soils components as *cobbles, gravels, sands,* and *fines (Table 36)*. In *Table 36*, silt and clay can both fall into particle sizes finer than a #200 sieve. A #200 sieve gets its name from the fact that it has 200 openings per lineal inch in each direction; thus it has 40,000 square openings per square inch. This is the finest sieve used in defining grain size distribution, and a particle must have a diameter of 0.074 mm or smaller in order to pass through this sieve. A #200 sieve in good condition will hold water.

Component	*Size Range*
Cobbles	Above 3 in.
Gravels	3 in. to #4 sieve
Coarse Gravels	3 in. to 3/4 in.
Fine Gravels	3/4 in. to #4 sieve
Sands	#4 sieve to #200 sieve
Coarse Sand	#4 sieve to #10 sieve
Medium Sand	#10 sieve to #40 sieve
Fine Sand	#40 sieve to #200 sieve
Fines	Finer than #200 sieve
Silts or Clays	

Table 36 *Soil components defined by size.*

10.9.1.2 — Moisture content

The moisture content of a soil is the ratio of weight of water to the weight of the solids of a soil sample, expressed as a percentage. (It is *not* weight of water to total weight). Soils have three fundamental components: solids, water, and air. Moisture content is significant in specifying density of soils during compaction. Of equal importance, moisture content assists us in understanding the nature, properties, and behavior of plastic soils.

10.9.1.3 — Atterberg limits and plasticity

The field of soil mechanics recognizes standardized tests to determine when a soil acts as a solid and when it acts as a liquid. This is contingent on the moisture content of the soil. The states of consistency of a soil are described and identified in *Figure 94* The percentage of moisture present in a soil when the properties indicate the soil is acting as a semi-solid is referred to as the plastic limit, while the percentage of moisture at which the soil starts to act as a liquid is referred to as the liquid limit. These terms are referred to as the Atterberg limits of a soil. The difference between the plastic limit and the liquid limit, called the plasticity index, is the range in which the soil acts as a plastic material.

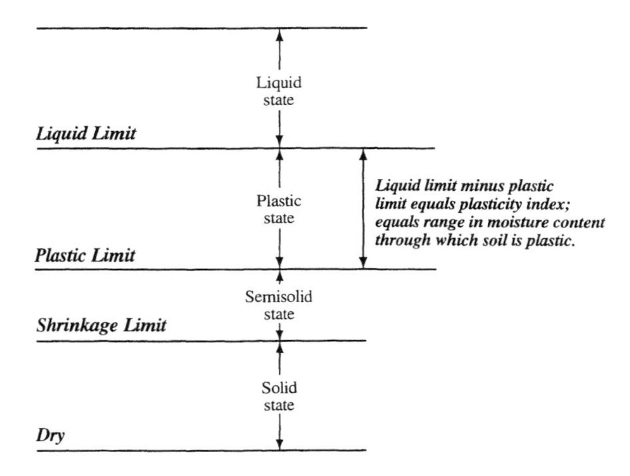

Figure 94 *Different states of soil consistency from* Reference 1. *Boundaries between the states, expressed in terms of soil moisture content, differ from soil to soil.*

Commentary:
The clays and the silts are the soils that are potentially dangerous as support materials. They can swell or they can collapse and they are not stiff (have low k *values).*

Clays tend to be more plastic than silts and are notably smaller in particle size. Sands, gravels, and cobbles, of course, do not possess these properties.

Having a fundamental knowledge of grain size, moisture content, and plasticity index is essential to proper subgrade design and to understanding the three basic soil classification systems.

10.9.2 — Soil classification systems

Three soil classification systems commonly used in the United States for floor design are:

- The Unified Soil Classification
- The AASHTO classification
- The FAA classification systems

Brief descriptions of the three systems follow.

Major Divisions (1)	(2)	Letter (3)	Name (4)	Value as Foundation When not Subject to Frost Action (5)	Value as Base Directly under Wearing Surface (6)	Potential Frost Action (7)	Compressibility and Expansion (8)	Drainage Characteristics (9)	Compaction Equipment (10)	Unit Dry Weight (pcf) (11)	Field CBR (12)	Subgrade Modulus k (pci) (13)
Coarse-grained soils	Gravel and gravelly soils	GW	Gravel or sandy gravel, well graded	Excellent	Good	None to very slight	Almost none	Excellent	Crawler-type tractor, rubber-tired equipment, steel-wheeled roller	125-140	60-80	300 or more
		GP	Gravel or sandy gravel, poorly graded	Good to excellent	Poor to fair	None to very slight	Almost none	Excellent	Crawler-type tractor, rubber-tired equipment, steel-wheeled roller	120-130	35-60	300 or more
		GU	Gravel or sandy gravel, uniformly graded	Good	Poor	None to very slight	Almost none	Excellent	Crawler-type tractor, rubber-tired equipment	115-125	25-50	300 or more
		GM	Silty gravel or silty sandy gravel	Good to excellent	Fair to good	Slight to medium	Very slight	Fair to poor	Rubber-tired equipment, sheepsfoot roller, close control of moisture	130-145	40-80	300 or more
		GC	Clayey gravel or clayey sandy gravel	Good	Poor	Slight to medium	Slight	Poor to practically impervious	Rubber-tired equipment, sheepsfoot roller	120-140	20-40	200-300
	Sand and sandy soils	SW	Sand or gravelly sand, well graded	Good	Poor	None to very slight	Almost none	Excellent	Crawler-type tractor, rubber-tired equipment	110-130	20-40	200-300
		SP	Sand or gravelly sand, poorly graded	Fair to good	Poor to not suitable	None to very slight	Almost none	Excellent	Crawler-type tractor, rubber-tired equipment	105-120	15-25	200-300
		SU	Sand or gravelly sand, uniformly graded	Fair to good	Not suitable	None to very slight	Almost none	Excellent	Crawler-type tractor, rubber-tired equipment	100-115	10-20	200-300
		SM	Silty sand or silty gravelly sand	Good	Poor	Slight to high	Very slight	Fair to poor	Rubber-tired equipment, sheepsfoot roller, close control of moisture	120-135	20-40	200-300
		SC	Clayey sand or clayey gravelly sand	Fair to good	Not suitable	Slight to high	Slight to medium	Poor to practically impervious	Rubber-tired equipment, sheepsfoot roller	105-130	10-20	200-300
Fine-grained soils	Low compressibility LL<50	ML	Silts, sandy silts, gravelly silts or diatomaceous soils	Fair to poor	Not suitable	Medium to very high	Slight to medium	Fair to poor	Rubber-tired equipment, sheepsfoot roller, close control of moisture	100-125	5-15	100-200
		CL	Lean clays, sandy clays, or gravelly clays	Fair to poor	Not suitable	Medium to high	Medium	Practically impervious	Rubber-tired equipment, sheepsfoot roller	100-125	5-15	100-200
		OL	Organic silts or lean organic clays	Poor	Not suitable	Medium to high	Medium to high	Poor	Rubber-tired equipment, sheepsfoot roller	90-105	4-8	100-200
	High compressibility LL>50	MH	Micaceous clays or diatomaceous soils	Poor	Not suitable	Medium to very high	High	Fair to poor	Rubber-tired equipment, sheepsfoot roller	80-100	4-8	100-200
		CH	Fat clays	Poor to very poor	Not suitable	Medium	High	Practically impervious	Rubber-tired equipment, sheepsfoot roller	90-110	3-5	50-100
		OH	Fat organic clays	Poor to very poor	Not suitable	Medium	High	Practically impervious	Rubber-tired equipment, sheepsfoot roller	80-105	3-5	50-100
Peat and other fibrous organic soils		PT	Peat humus, and other	Not suitable	Not suitable	Slight	Very high	Fair to poor	Compaction not practical			

Table 37 *Unified Soil Classification, from* Reference 2.

10.9.2.1 — Unified Soil Classification

The Unified Soil Classification used in this book is the system most recognized and referred to by geotechnical firms in their reports. Originally developed by Casagrande, the system is used by the United States Army Corps of Engineers *(Reference 2)*. A modified version of the same system adopted by the American Society for Testing and Materials (ASTM) has been designated ASTM D 2487 *(Reference 3)*. *Table 37*, taken from a Corps of Engineers publication *(Reference 2)*, shows the major divisions of soils with names commonly used indicating their primary characteristics.

When a combination of soils is encountered, the names and letters from the list below are combined; for example, SC means a sand, clay mixture. This can be further refined by using terms such as "sandy clay" or "clayey sand," depending on which item dominates the mixture.

G - Gravel   *W* - Well Graded
S - Sand   *P* - Poorly Graded
M - Silt   *U* - Uniformly Graded
C - Clay   *L* - Low Liquid Limit
H - High Liquid Limit

Table 37 provides information used in the design examples of this book. However, the tabular data are only guideline information, not a substitute for an appropriate soils report by a geotechnical engineer. This information should be used to help the floor designer both in interpreting the geotechnical report and in requesting pertinent data from the soils engineer for the project at hand.

General classification	***Granular materials*** ***(35% or less passing No.200)***			***Silt-clay materials*** ***(More than 35% passing No. 200)***			
Group classification	**A-1**	**A-3**	**A-2**	**A-4**	**A-5**	**A-6**	**A-7**
Sieve analysis, percent passing							
No. 10							
No. 40	50 max	51 min					
No. 200	26 max	10 max	35 max	36 min	36 min	36 min	36 min
Characteristics of fraction passing							
No. 40: Liquid limit				40 max	41 min	40 max	41 min
Plasticity index	6 max	NP		10 max	10 max	11 min	11 min
Group index			4 max	8 max	12 max	16 max	20 max
General rating as subgrade	Excellent to good			Fair to poor			

(Subgroups)

General classification	***Granular materials*** ***(35% or less passing No.200)***							***Silt-clay materials*** ***(More than 35% passing No. 200)***			
Group classification	**A-1**		**A-3**	**A-2**				**A-4**	**A-5**	**A-6**	**A-7**
	A-1-a	**A-1-b**		**A-2-4**	**A-2-5**	**A-2-6**	**A-2-7**				**A-7-5, A-7-6**
Sieve analysis, percent passing											
No. 10	50 max										
No. 40	30 max	50 max	51 min								
No. 200	15 max	25 max	10 max	35 max	35 max	35 max	35 max	36 min	36 min	36 min	36 min
Characteristics of fraction passing											
No. 40: Liquid limit				40 max	41 min	40 max	41 min	40 max	41 min	40 max	41 min
Plasticity index	6 max		NP	10 max	10 max	11 min	11 min	10 max	10 max	11 min	11 min
Group index	0		0	0		4 max		8 max	12 max	16 max	20 max
Usual types of significant constituent materials	Stone fragments, gravel, and sand		Fine Sand	Silty or clayey gravel and sand				Silty soils		Clayey soils	
General rating subgrade	Excellent to good							Fair to poor			

Table 38 *AASHTO classification of highway subgrade material from* Reference 33.

Commentary:
Many floor designers are forced into the situation where they must use tabulated information based on soil assumptions. This can be risky. A thorough site investigation is strongly recommended.

10.9.2.2 — AASHTO classification

The American Association of State Highway and Transportation Officials has developed through the years a standard specification for highway materials and methods of sampling and testing. This broad classification of soils for highway subgrade materials is shown in *Table 38*. Consult *Reference 33* if further information is needed on the AASHTO classification.

10.9.2.3 — FAA classification

The Federal Aviation Administration *(Reference 34)* classifies soils into thirteen groups designated as E-1 through E-13. These classifications, based on sieve analyses and Atterberg limits, are shown in *Table 39* and *Figure 95*. They can be quite helpful when the FAA system is encountered. If further information is needed on the FAA classification, see *Reference 34*.

	Mechanical Analysis					
		Material Finer Than No. 10 Sieve				
Soil Group	Retained on No. 10 Sieve (%)	Coarse Sand, Pass No. 10 Ret. No. 40 (%)	Fine Sand, Pass No. 40 Ret. No. 200 (%)	Combined Silt and Clay, Pass No. 200 (%)	Liquid Limit	Plasticity Index
E-1	0–45	40+	60–	15–	25–	6–
E-2	0–45	15+	85–	25–	25–	6–
E-3	0–45	–	–	25–	25–	6–
E-4	0–45	–	–	35–	35–	10–
E-5	0–55	–	–	45–	40–	15–
E-6	0–55	–	–	45+	40–	10–
E-7	0–55	–	–	45+	50–	10–30
E-8	0–55	–	–	45+	60–	15–40
E-9	0–55	–	–	45+	40+	30–
E-10	0–55	–	–	45+	70–	20–50
E-11	0–55	–	–	45+	80–	30+
E-12	0–55	–	–	45+	80+	–
E-13	Muck and peat — field examination					

Table 39 *FAA soil classification system, from* Reference 34.

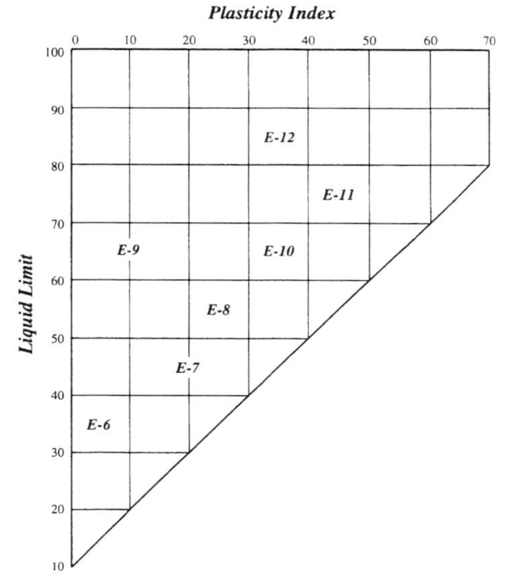

Figure 95 *FAA classification chart for fine-grained soils.*

10.9.3 — Determining the modulus of subgrade reaction

One of the most common methods for designing slabs on grade involves the use of a constant associated with the reaction of the soil to load. This constant is known as Westergaard's modulus of subgrade reaction or coefficient of subgrade reaction and is commonly referred to simply as k. It is expressed as the ratio of load per unit area on soil to the corresponding deformation. The units of k are pounds per square inch of pressure per inch of deflection; therefore, it is actually pounds per square inch per inch, but it is commonly written as pounds per cubic inch or pci. The standard method for field measurement of k is with the use of a 30-inch-diameter bearing plate. The ratio of the load pressure on the plate to a measured deflection close to 0.05 inch is taken as the k value. The field procedure is described in several soils and foundation books, including *Reference 1*.

This 30-inch plate test is relatively expensive and is commonly not used for industrial floors. In view of this, other procedures may be used to obtain a reasonable value for k. In addition to alternate procedures, there are general relationships that are available when limited geotechnical data are available.

When specified tests have not been performed to obtain a value for k, a value may be secured by consulting the soil description as found in the Unified Soil Classification chart *(Table 37)*. An example of this procedure follows.

A soil has been described for the site in question as a sandy clay. The liquid limit of the soil is 35. This is one of the Atterberg limits determined by a soils laboratory test. Using Table 37, *find the soil CL with LL (liquid limit) < 50, read the subgrade modulus (*k*) as ranging from 100 to 200 pci. With no additional input information, a value of k = 100 pci is recommended.*

Commentary:
The thickness selection is not overly sensitive to a specific value of k. *Thickness varies very close to the 8th root of relative values.*

More accurate alternative ways to obtain a k value may be the 12-inch-diameter plate test or use of the California bearing ratio (CBR). These values can be directly related to k as shown in *Figure 96*.

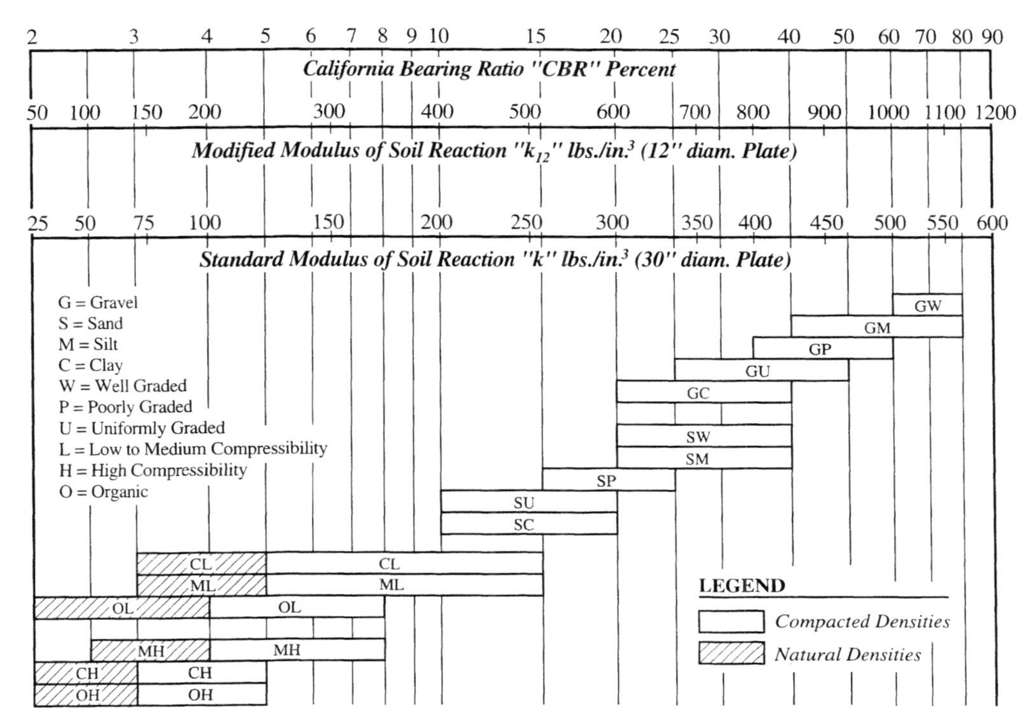

Figure 96 *Interrelationships of soil classifications and strength criteria, from* Reference 8.

A 12-inch-diameter plate test is much more economical than a 30-inch-diameter plate test and gives the owner more tests for the same expenditure. The pressure jack used with the 12-inch plate test can be reacted with the rear of a bulldozer, a water truck, or other similar available construction equipment. As indicated in *Figure 96*, the value from the 12-inch plate test, k_{12}, is double the value from the 30-inch plate test, k_{30}. Only the value of k_{30} is to be used in the design charts and tables in this book *(Reference 8)*.

Another method that can be used and related to a value for k is the on-site determination of the modulus of elasticity of the soil, E_s. All of these alternative procedures can be discussed with the geotechnical engineer as to the relative cost and practicality.

Commentary:
These potential problems apply to the construction process as well as to the final state of the slab when in place. Even soil with less than 45 percent of its material passing the #200 sieve may still be a problem soil. The material must be stable for construction activities such as passage of ready-mix trucks.

Highly compressible soils may allow a floor slab to settle or sink over a long period. Pumping and/or rocking may be amplified. A granular base is recommended to reduce pressures on the lower and softer soils.

Soils with a plasticity index (PI) *starting at between 7* (Reference 17) *and 13 begin to become potentially dangerous. These fine-grained soils can destroy an otherwise good floor. The expansive and unstable soils must be removed or have their characteristic properties altered.*

10.9.4 — Fine-grained soils

Most difficulties and areas of concern center around those materials designated as fine-grained soils. These soils, in natural state or used as compacted fill, are often associated with either high degrees of compressibility or with instability resulting from volume changes. As a general rule *(see Table 39)*, fine-grained soils can be identified as those soils having over 45 percent of the material (by weight) passing the #200 sieve. It is of interest that in the same soil there may be from 0 to 55 percent of that material retained on a #10 sieve *(Reference 1)*. These soils are designated as *ML, CL, OL, MH, CH,* or *OH* by the Unified Soil Classification. The liquid limit and the plasticity index help classify their characteristics relative to volume change, thereby classifying them as potentially compressible or expansive.

The important thing to remember when fine-grained soils are encountered is that more testing, other than a standard penetration test or a simple plate test, is warranted.

High compressibility soils are fine-grained soils which have liquid limit values greater than 50 percent. This usually indicates a clay material which is subject to substantial volume changes. Such clays can be defined for the purpose of applications here as having a consistency from *medium* to *soft*. This means that this clay has an unconfined compression strength of less than 1 ton per square foot. As mentioned above, this requires more testing than merely a simple standard penetration test.

Expansive soils are clays that experience a large volume change when their moisture content changes. Potentially expansive soils tend to have a high plasticity index, which indicates a large amount of clayey particles present in the soil and warrants special design considerations. This further emphasizes the need for competent soils investigations when fine-grained soils are encountered.

10.9.5 — Coarser materials: sands and gravels

The coarser materials, commonly known as sands and gravels, are not of great concern in the design of slabs on grade. They are typically considered stable and strong when encountered as natural soil or used as compacted soil. They are commonly used as base courses placed on top of the subgrade. The Unified Soil Classification system *(Reference 2)* classifies materials with more than 50 percent by weight larger than the #200 sieve as sands *(S)* and gravels *(G)*. Sands are defined as having more than 50 percent by weight smaller than the #4 sieve; gravels have more than 50 percent by weight larger than the #4 sieve. The #4 sieve has four openings per lineal inch in each direction. This is 16 square openings per square inch, with a clear opening of 0.187 inches in U.S. standard sieves. These soils are designated *SC, SM, SP, SW, GC, GM, GP* and *GW* in the Unified system.

These coarser materials generally drain reasonably well and can serve as a capillary break. How well they do this depends on the gradation of the material. The material should be generally uniformly graded (with few fines) for drainage and somewhat more gap-graded to serve as a capillary break. They can be effective as a base placed immediately beneath the slab and directly on the subgrade where they serve to stabilize the support system. This is advantageous for the construction process as well as for the long-term behavior of the floor slab. Depending on compaction and thickness, such a base can effectively increase the magnitude of the design value of k. On the other hand, if not smoothed on its top surface, the coarser base material can develop a high coefficient of friction. This excess drag can be quite detrimental where shrinkage-compensating concrete or larger joint spacings are used.

CHAPTER 11
TROUBLESHOOTING AND EVALUATING EXISTING FLOOR SLABS

11.1 — Introduction

A substantial part of a concrete floor designer's work involves troubleshooting existing floors in distress. Reasons for distress include the aging process, poor maintenance, design errors, construction errors, and changes in floor use. The investigator or designer also may have to evaluate an existing floor because of a contemplated change in use or owner.

The floor to be investigated may still be under construction; it may have been in service from a few weeks to several years, or it may not have been used at all. The investigator, who may or may not be the original designer, may have to analyze possible construction deficiencies as well as to check for design or detail errors. An independent investigator is frequently the best choice as the troubleshooter/evaluator.

The process of evaluating a floor slab consists of four parts:

- A walk-through visual inspection.
- A measurement and numerical session, during which quantitative information is obtained.
- Identification of repair or maintenance options.
- Reporting of results and opinions, which is always essential and important.

The four parts are distinctly different with respect to time required, cost, and objectives, but all are clearly interrelated. Depending on the situation, not all of the options discussed in this section are always necessary on a given job.

Commentary:
It is not possible to tell the reader everything that could be done, precisely how to do it, nor how much importance to attach to each item. All that can be done here is to point out options and give some general guidance.

11.2— Walk-through visual inspection

Without exception, the first step is a walk-through inspection. This gives the investigator a first look at whatever triggered the request for the evaluation, including cracks, edge chipping, settlement, bumpy floor surfaces, or other apparent distress. As much of the floor as possible should be looked at and carefully inspected.

The walk-through inspection should be predominantly visual with some preliminary measurements made and recorded, including crack width at the top surface, verification of joint spacings, joint types, and location of surface spalls. The investigator should have a floor plan in possession while making the visual inspection and observe the actual loading.

Although a complete list of potential defects is not given, some of the key features to be observed in the visual inspection are shown in the following list:

- Joints: Note edges, openings, curling, fillers.
- Corners, where joints intersect: Note cracks, curling, edge deterioration.

Evaluations cannot be done by telephone, nor from photographs alone. A site visit by the investigator is essential because it almost always reveals conditions different from or in addition to those reported initially.

Commentary:
The walk-through inspection will have to be done at least twice to verify and compare conditions. The investigator should take his time, make appropriate notes, and, with permission, take photographs.

For such an inspection, floor plan drawing(s), joint detail drawing(s), and a crack width indicator will prove helpful. See Figure 97.

- Surface conditions: Note spalling, scaling, pop-outs, and other defects.
- Surface flatness: Note any bumpiness.
- Random cracks: Note direction, position, openings, edge deterioration.
- Evidence of overload: Note cracking, faulting.
- Evidence of settlement: Note floor and wall cracking and lack of floor levelness.
- Subgrade problems: Note settlement, heaving, lack of floor levelness.

Figure 97 *A graphical indication of the width of the crack opening at the top surface of the floor. (Taken from a 2x3-inch transparent plastic card from Construction Technology Laboratories, Inc., 5420 Old Orchard Rd., Skokie, IL 60077)*

11.2.1—Joints

Joints are the preplanned separations in concrete slabs, such as contraction, construction, and isolation joints. Slabs tend to curve or *curl* at joints because of differential shrinkage—drying out faster on top than on the bottom. This may leave one edge of the slab higher than the adjoining one (commonly described as faulting), frequently producing a thumping noise as vehicles cross the joint. The investigator can also detect a vibration or movement by placing a hand on the slab while a vehicle crosses the joint. Other joint distress includes chipping of edges due to traffic, and loss of joint filler material when joints open too much.

11.2.2—Corners

Corners are the intersections of preplanned joints. Corners have the same problems as do joints between corners. The slabs at a corner can curl, showing faulting and allowing edge chipping and thumping. The slab corners can fail under load and break.

11.2.3—Surface conditions

A number of defects can be present on the concrete surface because of deficiencies in materials or workmanship. These defects, which are largely non-structural, include plastic shrinkage cracks, surface contamination due to foreign materials, pop-outs, mortar flakes, spalling and scaling.

A general lack of flatness is easily seen by light reflection across the floor and is frequently due to deficiencies in finishing techniques. Bumpiness, caused by a lack of flatness, is also observed by viewing a vehicle, preferably unloaded, moving across the floor.

11.2.4—Random cracks

Random cracks, particularly those due to shrinkage restraint, are wider at the top surface and tighter as the crack progresses down to the bottom. The crack, therefore, has more aggregate interlock than the top indicates.

The vast majority of random cracks are due to concrete shrinkage. They occur between preplanned joints and, although it should be verified, completely penetrate the depth of the slab. Random cracks are caused by subgrade settlement and heaving, excessive loading, and restraint to drying shrinkage.

These cracks frequently occur midway in a floor panel and subdivide it into two or four segments. They are approximately parallel to the joints and pass through the joints. Curl can occur at the random crack, causing faulting and edge chipping. All defects are made worse as the crack width widens.

11.2.5—Evidence of overload

General cracking can occur because of superimposed loadings. This cracking can be severe if the loading significantly exceeds the floor capacity or if the number of load applications is too high (fatigue). Faulting and edge deterioration are subsequent defects caused by overload cracking.

Commentary:
When cracking is evident, floors that are too thin are frequently suspected as the cause. The floor thickness is always part of the problem. It may or may not be as thick as was specified. If the floor thickness is to be physically measured, a sufficient number of randomly located cores or drill holes must be used to make the results statistically valid.

11.2.6—Evidence of settlement

Settlement of the floor causes a difference in elevation, which is most evident by a lack of levelness. Usually caused by consolidation of the supporting soils, it results in visible faulting at walls, footings, and other items fixed with respect to vertical motion. Cracks due to settlement can look like almost any other type of floor cracking, although they are more random in location.

11.2.7—Subgrade problems

Common subgrade deficiencies are softness (unstable for construction), weakness (low subgrade modulus), expansive soil materials (heaving), roughness (high friction), and uneveness (poor tolerance). Floor problems that can be caused by a deficient subgrade include settlement, heaving, random cracking, faulting, excess curling, and others.

11.3—Measurement and numerical analysis

This analysis involves two parts, both of which place numerical values on selected items. One part is the process of measuring certain physical features that may have been specified to see if they meet those specifications. Surface flatness and floor slab thickness are examples. The other part is determining what maximum loads and forces the floor is capable of sustaining. Where future loadings, including vehicle tire materials, are less stringent than originally planned, this numerical work may be unnecessary. If a more severe use is anticipated, then this step is quite important.

11.3.1—Meeting specifications

Most major building floors will be constructed according to drawings and specifications. One part of the evaluation is to determine if the floor, as built, meets those written contract specifications. The easiest areas to check are the joints, including spacings, positions, and types. Checking concrete strength is not difficult using existing cylinder or beam strength results or after concrete cores are obtained and tested. Floor surface flatness and floor thickness are more difficult. Not all of these features are always actually specified in pre-construction documents.

After the numerical measurements and numbers are obtained, they must be analyzed. Averages are frequently used, but a statistical analysis of the data will give a more realistic picture of what the numbers represent and how they vary for a given floor. Enough data must be obtained for the method selected to be accurate. See Section 11.6 for a list of helpful references.

Floors are frequently regarded as a part of the building where construction costs can be reduced. Shortcuts are sometimes taken, but owners and developers expect the quality of the final product to remain high. Virtually no floor totally meets all specification values over 100 percent of its area.

11.3.2—Determining maximum load capacity

Determining the maximum load capacity for various types of loads is sometimes necessary. The common sources of loading are:

- Lift truck axle loads, including axle loads for trucks or semi-trailers where appropriate. See the design procedures in *Chapter 3*.

A substantial change in floor usage, such as much heavier lift truck vehicles, is likely to bring about increased maintenance because of increased floor damage.

- Uniform storage loads as limited by the strength of the intermediate aisle. See the design procedures in *Chapter 4*.
- Post loads, including base plates, from storage racks. See the design procedures in *Chapter 4*.
- Anticipated new walls or columns involved in possible remodeling. See design procedures in *Chapter 5*.

Commentary:
Realistic numerical values that represent the actual floor at the time and age of the evaluation, such as concrete modulus of rupture, subgrade modulus, etc., are not easy to determine. Testing for the needed values is recommended. Note that it is not always appropriate to use only the specified minimums, nor is it always appropriate to use overly generous numbers.

11.3.3—Input for calculations

Determining the slab's load capacity is not difficult once the input values for concrete properties and load and site conditions are set. The difficulty is in determining appropriate values for an existing or aged concrete floor slab. These values fall into two categories: material and site properties, and factors for analysis as set by the investigator.

Material and site properties needed are the concrete's modulus of rupture (*MOR*) and the subgrade modulus, k. Factors set by the investigator include the nature of loading, such as a lift truck and its wheel contact area, with as narrow a range for the value of applied force as possible, joint load transfer factors, if appropriate, and the most suitable safety factor.

One effective approach is to assume the above values as accurately as is practical and then solve for a range of safety factors. The floor is then evaluated for load capacity by judging a range of forces, such as axle loads for lift trucks, compared to a range of safety factors. An example of this procedure is in the next section.

11.3.4—Example of load limit calculation

One numerical example illustrates the procedure, which solves the design problem in reverse to fine a range of safety factors. Acceptability of the loading on the existing floor is based on the investigator's opinion of these safety factors. Their range occurs because of a spread in the numbers selected for input values. The spread exists because exact vehicle data and some material properties are uncertain.

This example checks the load capacity of a floor for the axle weight of a lift truck. The lift truck is assumed appropriate for a Class 6 floor (see *Table 32*, Section 10.4.1) to which this example generally applies.

Input values not specifically known, or not set by an intended facility use, should have a reasonable range of magnitude other than a single value. These are determined by tests, by variations known to exist for the situation, or by assumption. The variation of input numbers causes a range of output values.

From materials and site:
  Modulus of rupture: Tested to be 732 psi
  Subgrade modulus: Assumed to be 150 to 200 pci
  Floor slab: Original plans called for 7 inch thickness

From lift truck data (See ACI 360R-92, Table 4.2):
  Axle load total: 20,200 to 23,800 pounds
  Wheel spacing: 37 to 45 inches
  Wheel contact area: 13 to 37 square inches

To do the floor analysis using a chart (in this case, *Figure A.7* from PCA for a two-wheel single-axle load), it is necessary to select representative input values for slab thickness and for subgrade modulus.

Subgrade modulus: Select k from available soil data. Plate tests may not be practical because too many are required for good results. The analysis is not sensitive to nominal changes in subgrade modulus (See Section 10.5.1). Use an average value unless better data are known. Use k = 175 pci.

Although the original plans called for a 7-inch thick floor, it is not logical to ignore the variations in this thickness due to lack of floor flatness (waviness) and subgrade uneveness.

If the floor has a flatness, F_F of 32 (assumed), the gap under a 10-foot straightedge would be about 3/16 inch (see Section 10.4.3). To include this effect, it is suggested that ± one-half of this value, ± 0.094 inches, be used to calculate the minimum and maximum thicknesses.

If the floor is placed on a subgrade whose top surface varies by ± 3/8 inch, based on a tolerance of +0/ − 3/4 inch, it is suggested that ± 0.375 inch be used to calculate the minimum and maximum thicknesses.

Floor thickness: Use two thicknesses.
Minimum thickness: t = 6.5 inches
(t = 7 − 0.094 − 0.375 = 6.531)
Maximum thickness: t = 7.5 inches
(t = 7 + 0.094 + 0.375 = 7.469)

Axle Load: Assume the average maximum axle load, which is 1/2 × (20,200 + 23,800) = 22,000 pounds. Expected facility use may make it more logical to use a different load magnitude.

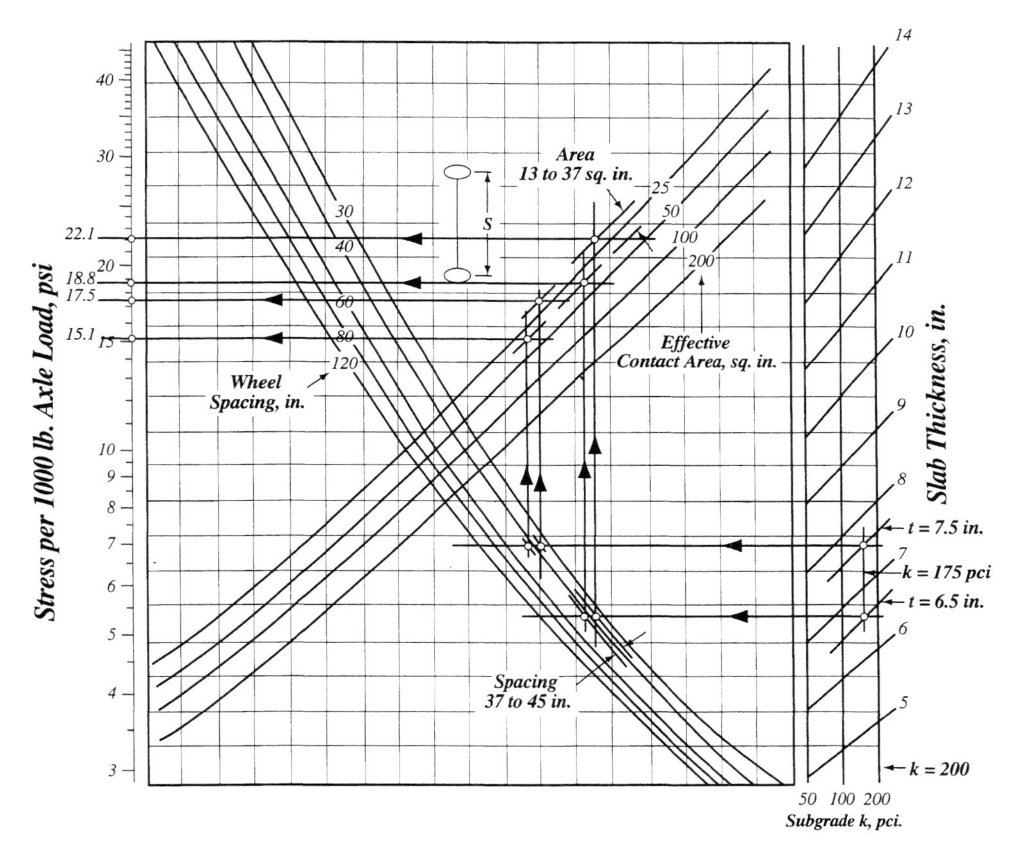

Figure 98 *Use of PCA chart to determine slab stress for a given lift truck axle load.*

The floor analysis is done graphically in *Figure 98* as follows:

- Enter *Figure 98* on the right-hand side, drawing a vertical line corresponding to k = 175 pci.
- Draw a diagonal line for each floor thickness, 6.5 and 7.5 inches, intersecting the vertical subgrade modulus line.
- Draw a horizontal line to the left from the intersection of each thickness line with the subgrade modulus line, to the wheel spacing curves. Due to the two wheel spacings, each horizontal line has two intersections, one at 37 inches and one at 45 inches.
- Draw four vertical lines, two for each thickness, to the effective wheel contact area curves. Due to the two contact areas, each vertical line has two intersections, one at 13 square inches and one at 37 square inches.

Commentary:

The sequence of lines is critical. It is the reverse of the design (thickness determination) procedure. The lines must be drawn from thickness/subgrade modulus to wheel spacing, to contact area, and then to stress per 1000 pounds of axle load. Any other sequence gives incorrect results.

- A total of eight horizontal lines can be drawn to the left from the above intersections with contact areas to the stress (psi per 1000 pounds axle load) axis. Although this decision is up to the investigator, it is suggested that only four lines be drawn (as shown in *Figure 98*) which correspond to the highest and lowest stresses for the thinnest slab and for the thickest slab.
- The stress magnitudes are 22.1 and 18.8 psi per 1000 pound axle load for the thinnest slab assumption (6.5 inches) and 17.5 and 15.1 psi per 1000 pound axle load for the thickest slab assumption (7.5 inches).
- Multiply the stress values (psi/1000 pound axle load) by the axle load to obtain the the expected bending stress in the floor slab:

  $22.1 \times 22 = 486$ psi
  $18.8 \times 22 = 413$ psi
  $17.5 \times 22 = 385$ psi
  $15.1 \times 22 = 332$ psi
- Divide the modulus of rupture (732 psi) by each stress to obtain the safety factor:

  $732 / 486 = 1.51$ }6.5-inch slab
  $732 / 413 = 1.77$
  $732 / 385 = 1.90$ }7.5-inch slab
  $732 / 332 = 2.20$
- Summary: The average safety factor for the thinnest slab sections is

  $(1.51 + 1.77) / 2 = 1.64$

  The average for the thickest slab sections is

  $(1.90 + 2.20) / 2 = 2.05$

  The overall average safety factor is 1.84.
- The decision here is not automatic. It is a judgement call based on a careful look at safety factors. Note that the safety factor accounts for the effect of fatigue caused by many load applications, as well as for the actual load/stress condition. Lower safety factors are acceptable for infrequent loadings. Higher safety factors are more appropriate for frequent load applications. In *Reference 9*, PCA gives a general relationship between the number of load applications and the stress ratio, which is the inverse of the safety factor. (If the safety factor is 1.84, the stress ratio is $1/1.84 = 0.54$)
- The floor slab is probably acceptable for this loading.

Commentary:
The resultant safety factors are not particularly sensitive to nominal variations in wheel spacing, wheel contact area, or subgrade modulus, but they are very sensitive to variations in floor thickness. The actual floor thickness is never constant.

Since the accuracy of the entire design and construction process is not at all exact, it is the authors' opinion that any safety factor less than 1.2 is unacceptable and indicates the probability of early cracking, maintenance, and repair.

11.4—Repair work options

A complete discussion on repair and maintenance is beyond the scope of this design book. Nonetheless, the following comments should help investigators outline feasible alternatives.

11.4.1—Describing floor distress

Describing floor distress in terms of appearance, function, or maintenance helps in the decision process when repair and maintenance are considered. Floor defects can be described as:

- Cosmetic: Only the appearance is affected. One example would be a hairline shrinkage crack which is tight without faulting, but is easily seen. Cosmetic defects may become maintenance problems.
- Maintenance related: Repair work may be appropriate to enhance both short-term and long-term floor use. One example would be an opened contraction joint (possibly in a food plant) requiring a joint filler.

Not all distress is due to errors or mistakes in design or construction. For example, some shrinkage cracking and some curling are common, especially in low-cost or fast-track floors. Therefore, some floor maintenance is always necessary. Another example is floor thickness, which may vary from design thickness according to tolerances for the subgrade top surface and the concrete top surface.

• Functional: The defects are so severe that they impact and impair the use of the floor and the building value. One example would be an extremely rough and bumpy floor which was intended for rapid transfer vehicles. In this case, grinding or another more radical repair may be appropriate.
• A combination of two or three defects: Appearance, use, and building value are all impacted. One example would be substantial chipping along a joint transverse to a prominent traffic aisle, requiring joint repair.

11.4.2—Evaluating the severity of distress

Although describing floor distress can be helpful, it is also necessary to establish the general level of severity for the defects before making a repair or maintenance decision. Floor defects may be described as slight, nominal, or severe.

• Slight: Repair effort, even maintenance, may be impractical and unnecessary. In some cases, it may be impossible, such as where a random crack is tight without faulting.

Random cracks less than 0.01 inch wide are almost impossible to fill. Remember that the crack is usually tighter in lower portions of the slab. For an indication of the difficulty of filling the crack, try folding and inserting a new dollar bill, which has a thickness of about 0.004 inches. For random cracks or open joints with widths of 0.03 inches or greater (about the thickness of a credit card), repair is possible. When the crack exceeds 0.05 inches, about the thickness of a dime, repair becomes even more practical.

As the crack width increases, repair tends to become more necessary. For simple and practical field measurements, a dime is slightly more than 0.05 inches thick. Inserting it into the crack is an indicator of repair difficulty. The easier and farther it goes into the crack, the easier the repair.

• Nominal: Repair is feasible but it may or may not always be necessary. Viable options are the use of crack or joint fillers, surface grinding, or replacement of small portions of floor strips along joints or in some areas.
• Severe: The distress is quite bad. This would be obvious because of thumping by lift trucks passing over curl at a joint, excess joint chipping, and wide random crack width. If it has become detrimental to the floor's use or to its value, repair is frequently mandatory. The question then becomes what to repair and how much.

Commentary:
Crack or joint repair is always visible and almost always shows as much as the original defect. Frequently, the completed repair is more obvious than the defect. Complete removal and replacement of large areas of the floor are undesirable and are a last resort.

11.4.3 — Some common repair procedures

Certain floor defects seem to require repair more often than others do. From the point of view of evaluating an existing floor, the following statements should provide guidance.

11.4.3.1 — Joint openings

Slab sections, except for shrinkage-compensating concrete slabs or post-tensioned slabs, will shrink and pull away from all joints. In general, total concrete shrinkage will be approximately 3/16 inch for each 25 feet. (*Reference 35*) This effect takes a substantial length of time for completion. The actual time span depends on a number of variables, but certainly exceeds six months and may pass one year. The joints will then open and frequently require a joint filler. The need for a joint filler and the properties of that filler depend on floor use. Where hard-wheeled lift trucks or heavy trucking vehicles are used, the filler must be hard enough to accept the abuse of the wheel.

Note that the cost and plant delay of clearing the area to be repaired is often substantially more than the cost of the repair itself.

11.4.3.2 — Chipping at joints

Joint distress in the form of chipping on the edges along construction or contraction joints can be severe. It could be caused by a combination of things, including some shrinkage and curling, hard-wheeled vehicle traffic, and incomplete concreting at keyed joints. This usually requires repair, which almost always involves removal of the top of the concrete along the joint followed by replacement with a hard filler material.

11.4.3.3 — Random cracks

Random cracking is usually caused by drying shrinkage and its restraint. It could also be caused by excessive loading, by joints placed too far apart, by saw-cuts made too late, and occasionally by settlement or heaving. Drying shrinkage cracks are most common. They are irregular and generally divide floor panels into half and quarter panels. They look bad and can readily affect floor use.

The appropriate repair depends on the width of the crack and on the condition of the crack edge. It is not always possible to effectively repair a narrow crack. With some slight width, a crack filler can be appropriate. When the width is substantial, the crack may have to be routed and filled or cut out and reformed.

11.4.3.4 — Lack of floor flatness

The floor surface can be bumpy or rough because of curling and distress at the joint or because of particular finishing techniques which leave the overall surface rough. The extent of curling depends on the concrete mix, its curing, and the joint type. The overall floor surface roughness is caused by placement and finishing procedures and is evaluated in terms of Flatness (F_F), Waviness Index (*WI*), or other suitable indicators. (See Section 10.4.3).

Roughness at joints can be repaired either by grinding or by replacement of a strip of the top floor surface along the joint. This may be restricted to critical locations such as aisles.

Overall roughness, indicated by low values of F_F or high values of *WI*, is more difficult to repair. Local grinding is possible if the areas are not extensive. Replacement of selected sections of the floor is possible but not desirable.

Commentary:
Settlement or heaving depends on the nature of the supporting soil. Both are discussed in Chapters 7, 8, and 10. Settlement occurs with high compressibility soils, which are fine-grained with liquid limit values greater than 50 percent. Expansive soils are clays that experience large volume changes when their moisture content changes. Heaving can also be serious in freezer floors where insulation is inadequate. See Section 10.6.

11.4.3.5 — Settlement or heaving

Slight settlements can usually be corrected by applying a layer of liquid floor leveling material or by slab jacking. Very small areas of heaving usually can be repaired by grinding. Excessive areas of settlement or heaving are usually repaired only by removal and replacement of the affected panels. None of these alternatives is desirable and all will probably cause an interruption to some part of facility operation.

11.4.3.6 — Other distresses

Other defects may also require repair, including surface spalling, surface crazing, plastic shrinkage surface cracks, and surface dusting.

11.5—Reporting results

All investigations will require some form of reporting process. This is true whether the investigator is in-house or an outside consultant. Reporting can take the form of an informal conversation, a more formal oral report, a brief written report, or an extensive written report. Often, both an oral and a written report are requested.

If the client makes no specific request as to a format, the investigator should consider giving his report orally, and then asking if a more thorough written report is needed. The latter is more time-consuming and expensive, but is frequently necessary, especially in litigation situations. Always assume that the oral report will be widely quoted and the written report will be widely copied and circulated. Don't put anything in it that you do not want everyone to read.

11.5.1—Oral report

The most effective report is oral, typically at a conference involving the investigator and concerned parties. In this format, questions can be asked and information and opinions exchanged. In such a conference, the investigator should consider keeping careful notes of his stated opinions, observations, etc.

11.5.2—Written report

When a written report is required or requested, it can be in the form of a brief letter or a more extensive multi-page document. One of these formats is almost always necessary when measurements are made and evaluated, as well as when load or force capabilities are determined.

11.6—Selected information sources

Listed below are some publications that the authors have found helpful when involved in troubleshooting and floor evaluation. They are listed according to their source.

Commentary:
These references contain practical information, including descriptions, causes, remedies, and photographs. Four are noted as most helpful by including the phrase "Excellent" after the publication date.

- The Aberdeen Group
  426 South Westgate Street
  Addison, IL 60101

  "A proposed method for determining compliance with floor thickness specifications," Snell and Rutledge, *Concrete Construction*, January 1989.

  "A Troubleshooting Guide," 19 pp., 1987. (*Excellent*)

  "Concrete and Masonry Problem Clinic," 198 pp., 1990. (*Excellent*)

  "Concrete Floors: Design and Construction," 60 pp., 1986.

  "Concrete Testing: A Guide to Better Field Practice," 39 pp., 1985.

  "Cracks in Concrete: Causes and Prevention," 40 pp., 1989.

  "Surface Defects in Concrete," 31 pp., 1989.

- The American Concrete Institute
  P.O. Box 19150
  Detroit, MI 48219

  "Concrete Floor Flatness and Levelness," ACI Compilation No. 9.

  "Concrete Repair Basics," Seminar Course Material, SCM-24(91).

  "Control of Cracking in Concrete Structures," ACI 224R(80) (Revised 1984).

  "Design of Slabs on Grade," ACI 360R-92.

  "Guide for Concrete Floor and Slab Construction," ACI 302.1R-89.

  "Guide for Design and Construction of Concrete Parking Lots," ACI 330R-87.

  "Troubleshooting Concrete Construction," Seminar Course Material, ACI SCM-22(90). (*Excellent*)

The International Concrete Repair Institute
1323 Shepard Drive Suite D
Sterling, VA 20164-4428

Concrete Repair Bulletin, a bi-monthly magazine.

The Portland Cement Association
5420 Old Orchard Road
Skokie, IL 60077

"Concrete Floors on Ground," 40 pp., 1983, revised 1990.

"Fiber Reinforced Concrete," 54 pp., 1991. (*Excellent*)

RESOURCES NEEDED FOR DESIGN
APPENDICES

A.1 — Introduction

The following section is a collection of charts, tables, and equations needed for the design of a slab on grade. These resources were reproduced in a smaller size in the examples shown in the book's text. The charts and tables have been enlarged in the Appendices so that they may be easily read and used. The authors recommend the photocopying of the charts or tables as needed, so that they can be marked up according to the specific job being worked on.

A.2 — Soil Properties, Conditions and the Thornthwaite Moisture Chart

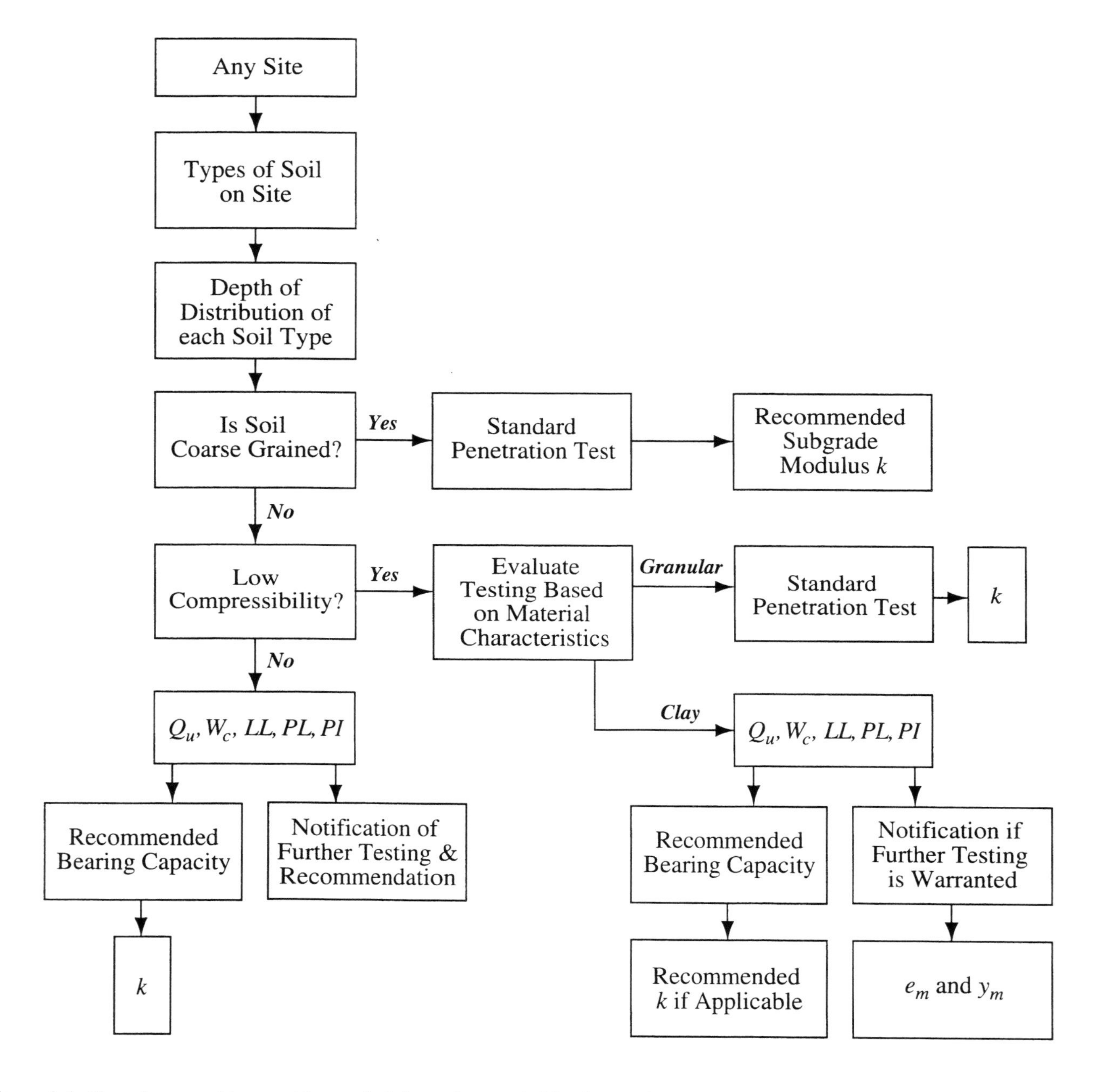

Figure A.1 *Flow chart provides a guide to soils information needed for design of slabs on grade.*

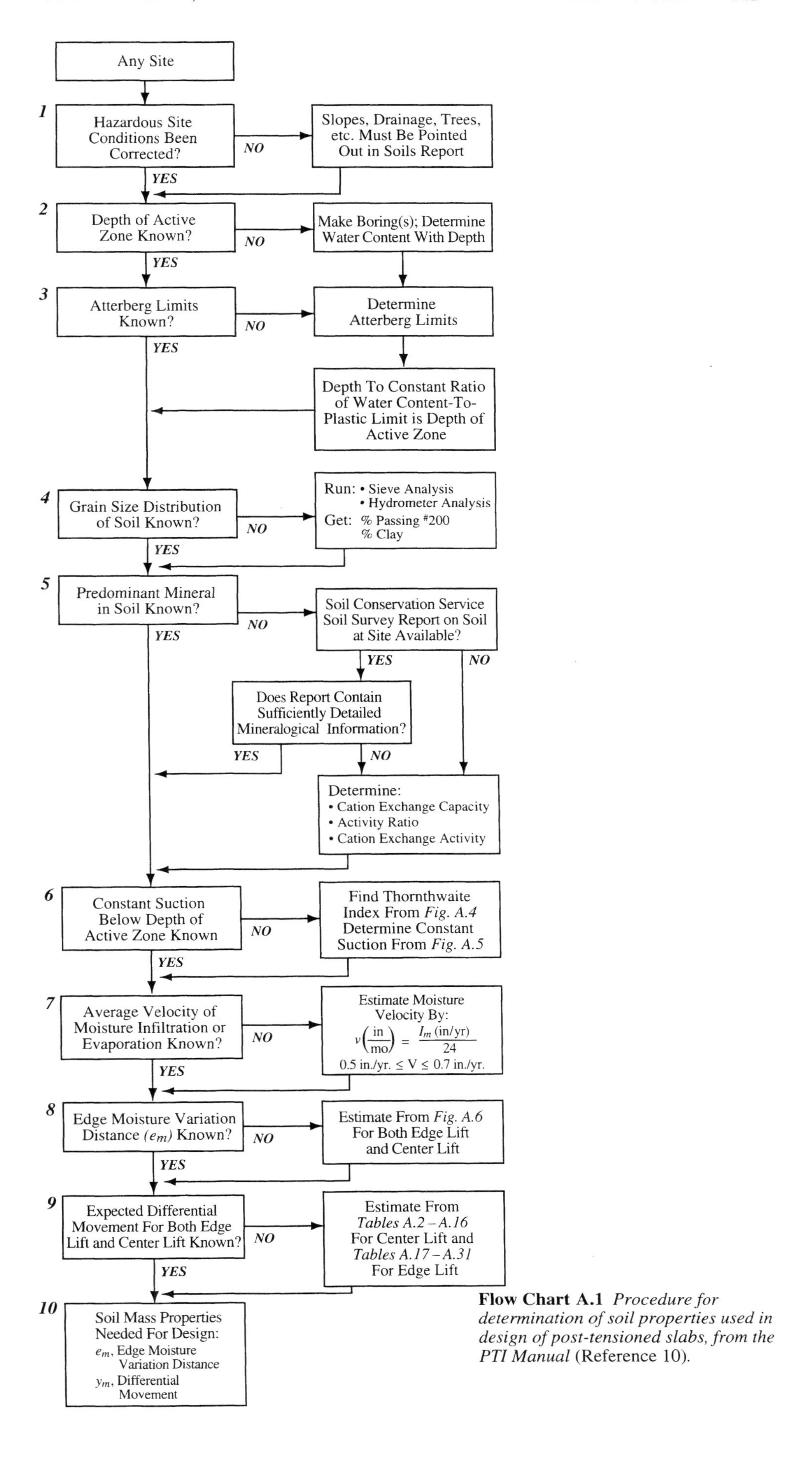

Flow Chart A.1 *Procedure for determination of soil properties used in design of post-tensioned slabs, from the PTI Manual* (Reference 10).

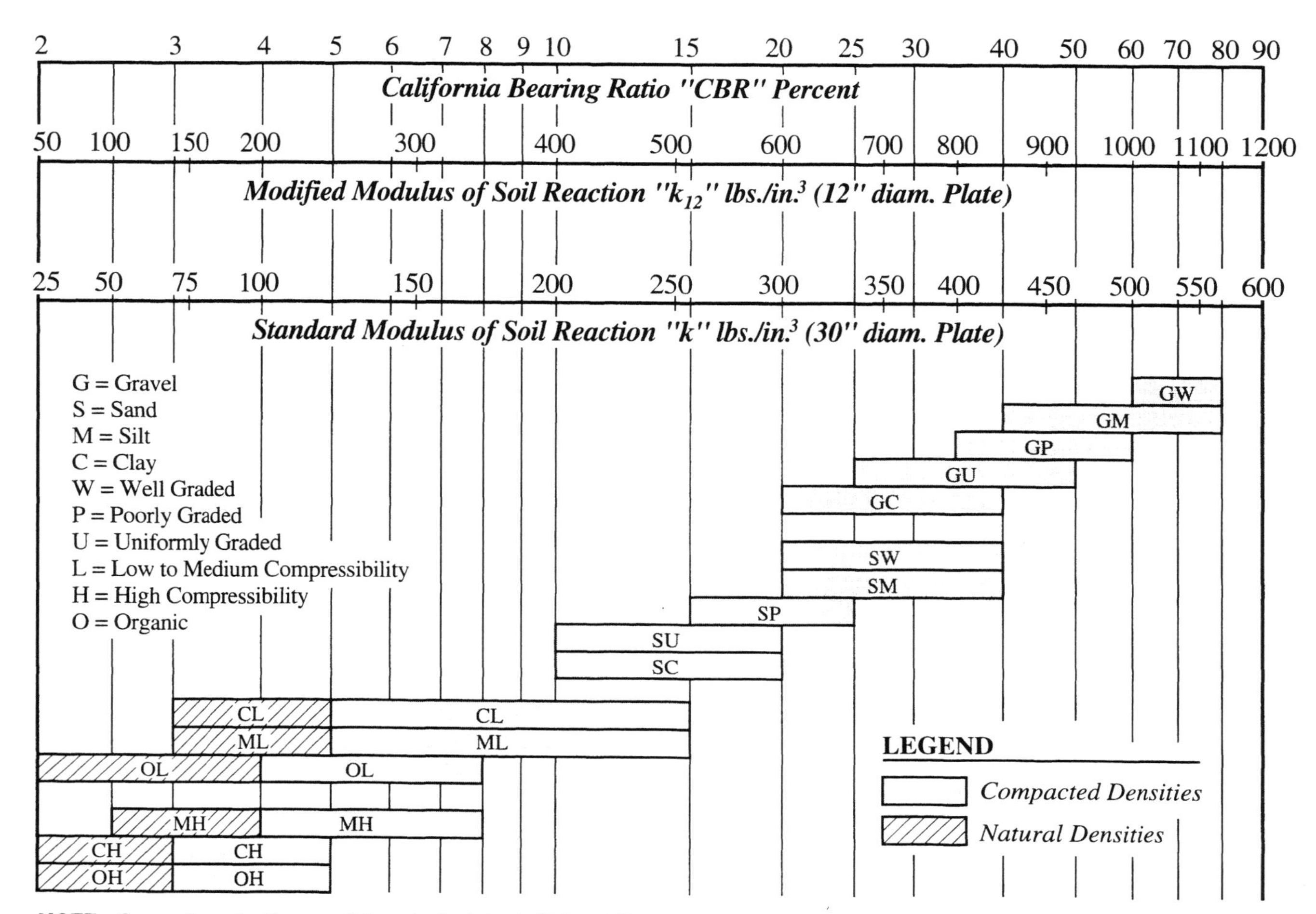

NOTE: *Comparison of soil type to "k" particularly in the "L" and "H" Groups, should generally be made in the lower range of the soil type.*

Figure A.2 *Interrelationships of soil classifications and strength criteria, from* Reference 8.

Table A.1 *Unified Soil Classification, from* Reference 2.

Major Divisions (1)	(2)	Letter (3)	Name (4)	Value as Foundation When not Subject to Frost Action (5)	Value as Base Directly under Wearing Surface (6)	Potential Frost Action (7)	Compressibility and Expansion (8)	Drainage Characteristics (9)	Compaction Equipment (10)	Unit Dry Weight (pcf) (11)	Field CBR (12)	Subgrade Modulus k (pci) (13)
Coarse-grained soils	Gravel and gravelly soils	GW	Gravel or sandy gravel, well graded	Excellent	Good	None to very slight	Almost none	Excellent	Crawler-type tractor, rubber-tired equipment, steel-wheeled roller	125-140	60-80	300 or more
		GP	Gravel or sandy gravel, poorly graded	Good to excellent	Poor to fair	None to very slight	Almost none	Excellent	Crawler-type tractor, rubber-tired equipment, steel-wheeled roller	120-130	35-60	300 or more
		GU	Gravel or sandy gravel, uniformly graded	Good	Poor	None to very slight	Almost none	Excellent	Crawler-type tractor, rubber-tired equipment	115-125	25-50	300 or more
		GM	Silty gravel or silty sandy gravel	Good to excellent	Fair to good	Slight to medium	Very slight	Fair to poor	Rubber-tired equipment, sheepsfoot roller, close control of moisture	130-145	40-80	300 or more
		GC	Clayey gravel or clayey sandy gravel	Good	Poor	Slight to medium	Slight	Poor to practically impervious	Rubber-tired equipment, sheepsfoot roller	120-140	20-40	200-300
	Sand and sandy soils	SW	Sand or gravelly sand, well graded	Good	Poor	None to very slight	Almost none	Excellent	Crawler-type tractor, rubber-tired equipment	110-130	20-40	200-300
		SP	Sand or gravelly sand, poorly graded	Fair to good	Poor to not suitable	None to very slight	Almost none	Excellent	Crawler-type tractor, rubber-tired equipment	105-120	15-25	200-300
		SU	Sand or gravelly sand, uniformly graded	Fair to good	Not suitable	None to very slight	Almost none	Excellent	Crawler-type tractor, rubber-tired equipment	100-115	10-20	200-300
		SM	Silty sand or silty gravelly sand	Good	Poor	Slight to high	Very slight	Fair to poor	Rubber-tired equipment, sheepsfoot roller, close control of moisture	120-135	20-40	200-300
		SC	Clayey sand or clayey gravelly sand	Fair to good	Not suitable	Slight to high	Slight to medium	Poor to practically impervious	Rubber-tired equipment, sheepsfoot roller	105-130	10-20	200-300
Fine-grained soils	Low compressibility LL<50	ML	Silts, sandy silts, gravelly silts or diatomaceous soils	Fair to poor	Not suitable	Medium to very high	Slight to medium	Fair to poor	Rubber-tired equipment, sheepsfoot roller, close control of moisture	100-125	5-15	100-200
		CL	Lean clays, sandy clays, or gravelly clays	Fair to poor	Not suitable	Medium to high	Medium	Practically impervious	Rubber-tired equipment, sheepsfoot roller	100-125	5-15	100-200
		OL	Organic silts or lean organic clays	Poor	Not suitable	Medium to high	Medium to high	Poor	Rubber-tired equipment, sheepsfoot roller	90-105	4-8	100-200
	High compressibility LL>50	MH	Micaceous clays or diatomaceous soils	Poor	Not suitable	Medium to very high	High	Fair to poor	Rubber-tired equipment, sheepsfoot roller	80-100	4-8	100-200
		CH	Fat clays	Poor to very poor	Not suitable	Medium	High	Practically impervious	Rubber-tired equipment, sheepsfoot roller	90-110	3-5	50-100
		OH	Fat organic clays	Poor to very poor	Not suitable	Medium	High	Practically impervious	Rubber-tired equipment, sheepsfoot roller	80-105	3-5	50-100
Peat and other fibrous organic soils		PT	Peat humus, and other	Not suitable	Not suitable	Slight	Very high	Fair to poor	Compaction not practical			

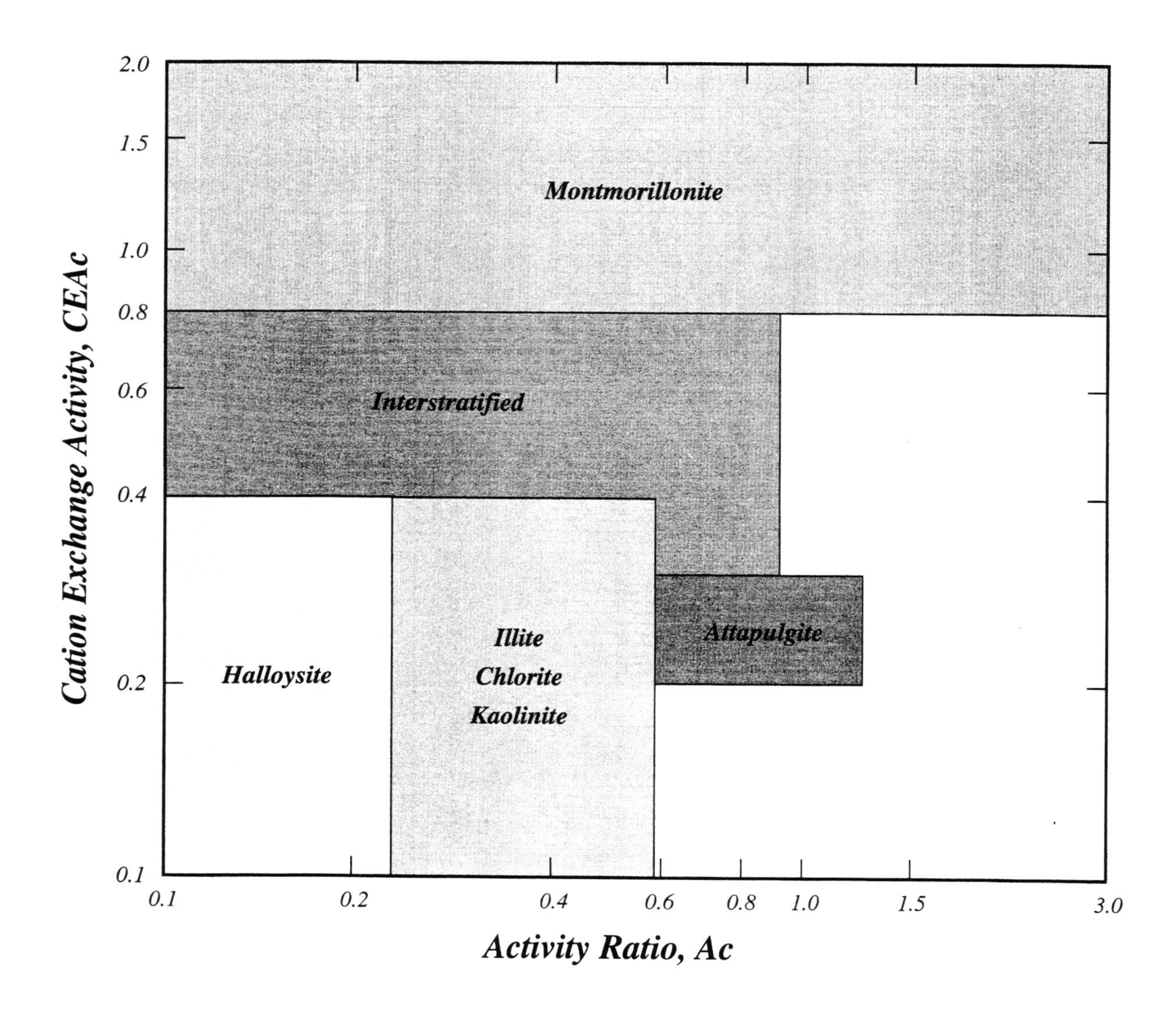

Figure A.3 *Clay type classification related to cation exchange activity and clay activity ratio, from the PTI Manual* (Reference 10).

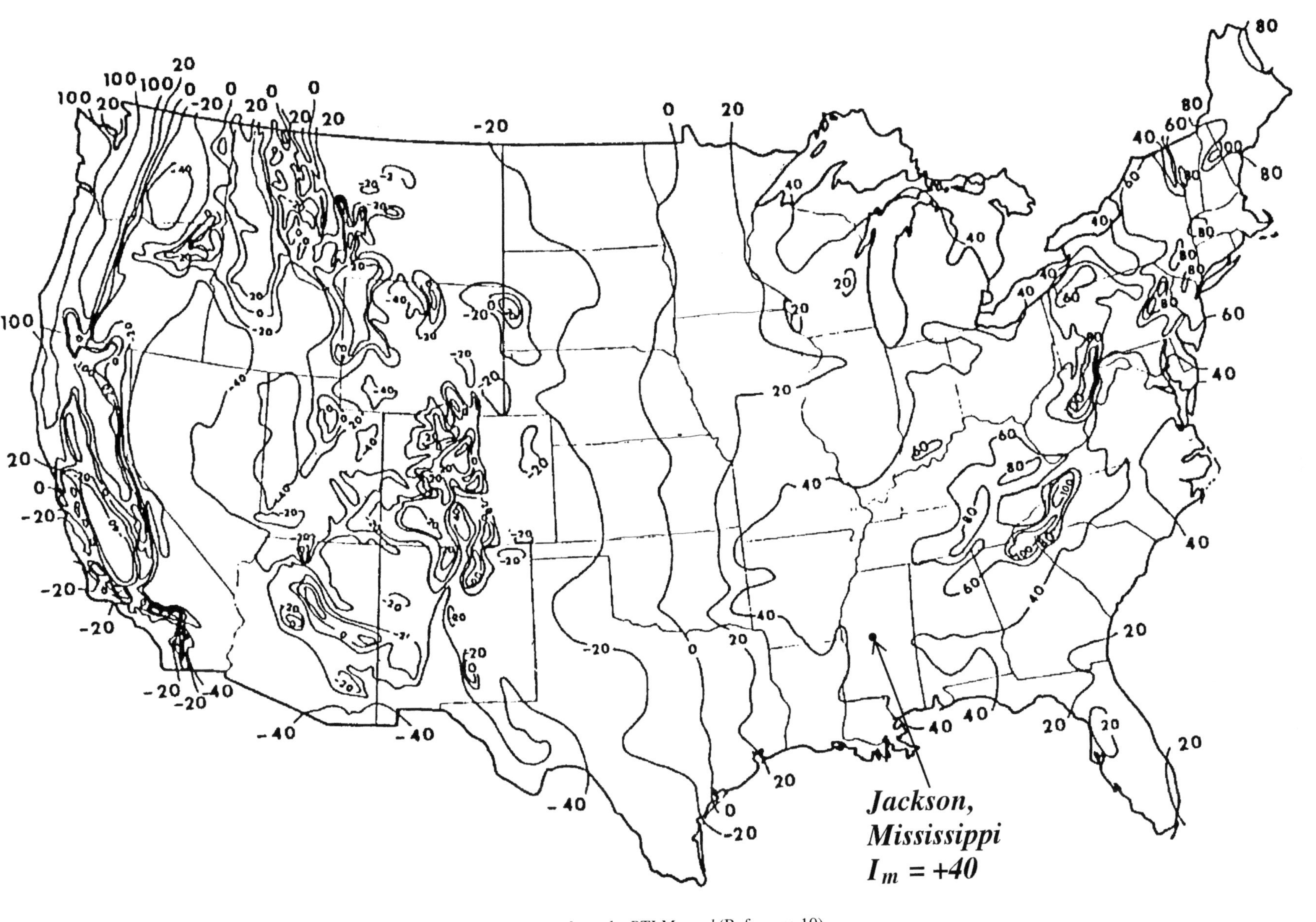

Figure A.4 *Thornthwaite moisture index distribution in the United States, from the PTI Manual* (Reference 10).

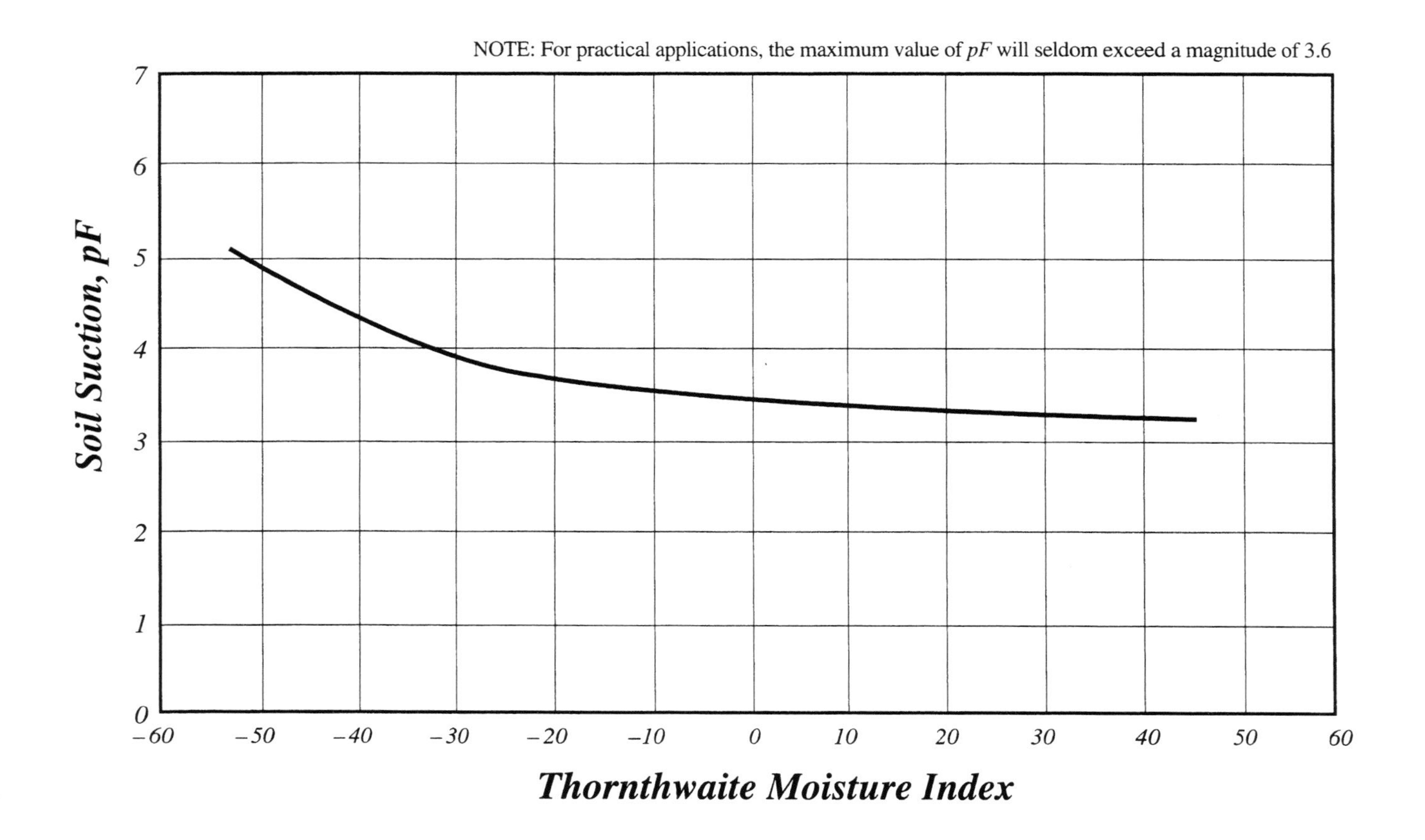

Figure A.5 *Variation of constant soil suction with Thornthwaite moisture index, from* Reference 10.

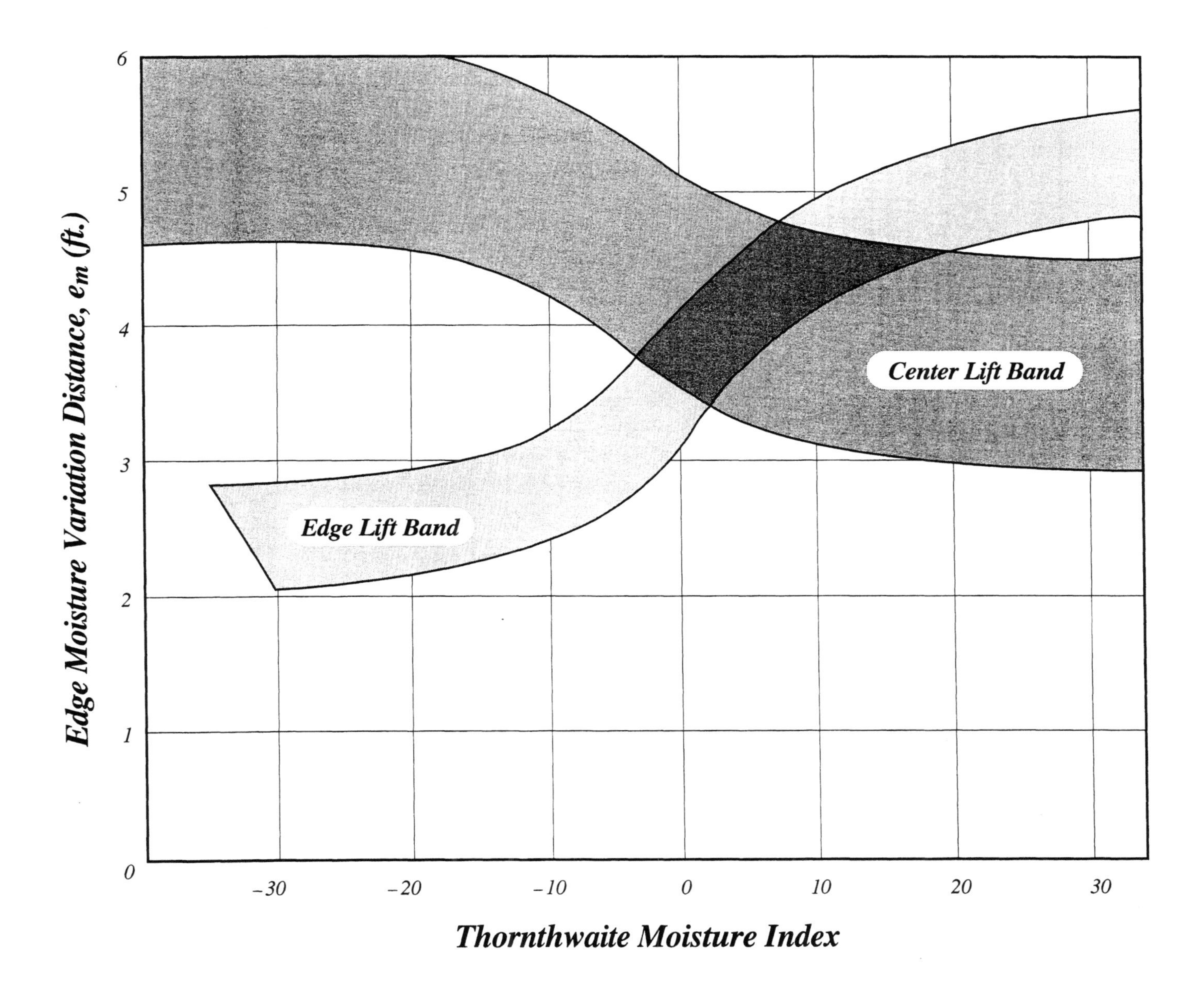

Figure A.6 *Approximate relationship between Thornthwaite index and edge moisture variation distance, from* Reference 10. *Note that extremely active clays may generate larger values of edge moisture variation than reflected by the above curves and related tables. Therefore these curves should be used only in conjunction with a site-specific soils investigation by knowledgeable geotechnical engineers.*

A.3 — Design Charts from Portland Cement Association, Wire Reinforcement Institute, Corps of Engineers, and American Concrete Institute

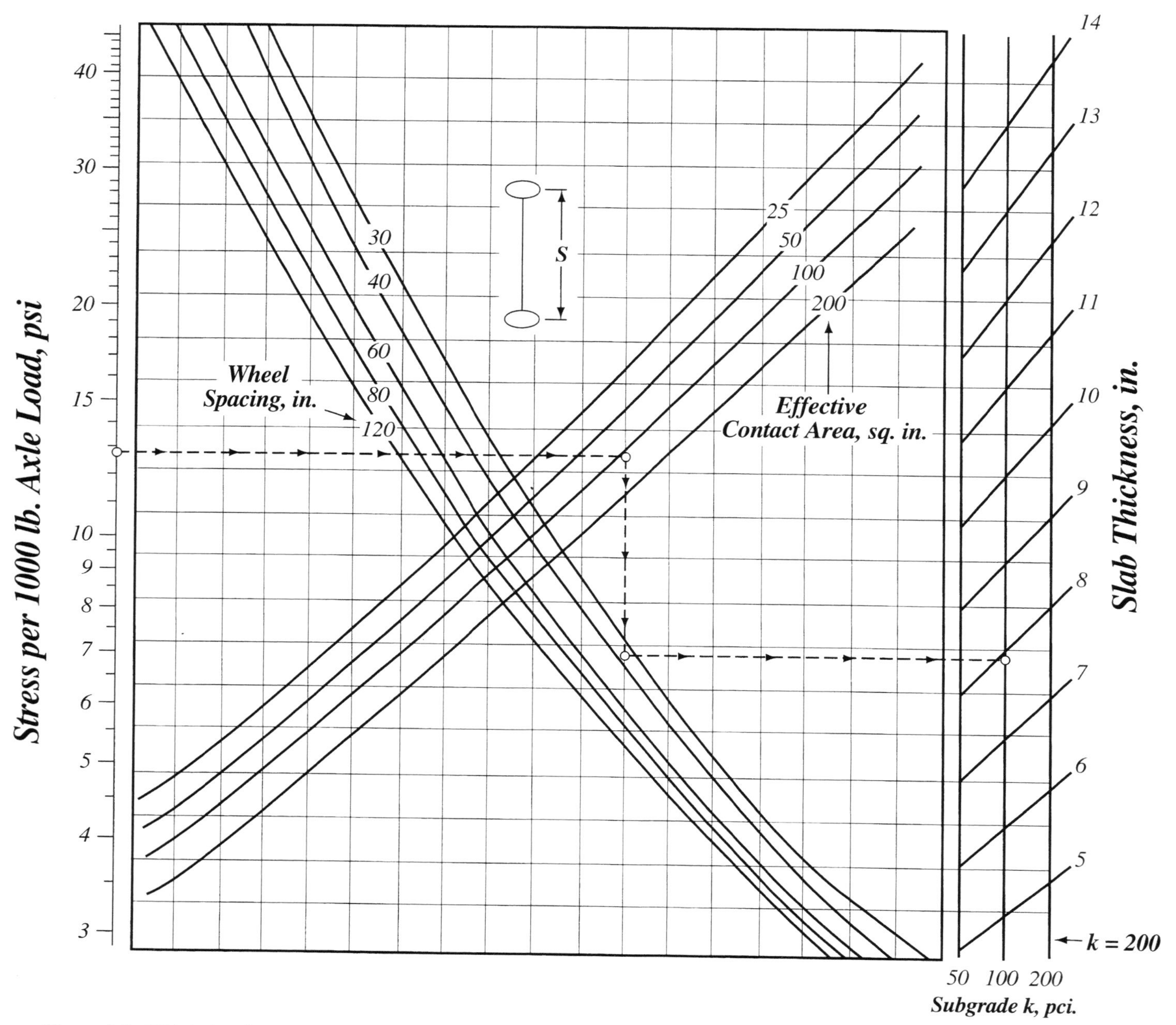

Figure A.7 *PCA design chart for selection of slab thickness for single axle loading.*

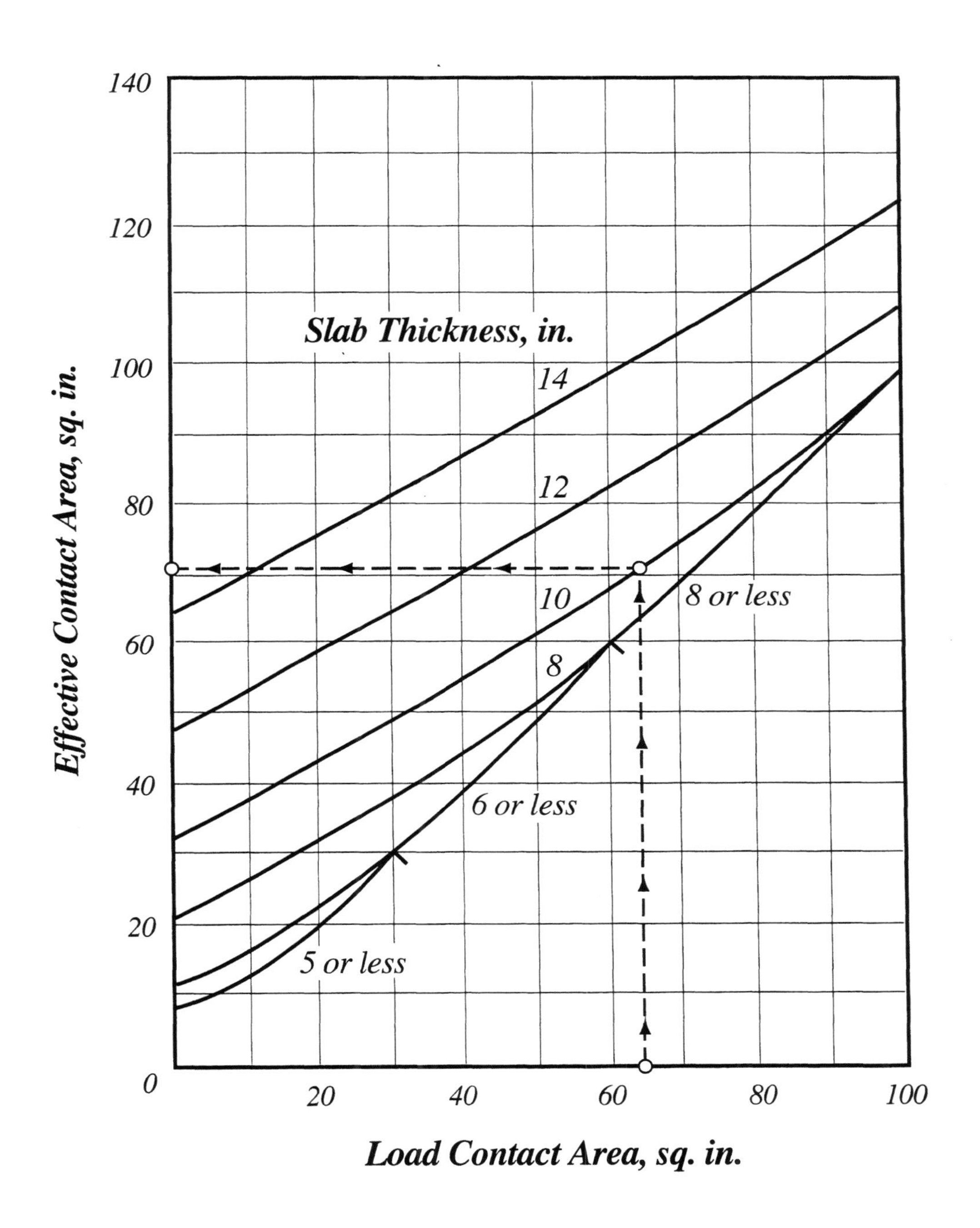

Figure A.8 *PCA chart for determining effective wheel contact area for concrete slabs thicker than 8 inches.*

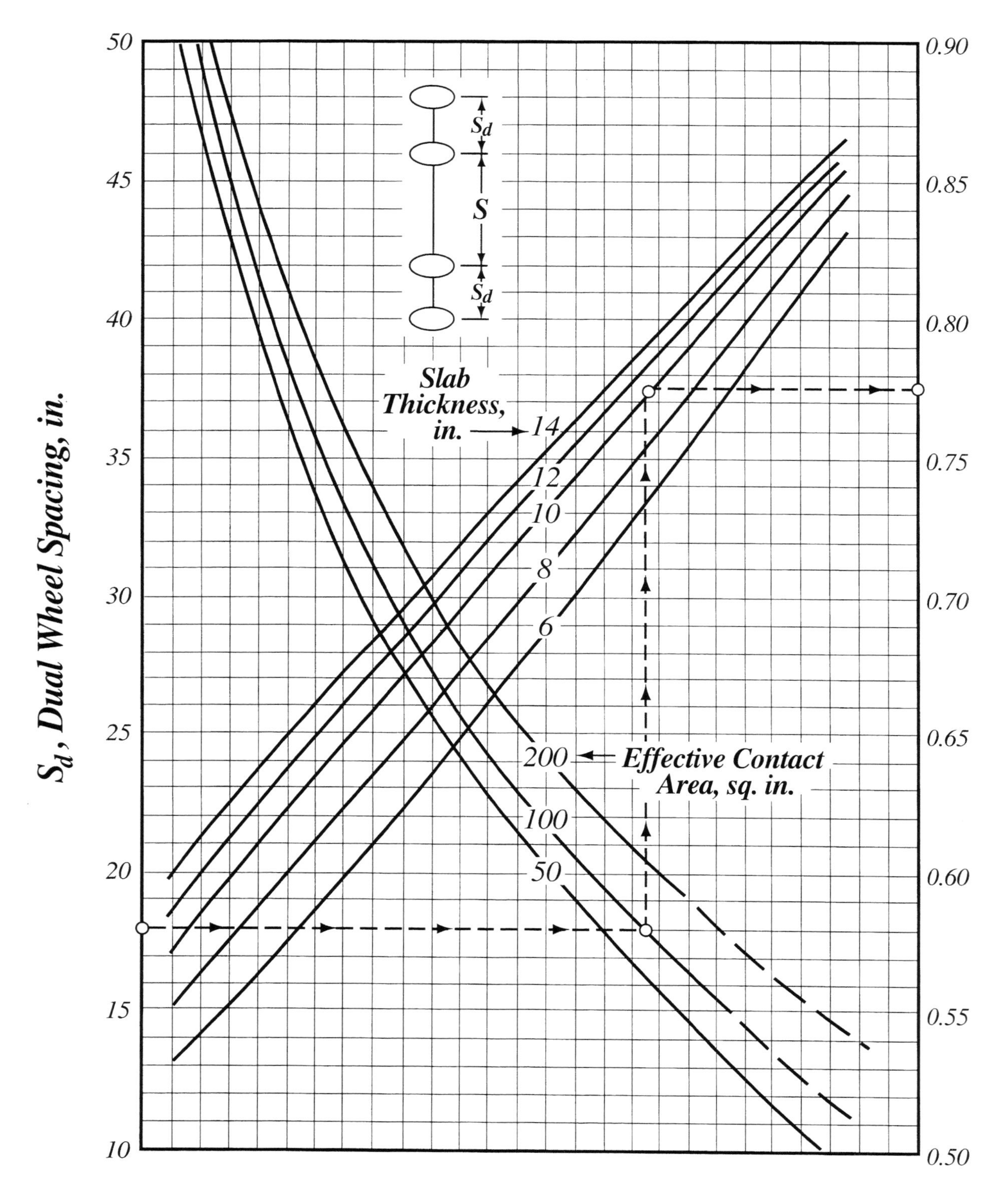

Figure A.9 *Reduction factor used with PCA charts when designing for dual-wheel loads.*

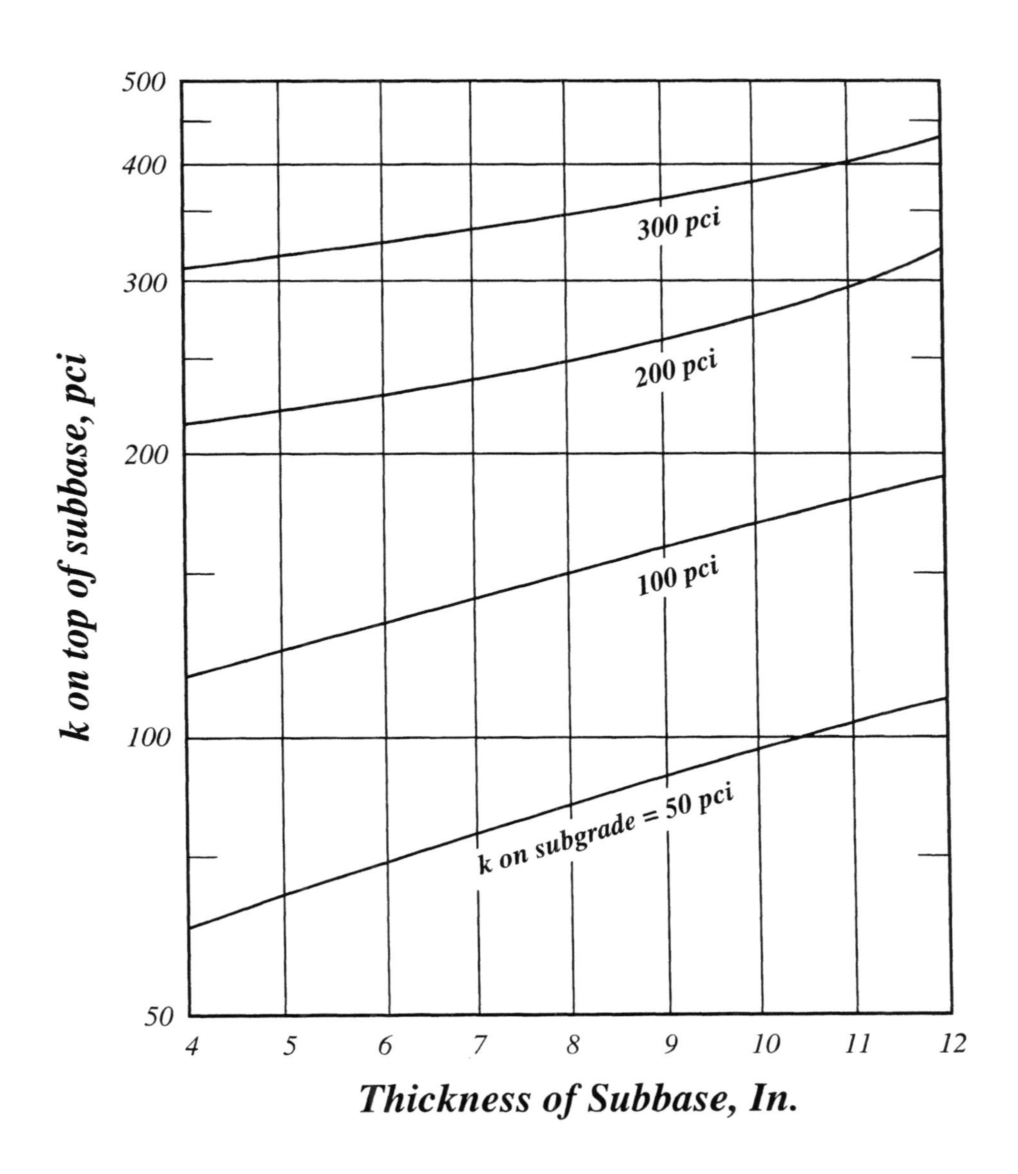

Figure A.10 *PCA chart for determining effect of granular subbase thickness on* k *value, from* Reference 14.

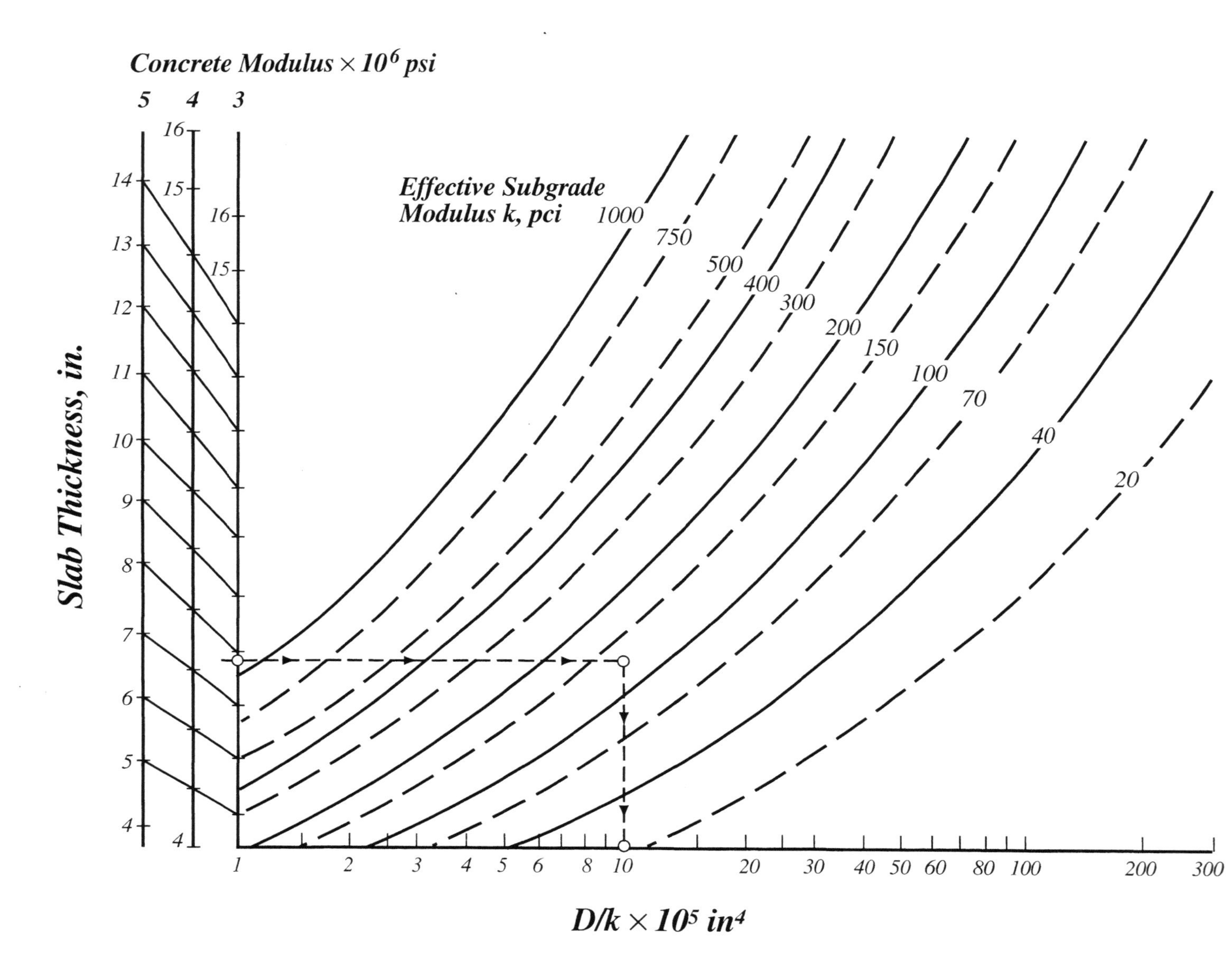

Figure A.11 *WRI chart for determination of the* D/k *value.*

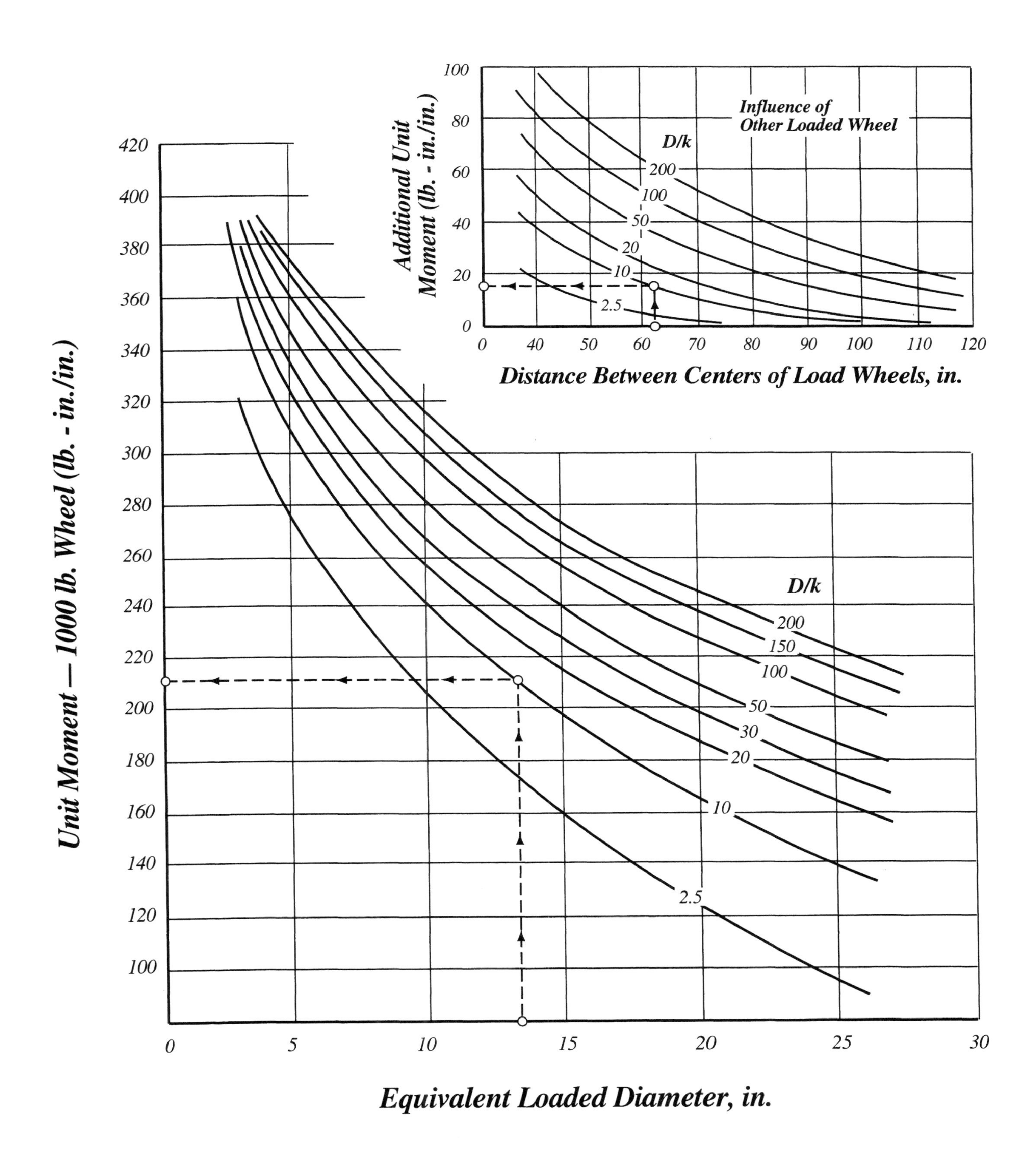

Figure A.12 *WRI determination of applied moment due to vehicle axle load requires two steps.*

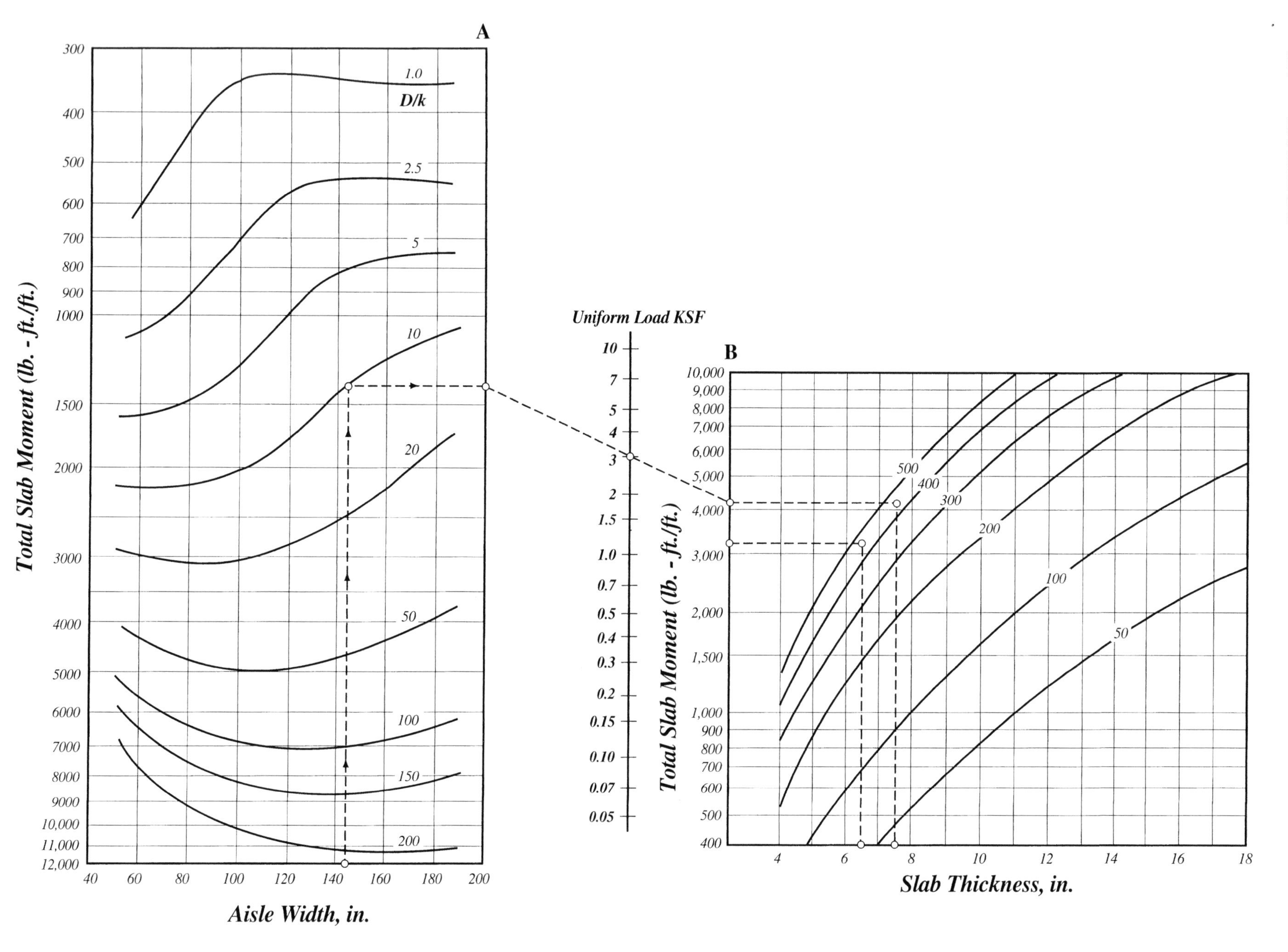

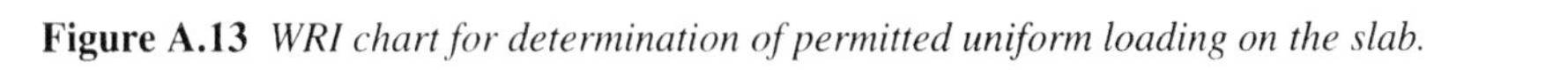

Figure A.13 *WRI chart for determination of permitted uniform loading on the slab.*

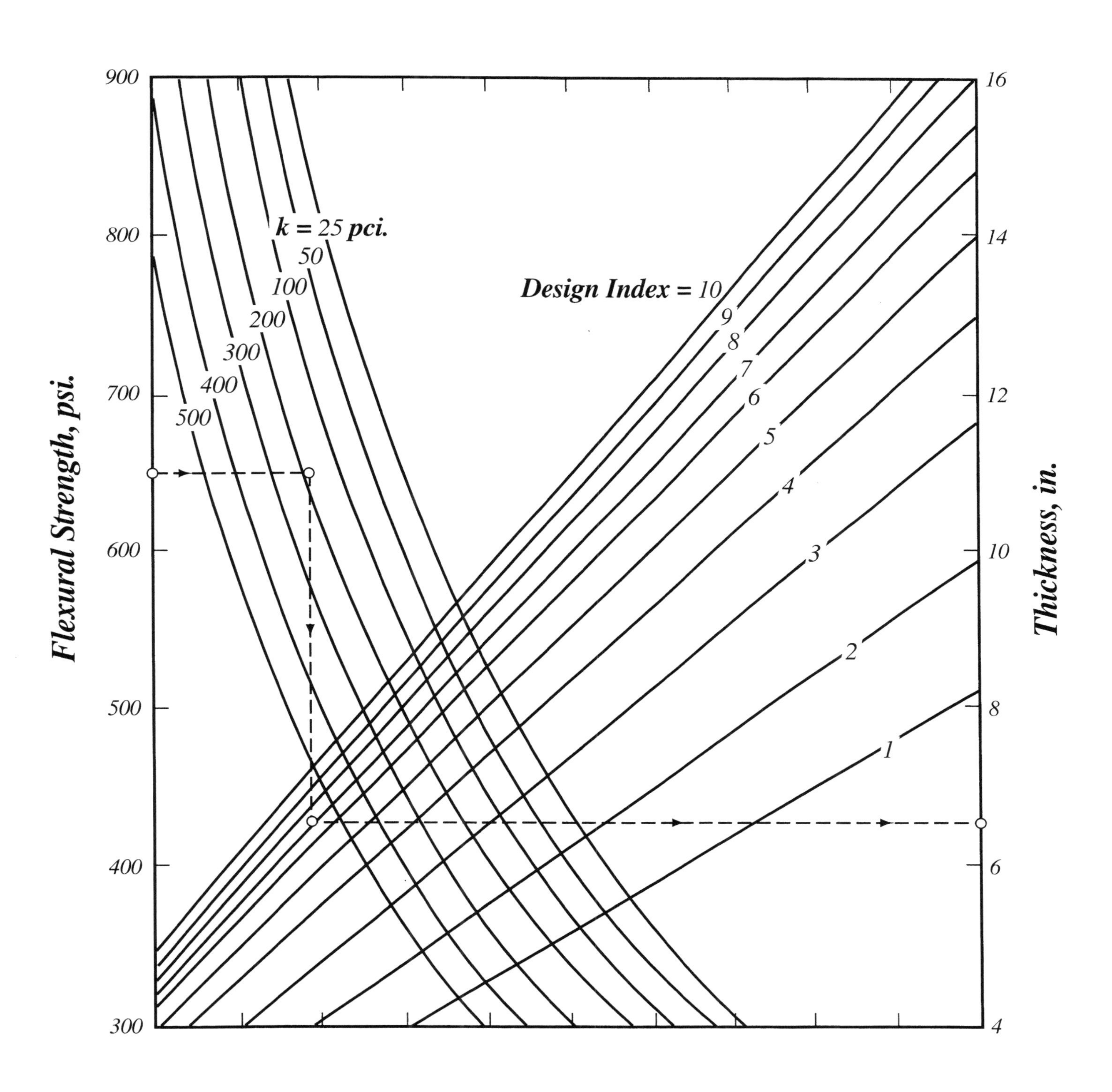

Figure A.14 *COE chart for slab thickness selection for relatively light lift truck loading.*

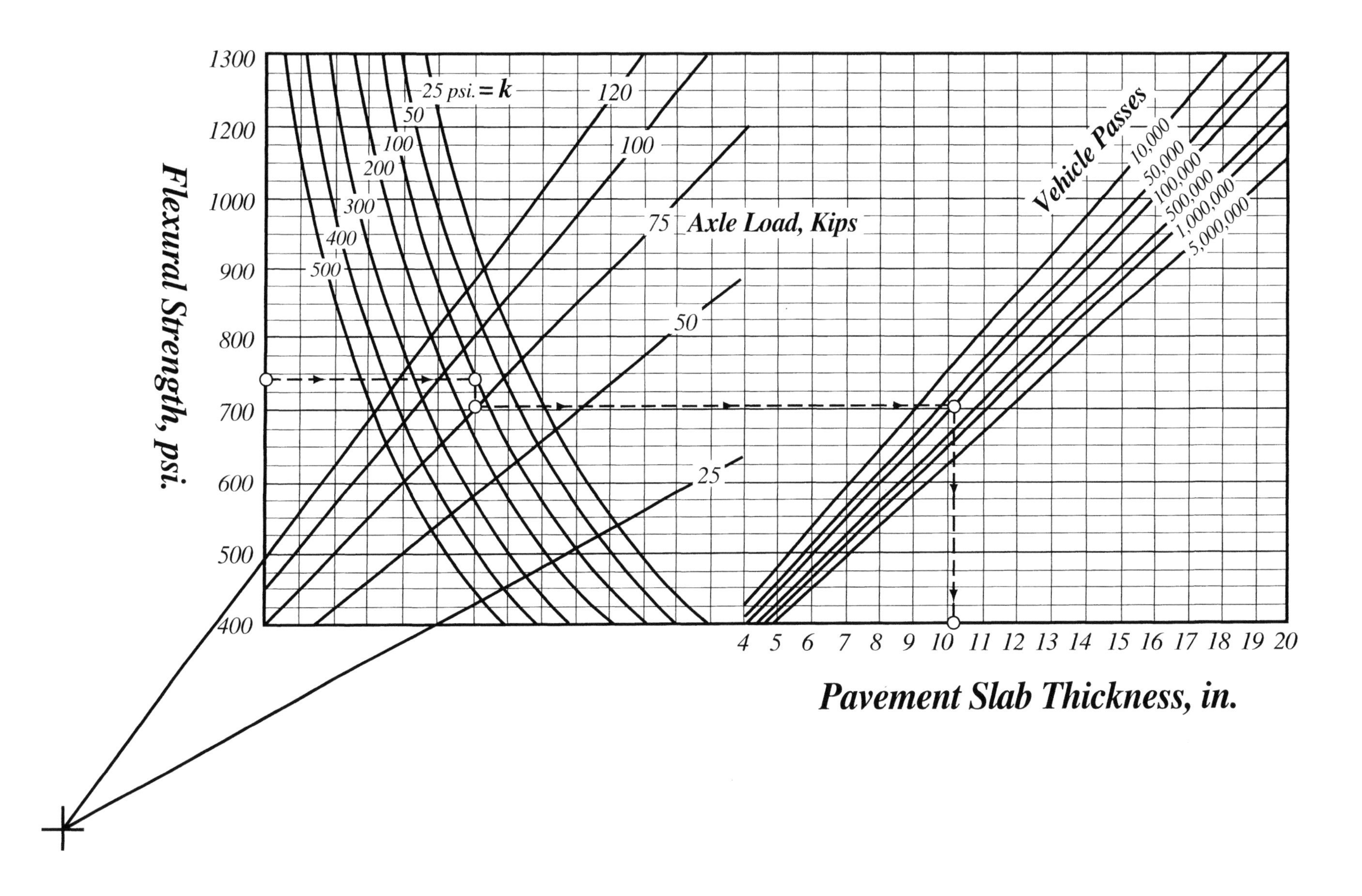

Figure A.15 *COE chart for slab thickness selection, heavy lift truck loadings.*

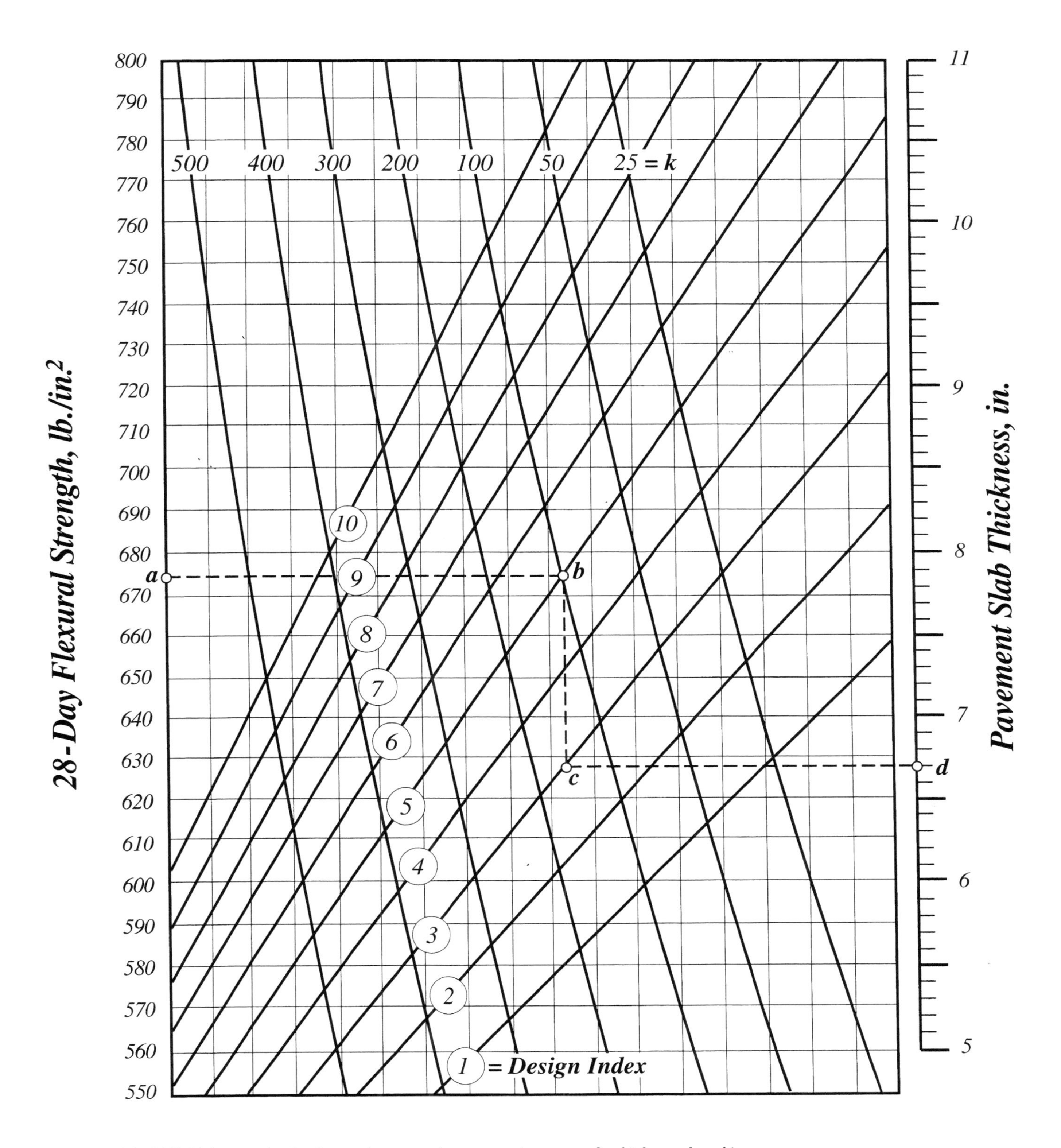

Figure A.16 *COE thickness selection for outdoor paved areas serving general vehicles and parking.*

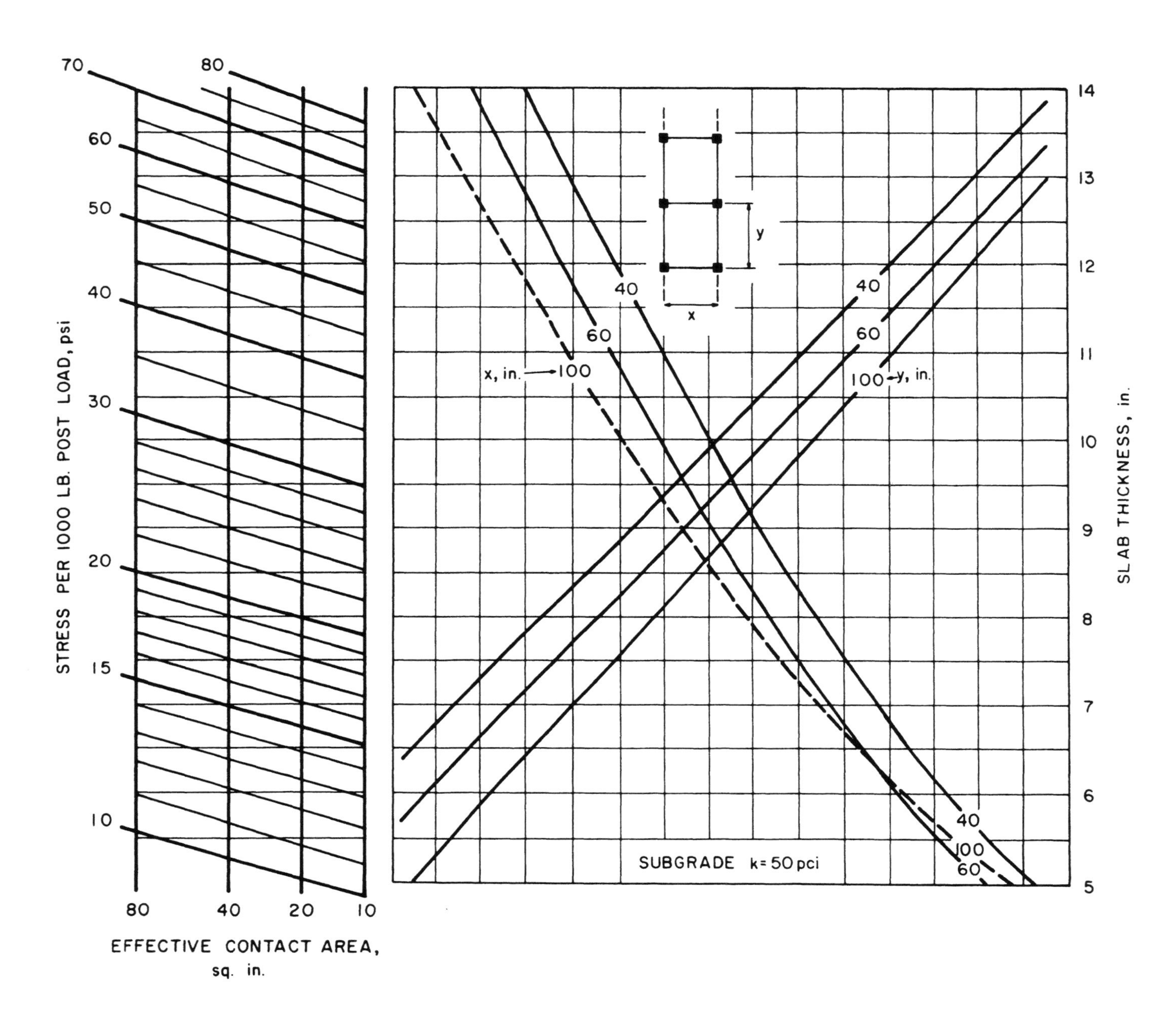

Figure A.17 *PCA chart for slab thickness selection when using post loading with subgrade* k = *50 pci.*

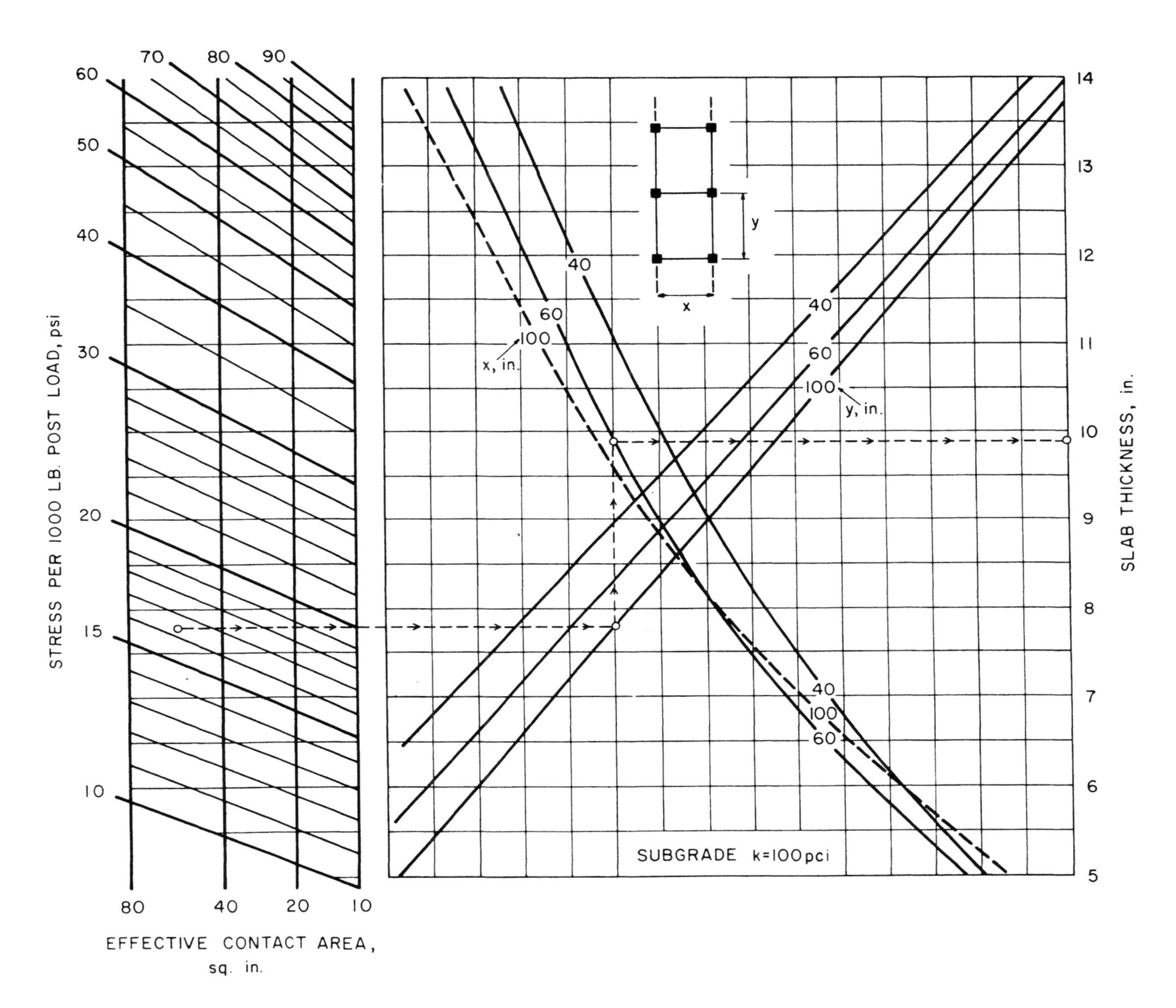

Figure A.18 *PCA chart for slab thickness selection when using post loading with subgrade* k = *100 pci.*

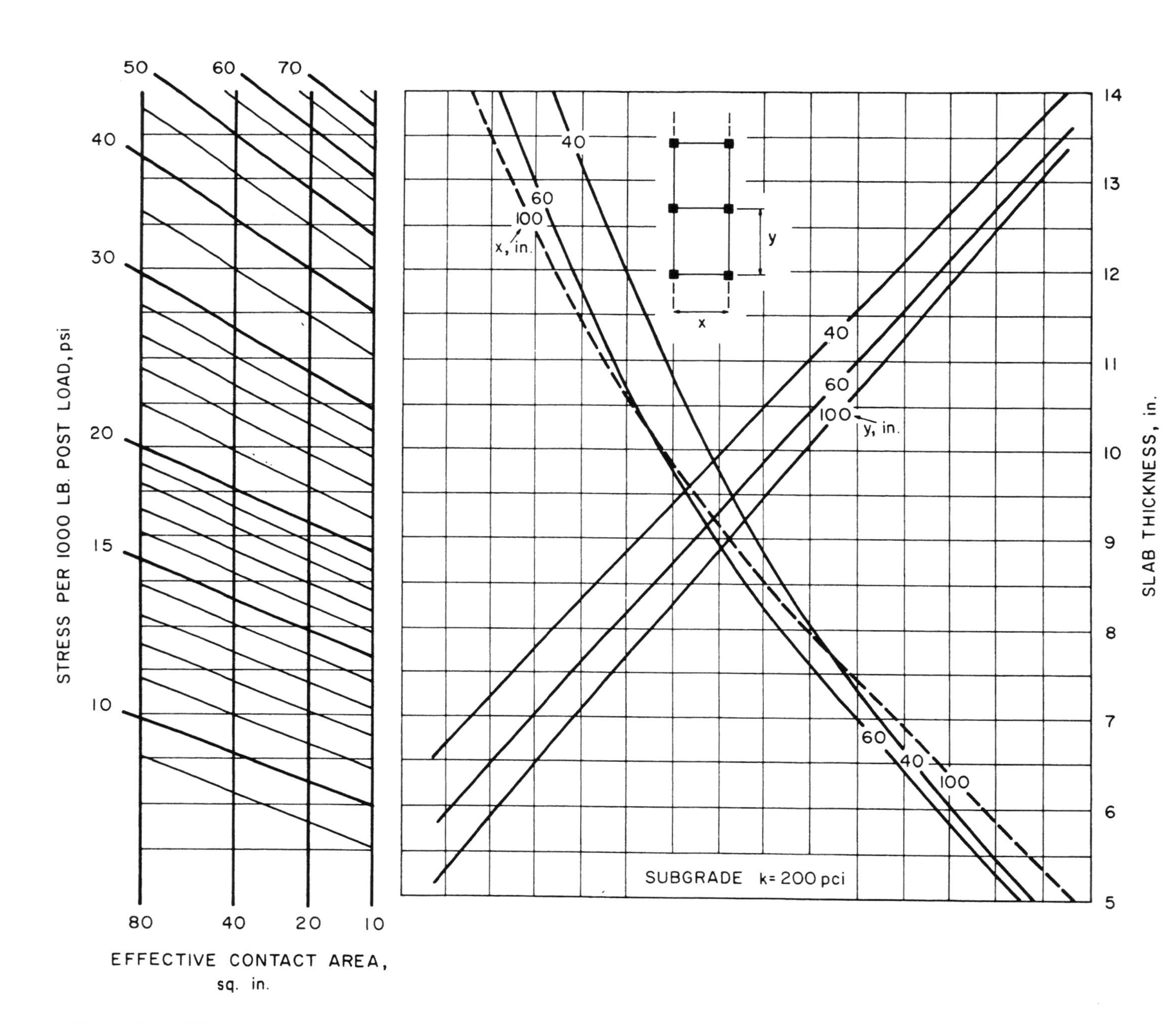

Figure A.19 *PCA chart for slab thickness selection when using post loading with subgrade* k = *200 pci.*

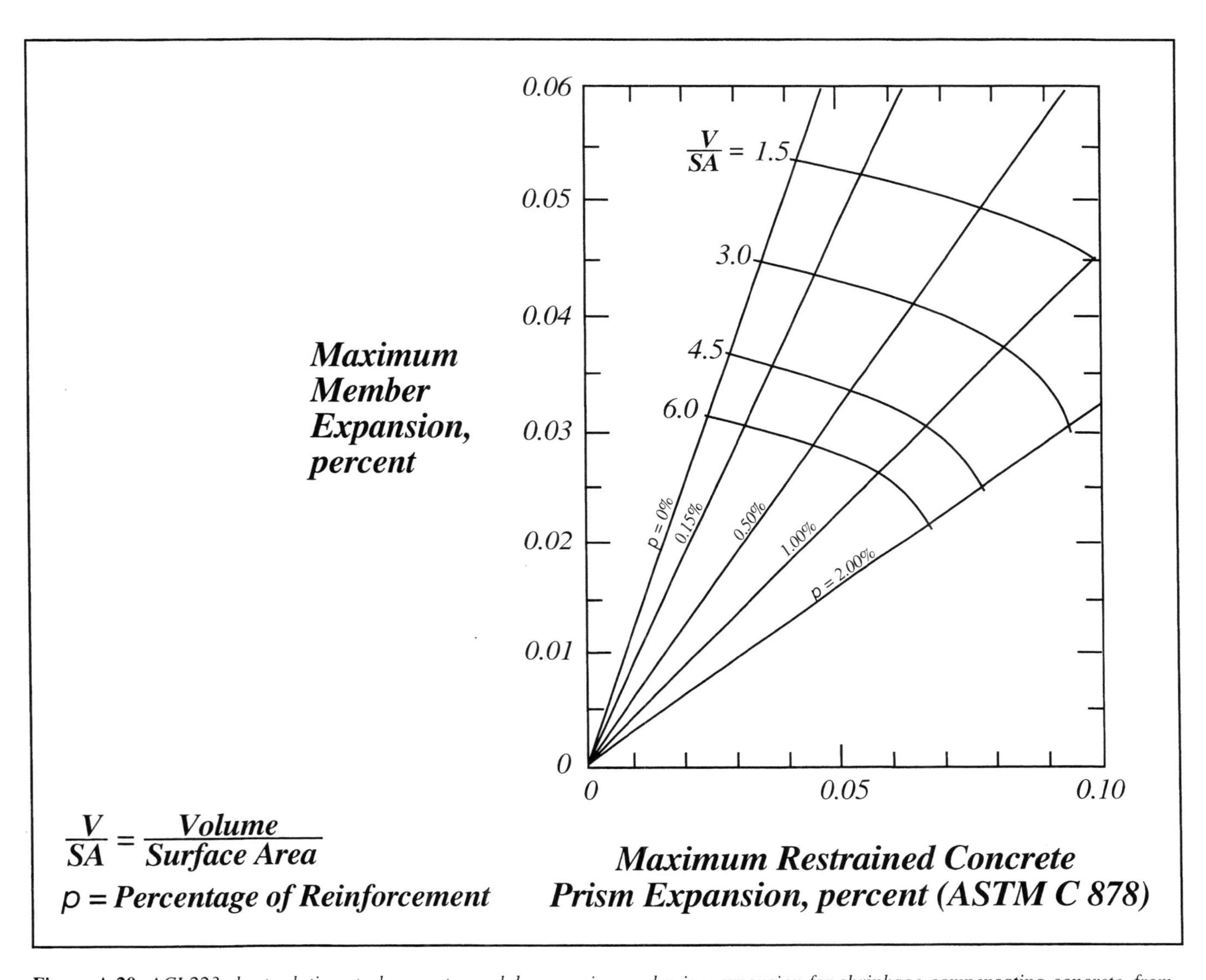

Figure A.20 *ACI 223 chart relating steel percentage, slab expansion, and prism expansion for shrinkage-compensating concrete, from* Reference 13.

A.4 — Post-Tensioned Slab Flow Charts, Equations Lists, and Post-Tensioning Institute Tables

A.4.1 — Sequential equations for design of slabs supported on plastic clays as presented in Chapter 7

Note: Although these equations are based on the PTI Manual design method *(Reference 7)*, they may not always be identical to the equations presented in that manual.

Trial section assumption; determining beam depth *d*

$$d = x^{1.176} \text{ for edge lift}$$

$$where\ x = \frac{L^{0.35} \times S^{0.88} \times e_m^{0.74} \times y_m^{0.76}}{12 \times \Delta_{allow} \times P^{0.01}}$$

Assume Δ_{allow} is $L/1700$, for this trial section determination only. In computing Δ_{allow} L is taken as the actual length or 6β, whichever is less. A common initial assumption for β is 10.

Slab on plastic clay, edge lift design, moment in the *long* direction

$$M_\ell = \frac{S^{0.10}(d\,e_m)^{0.78}\,y_m^{0.66}}{7.2 \times L^{0.0065}\,P^{0.04}}$$

Slab on plastic clay, edge lift design, moment in the *short* direction

$$M_s = d^{0.35} \times \frac{19 + e_m}{57.75} \times M_\ell$$

Slab on plastic clay, edge lift design, allowable differential deflection
First determine β:

$$\beta = \frac{1}{12}\sqrt[4]{\frac{E_c I}{E_s}}$$

Note that 6β is compared to the length under consideration, and the smaller of the two values is used for L in determining the allowable deflection:

$$\Delta_{allow} = 12\ (6\beta \text{ or } L)/800$$

Slab on plastic clay, edge lift design, expected differential deflection (both *long* and *short* directions)

$$\Delta = \frac{L^{0.35}\,S^{0.88}\,e_m^{0.74}\,y_m^{0.76}}{15.90\,d^{0.85}\,P^{0.01}}$$

Slab on plastic clay, edge lift design, expected shear force in *both* directions

$$V_s \text{ or } V_\ell = \frac{L^{0.07}\,d^{0.40}\,P^{0.03}\,e_m^{0.16}\,y_m^{0.67}}{3 \times S^{0.015}}$$

Slab on plastic clay, center lift design, moment in the *long* direction

$$M_\ell = A_o\left[B \times e_m{}^{1.238} + C\right]$$

$$\text{where } A_o = \frac{1}{727}\left[L^{0.013}\, S^{0.306}\, d^{0.688}\, P^{0.534}\, y_m{}^{0.193}\right]$$

$$\text{and for } 0 \le e_m \le 5,\ B = 1,\ C = 0$$

$$\text{For } e_m > 5,\ B = \frac{y_m{}^{-1}}{3} \le 1.0$$

$$C = \left[8 - \frac{P - 613}{225}\right]\left[\frac{4 - y_m}{3}\right] \ge 0$$

Slab on plastic clay, center lift design, moment in the *short* direction

$$M_s = \left[\frac{58 + e_m}{60}\right] \times M_\ell$$

Slab on plastic clay, center lift design, allowable differential deflection

$$\beta = \frac{1}{12}\sqrt[4]{\frac{E_c I}{E_s}}$$

$$\Delta_{allow} = 12\,(L \text{ or } 6\beta)/360$$

Use the smaller of L or 6β in calculating the allowable deflection.

Slab on plastic clay, center lift design, expected differential deflection, *both* directions

$$\Delta = \frac{[y_m L]^{0.205}\, S^{1.059}\, P^{0.523}\, e_m{}^{1.296}}{380 \times d^{1.214}}$$

Slab on plastic clay, center lift design, shear in *short* direction

$$V_s = \frac{1}{1350}\left[L^{0.19}\, S^{0.45}\, d^{0.20}\, P^{0.54}\, y_m{}^{0.04}\, e_m{}^{0.97}\right]$$

Slab on plastic clay, center lift design, shear in the *long* direction

$$V_\ell = \frac{1}{1940}\left[L^{0.09}\, S^{0.71}\, d^{0.43}\, P^{0.44}\, y_m{}^{0.16}\, e_m{}^{0.93}\right]$$

A.4.2 — Sequential equations for design of slabs supported on compressible clays as presented in Chapter 8

Note: Although these equations are based on the PTI Manual design method *(Reference 7)*, they may not all be identical to the equations presented in that manual.

Trial section assumption, beam depth *d*

$$d = x^{1.176} \text{ for edge lift}$$

$$where\ x = \frac{L^{0.35} \times S^{0.88} \times e_m^{0.74} \times y_m^{0.76}}{12 \times \Delta_{allow} \times P^{0.01}}$$

$$and\ \Delta_{allow} = \frac{12 \times (L \text{ or } 6\beta)}{1700}$$

$$\text{Assume } e_m = 0.67 \text{ and } y_m = \delta$$

Use the smaller of L or 6β when calculating deflection. An estimate of 6 feet for β is reasonable in this calculation.

Slab on compressible clay, edge lift design, moment in the *long* direction

$$M_{cs\ell} = \left[\frac{\delta}{\Delta_{ns}}\right]^{0.50} \times M_{ns\ell}$$

$$where\ M_{ns\ell} = \frac{d^{1.35}\ S^{0.36}}{80 \times L^{0.12} \times P^{0.10}}$$

$$and\ \Delta_{ns\ell} = \frac{L^{1.28}\ S^{0.80}}{133 \times d^{0.28} \times P^{0.62}}$$

Slab on compressible clay, edge lift design, moment in the *short* direction

$$M_{css} = \left[\frac{970 - d}{880}\right] \times M_{cs\ell}$$

Slab on compressible clay, edge lift design, allowable differential deflection, *long* and *short* directions

$\Delta_{allow} = 12\,(L \text{ or } 6\beta)/800$. Use L or 6β, whichever is smaller.

$$\beta = \frac{1}{12}\sqrt[4]{\frac{E_c I}{E_s \times \delta / \Delta_{ns}}}$$

Slab on compressible clay, edge lift design, expected differential deflection, *long* and *short* directions

$$\Delta_{cs} = \delta\ expZ$$

$$where\ Z = 1.78 - 0.103d - 1.65 \times 10^{-3}P + 3.95 \times 10^{-7}P^2$$

Note that $expZ$ = natural base e raised to the power Z. That is, $expZ = e^Z$.

Slab on compressible clay, edge lift design, shear in the *long* direction

$$V_{cs\ell} = \left[\frac{\delta}{\Delta_{ns\ell}} \right]^{0.30} \times V_{ns\ell}$$

$$where\ V_{ns\ell} = \frac{d^{0.90}\,[P \times S\,]^{0.30}}{550 \times L^{0.10}}$$

Slab on compressible clay, edge lift design, shear in the *short* direction

$$V_{css} = \left[\frac{116 - d}{94} \right] \times V_{cs\ell}$$

Slab on compressible clay, center lift design, moment in the *long* direction

$$M_\ell = A_o\,[\,B \times e_m^{\;1.238} + C\,]$$

$$where\ A_o = \frac{1}{727}\,[\,L^{0.013}\,S^{0.306}\,d^{0.688}\,P^{0.534}\,y_m^{\;0.193}\,]$$

$$and\,for\ 0 \le e_m \le 5,\ B = 1,\ C = 0$$

$$For\ e_m > 5,\ \ B = \frac{y_m^{\;-1}}{3} \le 1.0$$

$$C = \left[8 - \frac{P - 613}{225} \right] \left[\frac{4 - y_m}{3} \right] \ge 0$$

Slab on compressible clay, center lift design, moment in the *short* direction

$$M_s = \left[\frac{58 + e_m}{60} \right] \times M_\ell$$

Slab on compressible clay, center lift design, allowable differential deflection

$\Delta_{allow} = 12\,(L \text{ or } 6\beta)/360$. Use L or 6β, whichever is smaller.

$$\beta = \frac{1}{12} \sqrt[4]{\frac{E_c I}{E_s \times \delta / \Delta_{ns}}}$$

Slab on compressible clay, center lift design, expected differential deflection, *long* and *short* direction

$$\Delta = \frac{[y_m L]^{0.205}\,S^{1.059}\,P^{0.523}\,e_m^{\;1.296}}{380 \times d^{1.214}}$$

Slab on compressible clay, center lift design, shear in *short* direction

$$V_s = \frac{1}{1350}\,[\,L^{0.19}\,S^{0.45}\,d^{0.20}\,P^{0.54}\,y_m^{\;0.04}\,e_m^{\;0.97}\,]$$

Slab on compressible clay, center lift design, shear in the *long* direction

$$V_\ell = \frac{1}{1940}\,[\,L^{0.09}\,S^{0.71}\,d^{0.43}\,P^{0.44}\,y_m^{\;0.16}\,e_m^{\;0.93}\,]$$

A.4.3— Flow charts for post-tensioned slab design

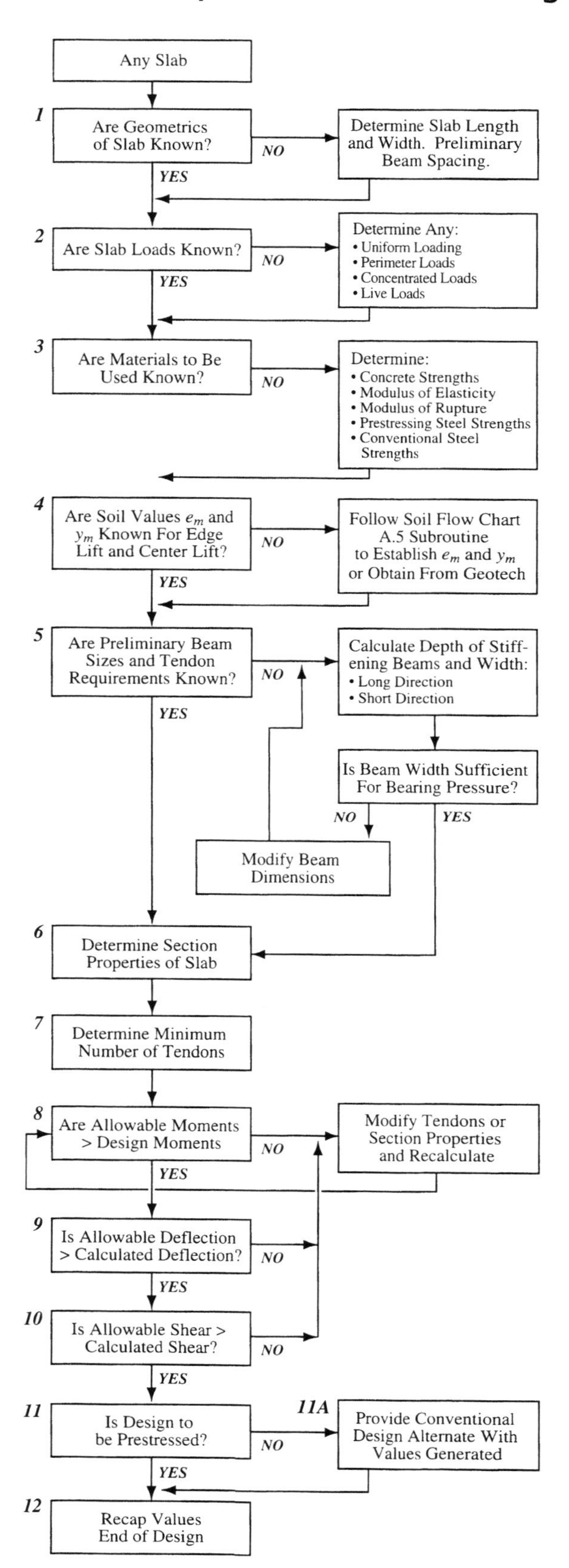

Flow Chart A.2 *Procedure for design of slabs on plastic clays.*

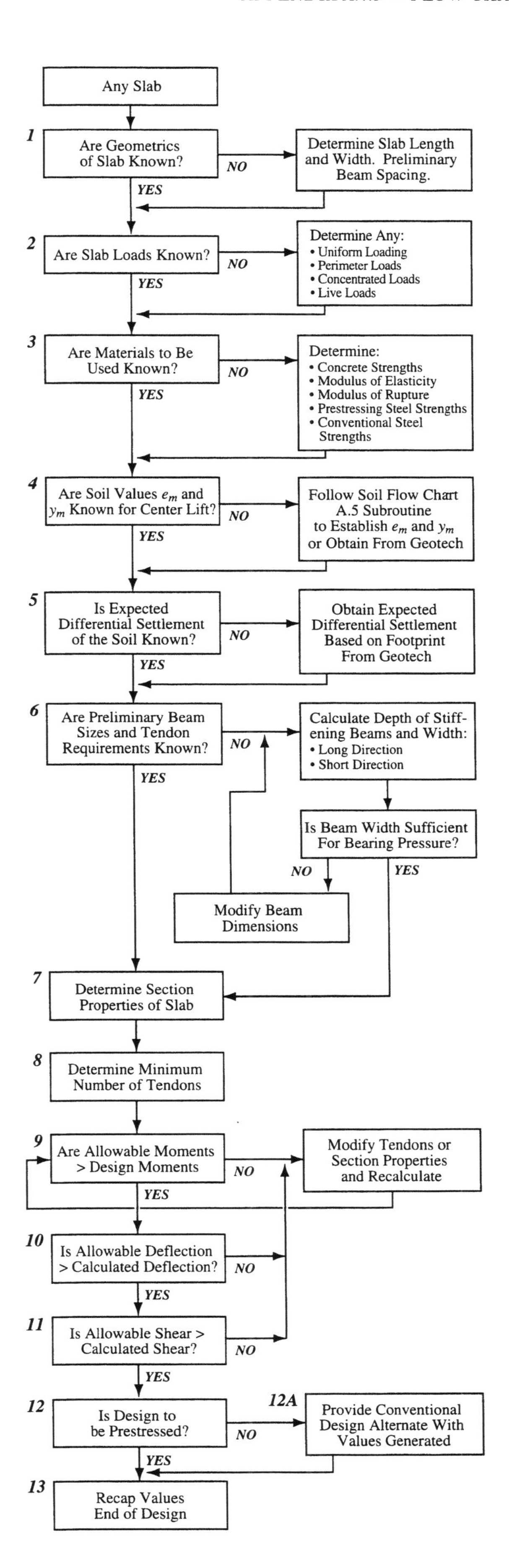

Flow Chart A.3 *Procedure for design of slabs on compressible clays.*

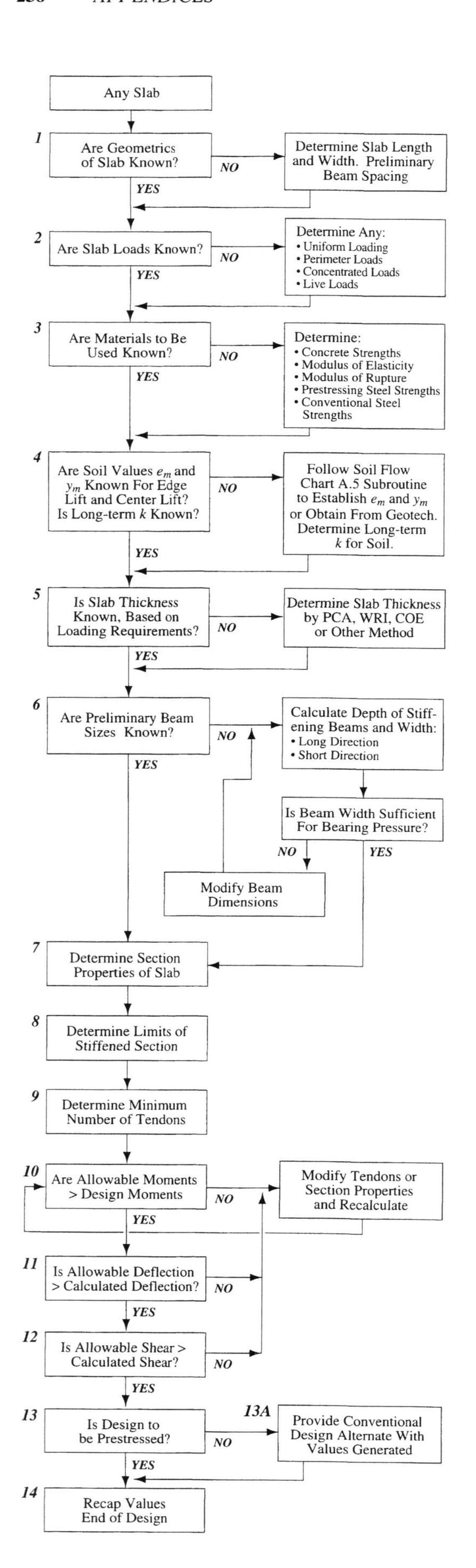

Flow Chart A.4 *Procedure for design of hybrid slabs on plastic clays.*

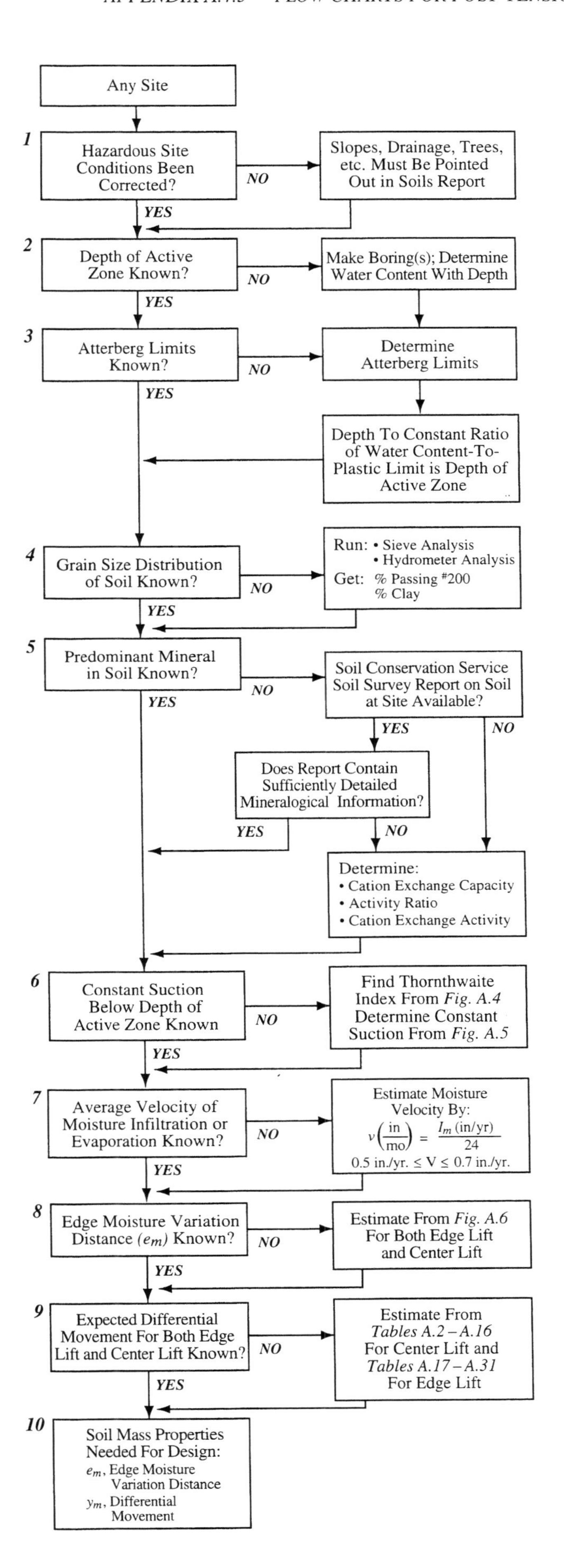

Flow Chart A.5 *Subroutine to Flow Charts A.2, A.3, and A.4 for determination of soil properties, from the PTI Manual* (Reference 10).

A.4.4— Post-Tensioning Institute Tables (Reference 10)

Percent Clay (%)	Depth to Constant Suction (FT)	Constant Suction (pF)	Velocity of Moisture Flow (inches /month)	DIFFERENTIAL SWELL (IN)							
				EDGE DISTANCE PENETRATION (FT)							
				1 FT	2 FT	3 FT	4 FT	5 FT	6 FT	7 FT	8 FT
30	3	3.2	0.1	0.001	0.001	0.002	0.002	0.003	0.004	0.004	0.005
			0.3	0.002	0.004	0.006	0.007	0.009	0.011	0.013	0.015
			0.5	0.003	0.005	0.008	0.011	0.015	0.018	0.021	0.025
			0.7	0.004	0.008	0.012	0.017	0.021	0.026	0.032	0.038
		3.4	0.1	0.001	0.002	0.003	0.004	0.006	0.007	0.008	0.009
			0.3	0.004	0.008	0.011	0.015	0.020	0.024	0.029	0.034
			0.5	0.006	0.012	0.019	0.026	0.034	0.044	0.054	0.067
			0.7	0.008	0.017	0.027	0.038	0.052	0.069	0.091	0.124
		3.6	0.1	0.003	0.006	0.009	0.012	0.015	0.019	0.022	0.026
			0.3	0.008	0.017	0.027	0.040	0.055	0.073	0.099	0.137
			0.5	0.014	0.030	0.050	0.077	0.117	0.192	0.370	0.881
			0.7	0.018	0.042	0.074	0.125	0.226	0.487	1.252	3.530
	5	3.2	0.1	0.001	0.003	0.004	0.005	0.007	0.008	0.010	0.012
			0.3	0.004	0.008	0.012	0.016	0.020	0.025	0.029	0.034
			0.5	0.007	0.013	0.020	0.028	0.035	0.043	0.051	0.060
			0.7	0.009	0.019	0.029	0.040	0.051	0.062	0.075	0.089
		3.4	0.1	0.003	0.005	0.008	0.011	0.014	0.017	0.020	0.023
			0.3	0.008	0.016	0.025	0.034	0.044	0.055	0.066	0.078
			0.5	0.014	0.028	0.043	0.060	0.079	0.100	0.125	0.153
			0.7	0.018	0.039	0.062	0.088	0.119	0.157	0.207	0.279
		3.6	0.1	0.006	0.013	0.020	0.027	0.034	0.042	0.050	0.059
			0.3	0.019	0.040	0.063	0.091	0.123	0.163	0.217	0.300
			0.5	0.030	0.067	0.112	0.171	0.258	0.413	0.776	1.797
			0.7	0.042	0.095	0.166	0.276	0.486	1.009	2.499	6.879
	7	3.2	0.1	0.002	0.004	0.007	0.009	0.012	0.015	0.017	0.020
			0.3	0.007	0.015	0.022	0.030	0.038	0.046	0.055	0.063
			0.5	0.012	0.025	0.038	0.051	0.065	0.080	0.095	0.111
			0.7	0.017	0.035	0.053	0.073	0.093	0.115	0.139	0.164
		3.4	0.1	0.005	0.010	0.016	0.021	0.026	0.031	0.037	0.042
			0.3	0.015	0.030	0.046	0.063	0.081	0.100	0.121	0.142
			0.5	0.024	0.050	0.079	0.110	0.144	0.184	0.228	0.281
			0.7	0.034	0.071	0.113	0.616	0.218	0.287	0.379	0.514
		3.58	0.1	0.011	0.022	0.033	0.045	0.057	0.069	0.082	0.096
			0.3	0.032	0.066	0.124	0.147	0.197	0.257	0.332	0.436
			0.5	0.051	0.110	0.182	0.272	0.396	0.596	1.006	2.098
			0.7	0.071	0.157	0.269	0.431	0.712	1.346	3.081	8.129

Table A.2 *Differential swell occurring at the perimeter of a slab for a center lift swelling condition in predominantly Kaolinite clay soil (30 percent clay).*

Percent Clay (%)	Depth to Constant Suction (FT)	Constant Suction (pF)	Velocity of Moisture Flow (inches /month)	DIFFERENTIAL SWELL (INCHES)							
				EDGE DISTANCE PENETRATION (FT)							
				1 FT	2 FT	3 FT	4 FT	5 FT	6 FT	7 FT	8 FT
40	3	3.2	0.1	0.001	0.001	0.002	0.002	0.004	0.005	0.005	0.006
			0.3	0.002	0.004	0.006	0.009	0.011	0.014	0.016	0.019
			0.5	0.004	0.008	0.012	0.016	0.020	0.024	0.029	0.034
			0.7	0.005	0.010	0.016	0.022	0.029	0.035	0.043	0.050
		3.4	0.1	0.002	0.004	0.005	0.007	0.008	0.009	0.011	0.014
			0.3	0.004	0.009	0.014	0.020	0.025	0.032	0.038	0.044
			0.5	0.007	0.016	0.074	0.037	0.046	0.058	0.073	0.090
			0.7	0.011	0.023	0.036	0.051	0.070	0.092	0.122	0.166
		3.6	0.1	0.003	0.007	0.011	0.015	0.020	0.024	0.029	0.034
			0.3	0.010	0.023	0.037	0.053	0.072	0.097	0.131	0.183
			0.5	0.018	0.040	0.066	0.102	0.157	0.256	0.495	1.181
			0.7	0.025	0.056	0.100	0.168	0.303	0.653	1.677	4.728
	5	3.2	0.1	0.002	0.004	0.006	0.007	0.009	0.011	0.013	0.015
			0.3	0.005	0.011	0.016	0.022	0.028	0.033	0.039	0.046
			0.5	0.009	0.018	0.027	0.037	0.047	0.057	0.068	0.080
			0.7	0.012	0.025	0.038	0.053	0.067	0.083	0.100	0.118
		3.4	0.1	0.003	0.007	0.011	0.015	0.019	0.023	0.027	0.031
			0.3	0.011	0.022	0.034	0.046	0.059	0.073	0.088	0.104
			0.5	0.018	0.037	0.148	0.081	0.106	0.134	0.167	0.206
			0.7	0.025	0.052	0.083	0.118	0.159	0.210	0.277	0.374
		3.6	0.1	0.008	0.017	0.027	0.036	0.046	0.057	0.068	0.079
			0.3	0.026	0.054	0.085	0.122	0.165	0.219	0.292	0.401
			0.5	0.041	0.090	0.150	0.229	0.346	0.553	1.040	2.408
			0.7	0.057	0.128	0.224	0.371	0.652	1.353	3.349	9.215
	7	3.2	0.1	0.003	0.006	0.010	0.013	0.017	0.020	0.023	0.027
			0.3	0.010	0.020	0.030	0.040	0.051	0.062	0.073	0.084
			0.5	0.016	0.033	0.051	0.069	0.087	0.107	0.127	0.148
			0.7	0.023	0.046	0.071	0.098	0.125	0.155	0.186	0.220
		3.4	0.1	0.006	0.013	0.020	0.027	0.034	0.041	0.048	0.056
			0.3	0.020	0.041	0.062	0.085	0.109	0.135	0.162	0.191
			0.5	0.033	0.069	0.107	0.148	0.194	0.246	0.306	0.377
			0.7	0.045	0.095	0.152	0.216	0.292	0.385	0.507	0.689
		3.58	0.1	0.014	0.029	0.044	0.060	0.076	0.093	0.110	0.129
			0.3	0.042	0.087	0.138	0.196	0.263	0.343	0.444	0.583
			0.5	0.069	0.148	0.244	0.365	0.531	0.799	1.348	2.791
			0.7	0.095	0.210	0.360	0.577	0.953	1.803	4.126	-

Table A.3 *Differential swell occurring at the perimeter of a slab for a center lift swelling condition in predominantly Kaolinite clay soil (40 percent clay).*

Percent Clay (%)	Depth to Constant Suction (FT)	Constant Suction (pF)	Velocity of Moisture Flow (inches /month)	DIFFERENTIAL SWELL (INCHES) EDGE DISTANCE PENETRATION (FT)							
				1 FT	2 FT	3 FT	4 FT	5 FT	6 FT	7 FT	8 FT
50	3	3.2	0.1	0.001	0.002	0.002	0.003	0.004	0.005	0.006	0.007
			0.3	0.003	0.005	0.008	0.011	0.014	0.018	0.021	0.024
			0.5	0.004	0.009	0.014	0.019	0.025	0.030	0.036	0.042
			0.7	0.006	0.013	0.020	0.028	0.036	0.044	0.053	0.063
		3.4	0.1	0.002	0.004	0.006	0.008	0.010	0.012	0.015	0.017
			0.3	0.006	0.012	0.018	0.025	0.032	0.040	0.048	0.057
			0.5	0.009	0.020	0.031	0.043	0.057	0.073	0.091	0.113
			0.7	0.013	0.028	0.044	0.064	0.086	0.115	0.153	0.207
		3.6	0.1	0.005	0.009	0.014	0.020	0.025	0.031	0.037	0.044
			0.3	0.014	0.029	0.046	0.067	0.091	0.123	0.165	0.230
			0.5	0.022	0.049	0.083	0.127	0.196	0.321	0.620	1.480
			0.7	0.031	0.070	0.124	0.210	0.380	0.818	2.103	5.926
	5	3.2	0.1	0.002	0.005	0.007	0.009	0.011	0.014	0.016	0.018
			0.3	0.007	0.014	0.020	0.027	0.035	0.042	0.050	0.057
			0.5	0.011	0.022	0.034	0.046	0.059	0.072	0.086	0.100
			0.7	0.015	0.031	0.048	0.066	0.084	0.104	0.125	0.148
		3.4	0.1	0.005	0.009	0.014	0.019	0.024	0.029	0.034	0.039
			0.3	0.013	0.027	0.042	0.058	0.074	0.092	0.110	0.130
			0.5	0.023	0.047	0.073	0.101	0.133	0.168	0.209	0.258
			0.7	0.031	0.066	0.105	0.148	0.200	0.264	0.347	0.469
		3.6	0.1	0.011	0.022	0.033	0.045	0.058	0.071	0.085	0.099
			0.3	0.031	0.066	0.106	0.151	0.206	0.274	0.365	0.502
			0.5	0.051	0.113	0.188	0.288	0.434	0.694	1.303	3.018
			0.7	0.071	0.160	0.281	0.465	0.817	1.696	4.196	-
	7	3.2	0.1	0.004	0.008	0.013	0.017	0.021	0.025	0.030	0.034
			0.3	0.012	0.024	0.037	0.050	0.064	0.077	0.091	0.106
			0.5	0.021	0.042	0.064	0.086	0.110	0.134	0.159	0.186
			0.7	0.028	0.058	0.090	0.122	0.157	0.194	0.233	0.276
		3.4	0.1	0.008	0.017	0.025	0.034	0.043	0.052	0.061	0.070
			0.3	0.025	0.051	0.078	0.107	0.137	0.169	0.203	0.240
			0.5	0.041	0.085	0.133	0.185	0.243	0.308	0.383	0.472
			0.7	0.057	0.120	0.191	0.272	0.366	0.483	0.636	0.864
		3.58	0.1	0.018	0.036	0.055	0.074	0.095	0.116	0.138	0.161
			0.3	0.053	0.111	0.174	0.246	0.330	0.430	0.557	0.732
			0.5	0.086	0.186	0.306	0.457	0.666	1.001	1.690	3.499
			0.7	0.119	0.263	0.452	0.723	1.194	2.260	5.172	-

Table A.4 *Differential swell occurring at the perimeter of a slab for a center lift swelling condition in predominantly Kaolinite clay soil (50 percent clay).*

Percent Clay (%)	Depth to Constant Suction (FT)	Constant Suction (pF)	Velocity of Moisture Flow (inches /month)	DIFFERENTIAL SWELL (INCHES) EDGE DISTANCE PENETRATION (FT)							
				1 FT	2 FT	3 FT	4 FT	5 FT	6 FT	7 FT	8 FT
60	3	3.2	0.1	0.001	0.003	0.004	0.005	0.006	0.007	0.008	0.010
			0.3	0.003	0.007	0.010	0.014	0.018	0.021	0.025	0.029
			0.5	0.006	0.012	0.018	0.024	0.030	0.037	0.044	0.052
			0.7	0.008	0.017	0.025	0.034	0.044	0.054	0.065	0.077
		3.4	0.1	0.002	0.005	0.007	0.010	0.012	0.015	0.017	0.020
			0.3	0.007	0.014	0.022	0.030	0.038	0.047	0.057	0.068
			0.5	0.012	0.024	0.038	0.053	0.069	0.088	0.110	0.136
			0.7	0.016	0.033	0.053	0.077	0.104	0.138	0.184	0.249
		3.6	0.1	0.006	0.012	0.018	0.024	0.031	0.038	0.045	0.053
			0.3	0.017	0.035	0.056	0.081	0.110	0.148	0.199	0.278
			0.5	0.026	0.059	0.099	0.154	0.236	0.386	0.745	1.779
			0.7	0.036	0.083	0.149	0.252	0.455	0.983	2.527	7.124
	5	3.2	0.1	0.003	0.005	0.008	0.011	0.014	0.016	0.019	0.022
			0.3	0.008	0.016	0.025	0.033	0.042	0.051	0.060	0.069
			0.5	0.013	0.027	0.041	0.056	0.071	0.087	0.103	0.120
			0.7	0.019	0.038	0.058	0.080	0.102	0.126	0.151	0.178
		3.4	0.1	0.006	0.011	0.017	0.023	0.029	0.034	0.041	0.047
			0.3	0.016	0.033	0.051	0.069	0.089	0.110	0.132	0.156
			0.5	0.028	0.056	0.087	0.122	0.160	0.202	0.252	0.310
			0.7	0.037	0.078	0.125	0.177	0.240	0.316	0.417	0.564
		3.6	0.1	0.013	0.027	0.040	0.055	0.070	0.086	0.102	0.120
			0.3	0.038	0.080	0.077	0.182	0.248	0.329	0.439	0.604
			0.5	0.062	0.135	0.226	0.345	0.521	0.834	1.566	3.628
			0.7	0.086	0.193	0.337	0.559	0.982	2.039	5.049	-
	7	3.2	0.1	0.005	0.010	0.016	0.021	0.026	0.031	0.036	0.041
			0.3	0.015	0.030	0.046	0.061	0.078	0.094	0.111	0.128
			0.5	0.025	0.050	0.077	0.104	0.132	0.161	0.192	0.224
			0.7	0.034	0.070	0.108	0.147	0.189	0.233	0.281	0.332
		3.4	0.1	0.010	0.020	0.030	0.041	0.052	0.062	0.073	0.085
			0.3	0.030	0.061	0.093	0.128	0.164	0.202	0.243	0.287
			0.5	0.050	0.103	0.160	0.223	0.292	0.371	0.461	0.568
			0.7	0.069	0.144	0.229	0.326	0.440	0.580	0.765	1.038
		3.58	0.1	0.021	0.043	0.066	0.089	0.114	0.139	0.166	0.194
			0.3	0.063	0.131	0.208	0.295	0.396	0.516	0.669	0.879
			0.5	0.103	0.223	0.367	0.549	0.800	1.203	2.031	4.205
			0.7	0.142	0.316	0.543	0.870	1.436	2.717	6.217	-

Table A.5 *Differential swell occurring at the perimeter of a slab for a center lift swelling condition in predominantly Kaolinite clay soil (60 percent clay).*

Percent Clay (%)	Depth to Constant Suction (FT)	Constant Suction (pF)	Velocity of Moisture Flow (inches /month)	DIFFERENTIAL SWELL (INCHES) EDGE DISTANCE PENETRATION (FT) 1 FT	2 FT	3 FT	4 FT	5 FT	6 FT	7 FT	8 FT
70	3	3.2	0.1	0.001	0.003	0.004	0.005	0.007	0.008	0.010	0.011
			0.3	0.004	0.008	0.012	0.016	0.021	0.025	0.029	0.034
			0.5	0.006	0.013	0.020	0.027	0.035	0.042	0.051	0.060
			0.7	0.010	0.019	0.029	0.040	0.051	0.063	0.076	0.089
		3.4	0.1	0.003	0.005	0.008	0.011	0.014	0.017	0.020	0.023
			0.3	0.008	0.017	0.026	0.035	0.045	0.056	0.067	0.079
			0.5	0.014	0.028	0.044	0.062	0.081	0.103	0.129	0.159
			0.7	0.019	0.039	0.063	0.090	0.122	0.162	0.215	0.292
		3.6	0.1	0.006	0.013	0.020	0.027	0.035	0.043	0.052	0.061
			0.3	0.019	0.041	0.065	0.094	0.128	0.172	0.232	0.324
			0.5	0.032	0.069	0.117	0.180	0.276	0.451	0.871	2.079
			0.7	0.043	0.098	0.174	0.294	0.532	1.148	2.952	8.322
	5	3.2	0.1	0.003	0.006	0.009	0.013	0.016	0.019	0.022	0.026
			0.3	0.010	0.019	0.029	0.039	0.049	0.059	0.070	0.080
			0.5	0.016	0.032	0.048	0.065	0.083	0.101	0.120	0.140
			0.7	0.022	0.044	0.068	0.093	0.119	0.146	0.176	0.208
		3.4	0.1	0.006	0.012	0.019	0.026	0.033	0.039	0.046	0.054
			0.3	0.018	0.038	0.059	0.081	0.104	0.128	0.155	0.183
			0.5	0.031	0.065	0.101	0.141	0.185	0.235	0.293	0.361
			0.7	0.043	0.092	0.146	0.207	0.280	0.370	0.487	0.659
		3.6	0.1	0.015	0.030	0.046	0.063	0.081	0.099	0.119	0.139
			0.3	0.045	0.094	0.149	0.213	0.290	0.385	0.512	0.706
			0.5	0.072	0.158	0.264	0.404	0.609	0.974	1.830	4.239
			0.7	0.100	0.225	0.394	0.653	1.147	2.381	5.893	-
	7	3.2	0.1	0.006	0.011	0.017	0.023	0.029	0.035	0.041	0.047
			0.3	0.018	0.035	0.053	0.072	0.090	0.110	0.129	0.149
			0.5	0.030	0.059	0.090	0.121	0.154	0.188	0.224	0.262
			0.7	0.040	0.082	0.126	0.172	0.221	0.273	0.328	0.388
		3.4	0.1	0.012	0.024	0.036	0.048	0.060	0.073	0.086	0.099
			0.3	0.035	0.071	0.109	0.149	0.192	0.237	0.284	0.336
			0.5	0.057	0.119	0.186	0.260	0.341	0.432	0.538	0.663
			0.7	0.080	0.168	0.267	0.381	0.514	0.678	0.893	1.213
		3.58	0.1	0.025	0.050	0.077	0.104	0.133	0.163	0.194	0.227
			0.3	0.073	0.154	0.244	0.345	0.463	0.604	0.782	1.028
			0.5	0.120	0.260	0.429	0.642	0.935	1.406	2.373	4.913
			0.7	0.166	0.369	0.634	1.016	1.677	3.175	7.263	-

Table A.6 *Differential swell occurring at the perimeter of a slab for a center lift swelling condition in predominantly Kaolinite clay soil (70 percent clay).*

Percent Clay (%)	Depth to Constant Suction (FT)	Constant Suction (pF)	Velocity of Moisture Flow (inches /month)	DIFFERENTIAL SWELL (INCHES) EDGE DISTANCE PENETRATION (FT) 1 FT	2 FT	3 FT	4 FT	5 FT	6 FT	7 FT	8 FT
30	**3**	**3.2**	**0.1**	**0.002**	**0.003**	**0.004**	**0.005**	**0.006**	**0.007**	**0.008**	**0.010**
			0.3	**0.003**	**0.007**	**0.010**	**0.014**	**0.017**	**0.021**	**0.025**	**0.029**
			0.5	**0.006**	**0.012**	**0.018**	**0.024**	**0.030**	**0.037**	**0.044**	**0.051**
			0.7	**0.008**	**0.016**	**0.024**	**0.033**	**0.043**	**0.053**	**0.064**	**0.075**
		3.4	0.1	0.003	0.005	0.007	0.010	0.012	0.015	0.017	0.020
			0.3	0.007	0.014	0.022	0.030	0.038	0.047	0.057	0.067
			0.5	0.011	0.024	0.037	0.052	0.068	0.087	0.109	0.135
			0.7	0.016	0.034	0.054	0.077	0.104	0.138	0.182	0.248
		3.6	0.1	0.005	0.011	0.017	0.023	0.030	0.037	0.044	0.051
			0.3	0.016	0.035	0.055	0.080	0.109	0.146	0.197	0.274
			0.5	0.027	0.058	0.098	0.152	0.234	0.382	0.737	1.760
			0.7	0.036	0.083	0.147	0.249	0.451	0.973	2.500	7.049
	5	3.2	0.1	0.003	0.006	0.008	0.011	0.014	0.016	0.019	0.022
			0.3	0.008	0.016	0.024	0.032	0.041	0.050	0.059	0.068
			0.5	0.013	0.027	0.041	0.055	0.070	0.086	0.102	0.119
			0.7	0.018	0.037	0.057	0.078	0.100	0.124	0.149	0.176
		3.4	0.1	0.005	0.011	0.016	0.022	0.028	0.034	0.040	0.046
			0.3	0.016	0.033	0.051	0.069	0.089	0.109	0.131	0.155
			0.5	0.027	0.055	0.086	0.120	0.157	0.200	0.248	0.306
			0.7	0.037	0.078	0.124	0.176	0.238	0.319	0.413	0.558
		3.6	0.1	0.012	0.025	0.039	0.053	0.068	0.084	0.100	0.118
			0.3	0.037	0.079	0.126	0.180	0.245	0.326	0.434	0.598
			0.5	0.062	0.134	0.224	0.342	0.516	0.825	1.551	3.591
			0.7	0.084	0.190	0.333	0.553	0.971	2.016	4.991	-
	7	3.2	0.1	0.005	0.010	0.015	0.020	0.025	0.030	0.035	0.041
			0.3	0.015	0.030	0.045	0.061	0.076	0.093	0.109	0.126
			0.5	0.025	0.050	0.076	0.103	0.131	0.160	0.190	0.221
			0.7	0.034	0.070	0.107	0.146	0.187	0.231	0.278	0.329
		3.4	0.1	0.010	0.020	0.030	0.041	0.051	0.062	0.073	0.084
			0.3	0.029	0.060	0.092	0.126	0.162	0.200	0.241	0.284
			0.5	0.048	0.102	0.158	0.221	0.288	0.367	0.456	0.562
			0.7	0.068	0.143	0.227	0.323	0.436	0.574	0.757	1.028
		3.58	0.1	0.021	0.042	0.065	0.088	0.113	0.138	0.164	0.191
			0.3	0.062	0.131	0.207	0.292	0.392	0.511	0.662	0.870
			0.5	0.103	0.221	0.363	0.543	0.792	1.191	2.010	4.162
			0.7	0.141	0.313	0.537	0.861	1.421	2.689	6.153	-

Table A.7 *Differential swell occurring at the perimeter of a slab for a center lift swelling condition in predominantly Illite clay soil (30 percent clay).*

Percent Clay (%)	Depth to Constant Suction (FT)	Constant Suction (pF)	Velocity of Moisture Flow (inches /month)	DIFFERENTIAL SWELL (INCHES) EDGE DISTANCE PENETRATION (FT)							
				1 FT	2 FT	3 FT	4 FT	5 FT	6 FT	7 FT	8 FT
40	3	3.2	0.1	0.002	0.003	0.005	0.007	0.008	0.010	0.012	0.014
			0.3	0.005	0.010	0.015	0.020	0.026	0.031	0.037	0.042
			0.5	0.008	0.016	0.025	0.034	0.043	0.052	0.063	0.073
			0.7	0.011	0.023	0.035	0.048	0.062	0.076	0.092	0.109
		3.4	0.1	0.003	0.006	0.010	0.013	0.017	0.020	0.024	0.028
			0.3	0.010	0.021	0.032	0.043	0.055	0.068	0.082	0.098
			0.5	0.017	0.034	0.054	0.075	0.099	0.126	0.157	0.194
			0.7	0.023	0.048	0.077	0.110	0.149	0.198	0.263	0.357
		3.6	0.1	0.008	0.017	0.025	0.034	0.044	0.054	0.064	0.075
			0.3	0.023	0.050	0.080	0.115	0.157	0.211	0.284	0.396
			0.5	0.039	0.085	0.142	0.220	0.338	0.552	1.065	2.542
			0.7	0.053	0.120	0.213	0.360	0.651	1.405	3.611	-
	5	3.2	0.1	0.004	0.007	0.011	0.015	0.019	0.023	0.027	0.031
			0.3	0.012	0.024	0.035	0.047	0.060	0.072	0.085	0.098
			0.5	0.019	0.039	0.059	0.080	0.101	0.124	0.147	0.172
			0.7	0.026	0.054	0.083	0.113	0.145	0.179	0.215	0.254
		3.4	0.1	0.007	0.015	0.023	0.031	0.040	0.048	0.057	0.065
			0.3	0.023	0.047	0.073	0.099	0.128	0.157	0.189	0.224
			0.5	0.039	0.080	0.124	0.173	0.227	0.288	0.358	0.442
			0.7	0.053	0.112	0.178	0.254	0.343	0.452	0.596	0.805
		3.6	0.1	0.018	0.037	0.057	0.077	0.099	0.121	0.145	0.170
			0.3	0.054	0.114	0.182	0.261	0.354	0.471	0.627	0.863
			0.5	0.089	0.194	0.323	0.494	0.745	1.192	2.239	5.184
			0.7	0.122	0.275	0.482	0.799	1.403	2.912	7.207	-
	7	3.2	0.1	0.007	0.014	0.021	0.029	0.036	0.043	0.051	0.058
			0.3	0.022	0.043	0.065	0.088	0.111	0.134	0.158	0.183
			0.5	0.035	0.072	0.109	0.148	0.188	0.230	0.274	0.320
			0.7	0.049	0.100	0.154	0.210	0.270	0.330	0.401	0.474
		3.4	0.1	0.015	0.030	0.045	0.060	0.075	0.091	0.106	0.122
			0.3	0.042	0.087	0.134	0.183	0.234	0.289	0.348	0.410
			0.5	0.070	0.146	0.228	0.318	0.417	0.528	0.657	0.811
			0.7	0.098	0.207	0.328	0.466	0.629	0.829	1.093	1.434
		3.58	0.1	0.030	0.062	0.094	0.128	0.163	0.199	0.236	0.277
			0.3	0.090	0.188	0.298	0.422	0.566	0.738	0.956	1.256
			0.5	0.147	0.318	0.524	0.784	1.143	1.719	2.902	6.008
			0.7	0.203	0.451	0.775	1.242	2.051	3.883	8.882	-

Table A.8 *Differential swell occurring at the perimeter of a slab for a center lift swelling condition in predominantly Illite clay soil (40 percent clay).*

Percent Clay (%)	Depth to Constant Suction (FT)	Constant Suction (pF)	Velocity of Moisture Flow (inches /month)	DIFFERENTIAL SWELL (INCHES) EDGE DISTANCE PENETRATION (FT)							
				1 FT	2 FT	3 FT	4 FT	5 FT	6 FT	7 FT	8 FT
50	3	3.2	0.1	0.002	0.004	0.006	0.008	0.010	0.013	0.015	0.017
			0.3	0.006	0.013	0.019	0.026	0.033	0.040	0.047	0.055
			0.5	0.010	0.021	0.033	0.042	0.056	0.069	0.082	0.096
			0.7	0.014	0.030	0.046	0.063	0.081	0.099	0.120	0.142
		3.4	0.1	0.004	0.009	0.013	0.018	0.022	0.027	0.032	0.037
			0.3	0.013	0.026	0.041	0.056	0.072	0.089	0.107	0.127
			0.5	0.022	0.045	0.070	0.098	0.129	0.164	0.205	0.254
			0.7	0.030	0.063	0.101	0.144	0.196	0.260	0.344	0.467
		3.6	0.1	0.011	0.021	0.033	0.045	0.057	0.070	0.084	0.098
			0.3	0.031	0.065	0.105	0.150	0.206	0.276	0.372	0.518
			0.5	0.050	0.111	0.185	0.287	0.441	0.721	1.391	3.322
			0.7	0.069	0.156	0.278	0.470	0.851	1.836	4.720	-
	5	3.2	0.1	0.005	0.010	0.015	0.020	0.026	0.031	0.036	0.041
			0.3	0.015	0.030	0.046	0.062	0.078	0.094	0.111	0.128
			0.5	0.025	0.051	0.077	0.104	0.133	0.162	0.193	0.225
			0.7	0.035	0.071	0.110	0.148	0.190	0.235	0.282	0.333
		3.4	0.1	0.010	0.021	0.031	0.042	0.053	0.064	0.075	0.086
			0.3	0.031	0.062	0.096	0.131	0.167	0.207	0.248	0.293
			0.5	0.051	0.104	0.163	0.227	0.298	0.377	0.469	0.579
			0.7	0.070	0.147	0.233	0.332	0.449	0.592	0.779	1.054
		3.6	0.1	0.025	0.049	0.075	0.102	0.130	0.160	0.191	0.224
			0.3	0.071	0.150	0.239	0.341	0.463	0.615	0.820	1.129
			0.5	0.116	0.253	0.423	0.646	0.974	1.558	2.927	6.778
			0.7	0.159	0.359	0.629	1.043	1.834	3.807	9.422	-
	7	3.2	0.1	0.009	0.018	0.028	0.037	0.047	0.056	0.066	0.076
			0.3	0.028	0.056	0.085	0.114	0.144	0.175	0.206	0.238
			0.5	0.046	0.094	0.143	0.193	0.246	0.301	0.358	0.418
			0.7	0.064	0.131	0.201	0.275	0.353	0.436	0.524	0.620
		3.4	0.1	0.019	0.038	0.057	0.077	0.097	0.117	0.138	0.159
			0.3	0.055	0.114	0.175	0.239	0.307	0.378	0.455	0.537
			0.5	0.092	0.192	0.299	0.416	0.546	0.692	0.860	1.061
			0.7	0.129	0.271	0.429	0.610	0.823	1.085	1.429	1.940
		3.58	0.1	0.040	0.081	0.123	0.167	0.213	0.261	0.310	0.362
			0.3	0.117	0.246	0.389	0.551	0.739	0.965	1.250	1.642
			0.5	0.193	0.417	0.685	1.026	1.495	2.248	3.795	7.856
			0.7	0.265	0.590	1.013	1.624	2.682	5.075	-	-

Table A.9 *Differential swell occurring at the perimeter of a slab for a center lift swelling condition in predominantly Illite clay soil (50 percent clay).*

Percent Clay (%)	Depth to Constant Suction (FT)	Constant Suction (pF)	Velocity of Moisture Flow (inches /month)	DIFFERENTIAL SWELL (INCHES) EDGE DISTANCE PENETRATION (FT) 1 FT	2 FT	3 FT	4 FT	5 FT	6 FT	7 FT	8 FT
60	3	3.2	0.1	0.003	0.006	0.008	0.011	0.014	0.016	0.019	0.022
			0.3	0.008	0.016	0.024	0.033	0.041	0.050	0.059	0.068
			0.5	0.013	0.027	0.040	0.055	0.070	0.085	0.102	0.119
			0.7	0.018	0.037	0.057	0.078	0.100	0.123	0.148	0.176
		3.4	0.1	0.006	0.011	0.017	0.022	0.028	0.034	0.040	0.046
			0.3	0.016	0.033	0.051	0.069	0.089	0.110	0.133	0.157
			0.5	0.027	0.055	0.087	0.121	0.159	0.203	0.253	0.314
			0.7	0.037	0.078	0.124	0.178	0.241	0.320	0.424	0.576
		3.6	0.1	0.013	0.026	0.040	0.055	0.070	0.086	0.103	0.121
			0.3	0.038	0.080	0.129	0.185	0.254	0.340	0.459	0.639
			0.5	0.062	0.136	0.229	0.355	0.544	0.890	1.659	4.104
			0.7	0.085	0.194	0.344	0.582	1.052	2.268	5.831	-
	5	3.2	0.1	0.006	0.012	0.018	0.025	0.031	0.038	0.044	0.050
			0.3	0.018	0.037	0.056	0.075	0.095	0.116	0.137	0.158
			0.5	0.031	0.062	0.095	0.129	0.164	0.200	0.238	0.277
			0.7	0.043	0.088	0.134	0.183	0.235	0.290	0.348	0.411
		3.4	0.1	0.012	0.025	0.038	0.051	0.065	0.078	0.092	0.106
			0.3	0.038	0.077	0.118	0.161	0.206	0.255	0.306	0.361
			0.5	0.062	0.129	0.201	0.280	0.367	0.466	0.579	0.714
			0.7	0.086	0.181	0.288	0.410	0.554	0.730	0.962	1.301
		3.6	0.1	0.030	0.061	0.093	0.126	0.161	0.197	0.236	0.276
			0.3	0.088	0.185	0.295	0.421	0.572	0.761	1.013	1.394
			0.5	0.144	0.313	0.522	0.797	1.203	1.924	3.615	8.371
			0.7	0.197	0.444	0.778	1.289	2.265	4.702	-	-
	7	3.2	0.1	0.012	0.023	0.035	0.047	0.058	0.070	0.082	0.094
			0.3	0.035	0.069	0.105	0.141	0.178	0.216	0.255	0.294
			0.5	0.057	0.116	0.176	0.239	0.304	0.372	0.442	0.516
			0.7	0.079	0.162	0.249	0.340	0.436	0.538	0.647	0.766
		3.4	0.1	0.023	0.047	0.071	0.095	0.120	0.145	0.170	0.196
			0.3	0.069	0.141	0.216	0.294	0.379	0.467	0.562	0.663
			0.5	0.114	0.237	0.369	0.514	0.674	0.855	1.063	1.310
			0.7	0.158	0.333	0.528	0.752	1.015	1.340	1.764	2.395
		3.58	0.1	0.050	0.100	0.153	0.207	0.263	0.322	0.383	0.447
			0.3	0.145	0.304	0.481	0.681	0.912	1.192	1.543	2.028
			0.5	0.238	0.514	0.846	1.266	1.846	2.776	4.687	9.702
			0.7	0.328	0.730	1.252	2.006	3.312	6.268	-	-

Table A.10 *Differential swell occurring at the perimeter of a slab for a center lift swelling condition in predominantly Illite clay soil (60 percent clay).*

Percent Clay (%)	Depth to Constant Suction (FT)	Constant Suction (pF)	Velocity of Moisture Flow (inches /month)	DIFFERENTIAL SWELL (INCHES) EDGE DISTANCE PENETRATION (FT) 1 FT	2 FT	3 FT	4 FT	5 FT	6 FT	7 FT	8 FT
70	3	3.2	0.1	0.003	0.006	0.009	0.013	0.016	0.019	0.022	0.026
			0.3	0.009	0.019	0.028	0.038	0.048	0.059	0.069	0.080
			0.5	0.016	0.032	0.048	0.066	0.083	0.102	0.121	0.141
			0.7	0.021	0.044	0.068	0.092	0.119	0.147	0.177	0.209
		3.4	0.1	0.007	0.013	0.020	0.027	0.034	0.041	0.047	0.055
			0.3	0.019	0.039	0.061	0.083	0.106	0.131	0.158	0.187
			0.5	0.032	0.066	0.103	0.144	0.190	0.242	0.302	0.374
			0.7	0.044	0.093	0.148	0.212	0.287	0.381	0.510	0.686
		3.6	0.1	0.016	0.032	0.049	0.066	0.084	0.103	0.124	0.145
			0.3	0.045	0.095	0.153	0.220	0.302	0.405	0.545	0.761
			0.5	0.074	0.162	0.273	0.423	0.648	1.061	2.047	4.885
			0.7	0.101	0.231	0.409	0.692	1.251	2.700	6.940	-
	5	3.2	0.1	0.008	0.015	0.022	0.030	0.037	0.045	0.035	0.060
			0.3	0.022	0.044	0.067	0.090	0.114	0.138	0.163	0.189
			0.5	0.037	0.074	0.113	0.153	0.195	0.238	0.283	0.330
			0.7	0.051	0.104	0.159	0.218	0.279	0.345	0.414	0.489
		3.4	0.1	0.015	0.030	0.046	0.062	0.077	0.094	0.110	0.127
			0.3	0.044	0.091	0.140	0.191	0.245	0.303	0.364	0.430
			0.5	0.073	0.153	0.239	0.330	0.437	0.554	0.689	0.850
			0.7	0.102	0.215	0.342	0.487	0.659	0.869	1.145	1.548
		3.6	0.1	0.035	0.071	0.109	0.149	0.191	0.234	0.279	0.327
			0.3	0.104	0.219	0.350	0.501	0.680	0.904	1.204	1.659
			0.5	0.170	0.372	0.620	0.948	1.431	2.290	4.302	9.964
			0.7	0.234	0.528	0.925	1.534	2.696	5.597	-	-
	7	3.2	0.1	0.014	0.027	0.041	0.055	0.069	0.083	0.097	0.112
			0.3	0.040	0.082	0.124	0.167	0.212	0.257	0.303	0.350
			0.5	0.068	0.138	0.210	0.284	0.362	0.442	0.526	0.614
			0.7	0.094	0.193	0.296	0.404	0.519	0.640	0.771	0.911
		3.4	0.1	0.028	0.056	0.084	0.113	0.142	0.173	0.202	0.233
			0.3	0.082	0.167	0.257	0.352	0.451	0.556	0.669	0.789
			0.5	0.136	0.282	0.439	0.612	0.803	1.018	1.265	1.560
			0.7	0.188	0.396	0.629	0.895	1.209	1.594	2.100	2.851
		3.58	0.1	0.058	0.118	0.181	0.246	0.313	0.383	0.456	0.532
			0.3	0.173	0.362	0.573	0.812	1.088	1.419	1.838	2.415
			0.5	0.283	0.612	1.007	1.507	2.197	3.304	5.579	11.549
			0.7	0.391	0.869	1.490	2.388	3.943	7.462	-	-

Table A.11 *Differential swell occurring at the perimeter of a slab for a center lift swelling condition in predominantly Illite clay soil (70 percent clay).*

Percent Clay (%)	Depth to Constant Suction (FT)	Constant Suction (pF)	Velocity of Moisture Flow (inches /month)	DIFFERENTIAL SWELL (INCHES) EDGE DISTANCE PENETRATION (FT)							
				1 FT	2 FT	3 FT	4 FT	5 FT	6 FT	7 FT	8 FT
30	3	3.2	0.1	0.001	0.003	0.004	0.005	0.007	0.008	0.009	0.011
			0.3	0.004	0.009	0.012	0.016	0.020	0.025	0.030	0.035
			0.5	0.006	0.013	0.020	0.027	0.035	0.043	0.051	0.060
			0.7	0.010	0.019	0.029	0.040	0.051	0.063	0.076	0.089
		3.4	0.1	0.003	0.005	0.008	0.011	0.014	0.017	0.020	0.023
			0.3	0.008	0.017	0.026	0.035	0.045	0.056	0.067	0.079
			0.5	0.014	0.028	0.044	0.062	0.081	0.102	0.128	0.159
			0.7	0.019	0.039	0.063	0.089	0.122	0.162	0.214	0.291
		3.6	0.1	0.006	0.013	0.020	0.027	0.035	0.043	0.052	0.061
			0.3	0.019	0.041	0.065	0.094	0.128	0.172	0.232	0.323
			0.5	0.031	0.069	0.116	0.179	0.275	0.450	0.868	2.073
			0.7	0.043	0.098	0.173	0.294	0.532	1.145	2.945	8.301
	5	3.2	0.1	0.003	0.006	0.010	0.013	0.016	0.019	0.022	0.026
			0.3	0.009	0.018	0.028	0.038	0.048	0.058	0.069	0.079
			0.5	0.016	0.032	0.048	0.065	0.083	0.101	0.120	0.140
			0.7	0.022	0.044	0.068	0.093	0.119	0.147	0.176	0.208
		3.4	0.1	0.006	0.013	0.019	0.026	0.033	0.039	0.046	0.054
			0.3	0.019	0.039	0.060	0.081	0.104	0.129	0.155	0.182
			0.5	0.032	0.065	0.102	0.141	0.186	0.235	0.293	0.361
			0.7	0.043	0.091	0.145	0.207	0.280	0.369	0.486	0.657
		3.6	0.1	0.015	0.030	0.047	0.063	0.081	0.099	0.119	0.139
			0.3	0.045	0.094	0.149	0.213	0.289	0.384	0.512	0.704
			0.5	0.072	0.158	0.264	0.403	0.607	0.972	1.826	2.514
			0.7	0.100	0.224	0.393	0.651	1.144	2.375	5.870	-
	7	3.2	0.1	0.006	0.012	0.017	0.023	0.029	0.035	0.041	0.047
			0.3	0.017	0.035	0.053	0.071	0.090	0.109	0.128	0.148
			0.5	0.029	0.059	0.089	0.121	0.154	0.188	0.223	0.261
			0.7	0.039	0.081	0.125	0.171	0.219	0.271	0.326	0.386
		3.4	0.1	0.012	0.024	0.036	0.048	0.061	0.074	0.086	0.099
			0.3	0.035	0.071	0.109	0.149	0.191	0.236	0.284	0.335
			0.5	0.058	0.120	0.187	0.261	0.342	0.433	0.537	0.662
			0.7	0.081	0.169	0.268	0.381	0.514	0.677	0.892	1.211
		3.58	0.1	0.024	0.050	0.076	0.104	0.132	0.162	0.193	0.225
			0.3	0.073	0.154	0.243	0.344	0.461	0.602	0.780	1.025
			0.5	0.120	0.259	0.427	0.639	0.932	1.402	2.367	4.900
			0.7	0.165	0.368	0.632	1.013	1.672	3.165	7.244	-

Table A.12 *Differential swell occurring at the perimeter of a slab for a center lift swelling condition in predominantly Montmorillonite clay soil (30 percent clay).*

Percent Clay (%)	Depth to Constant Suction (FT)	Constant Suction (pF)	Velocity of Moisture Flow (inches /month)	DIFFERENTIAL SWELL (INCHES) EDGE DISTANCE PENETRATION (FT)							
				1 FT	2 FT	3 FT	4 FT	5 FT	6 FT	7 FT	8 FT
40	3	3.2	0.1	0.002	0.004	0.006	0.008	0.010	0.012	0.014	0.016
			0.3	0.006	0.012	0.018	0.024	0.030	0.036	0.042	0.049
			0.5	0.009	0.019	0.029	0.040	0.050	0.062	0.074	0.086
			0.7	0.014	0.027	0.042	0.057	0.074	0.091	0.109	0.129
		3.4	0.1	0.004	0.008	0.012	0.016	0.021	0.025	0.029	0.034
			0.3	0.011	0.024	0.037	0.050	0.065	0.080	0.097	0.115
			0.5	0.019	0.040	0.063	0.088	0.116	0.148	0.185	0.229
			0.7	0.027	0.057	0.091	0.130	0.177	0.234	0.311	0.422
		3.6	0.1	0.010	0.019	0.030	0.040	0.051	0.063	0.076	0.089
			0.3	0.028	0.059	0.094	0.136	0.186	0.249	0.335	0.468
			0.5	0.046	0.100	0.168	0.260	0.399	0.652	1.258	3.004
			0.7	0.062	0.142	0.251	0.425	0.769	1.660	4.267	7.762
	5	3.2	0.1	0.004	0.009	0.013	0.018	0.022	0.027	0.032	0.037
			0.3	0.013	0.027	0.041	0.055	0.069	0.085	0.100	0.115
			0.5	0.023	0.046	0.069	0.094	0.120	0.147	0.174	0.203
			0.7	0.032	0.064	0.098	0.134	0.172	0.212	0.255	0.301
		3.4	0.1	0.010	0.019	0.028	0.038	0.048	0.058	0.068	0.079
			0.3	0.027	0.056	0.086	0.118	0.151	0.186	0.224	0.264
			0.5	0.045	0.094	0.147	0.205	0.269	0.341	0.424	0.522
			0.7	0.063	0.133	0.211	0.300	0.405	0.535	0.704	0.952
		3.6	0.1	0.021	0.044	0.067	0.092	0.117	0.144	0.172	0.201
			0.3	0.064	0.135	0.216	0.308	0.419	0.556	0.741	1.020
			0.5	0.105	0.229	0.382	0.584	0.880	1.408	2.645	6.127
			0.7	0.144	0.325	0.569	0.944	1.658	3.441	8.517	-
	7	3.2	0.1	0.009	0.018	0.026	0.035	0.043	0.052	0.061	0.070
			0.3	0.025	0.051	0.077	0.103	0.131	0.158	0.186	0.215
			0.5	0.042	0.085	0.129	0.175	0.223	0.272	0.324	0.378
			0.7	0.059	0.119	0.183	0.249	0.320	0.394	0.475	0.561
		3.4	0.1	0.017	0.035	0.052	0.070	0.088	0.106	0.125	0.144
			0.3	0.050	0.103	0.158	0.216	0.277	0.342	0.411	0.485
			0.5	0.084	0.173	0.270	0.377	0.494	0.626	0.778	0.960
			0.7	0.116	0.244	0.387	0.550	0.743	0.980	1.291	1.753
		3.56	0.1	0.033	0.066	0.101	0.136	0.173	0.211	0.251	0.292
			0.3	0.096	0.201	0.316	0.444	0.590	0.760	0.967	1.231
			0.5	0.158	0.339	0.553	0.815	1.160	1.668	2.583	4.748
			0.7	0.218	0.480	0.813	1.271	2.004	3.504	7.412	-

Table A.13 *Differential swell occurring at the perimeter of a slab for a center lift swelling condition in predominantly Montmorillonite clay soil (40 percent clay).*

Percent Clay (%)	Depth to Constant Suction (FT)	Constant Suction (pF)	Velocity of Moisture Flow (inches/month)	DIFFERENTIAL SWELL (INCHES) EDGE DISTANCE PENETRATION (FT) 1 FT	2 FT	3 FT	4 FT	5 FT	6 FT	7 FT	8 FT
50	3	3.2	0.1	0.003	0.005	0.008	0.010	0.013	0.016	0.018	0.021
			0.3	0.007	0.015	0.023	0.031	0.039	0.047	0.055	0.065
			0.5	0.012	0.026	0.038	0.052	0.066	0.081	0.097	0.113
			0.7	0.018	0.036	0.055	0.075	0.096	0.119	0.143	0.169
		3.4	0.1	0.006	0.011	0.016	0.022	0.027	0.033	0.038	0.044
			0.3	0.016	0.032	0.049	0.067	0.086	0.106	0.127	0.151
			0.5	0.026	0.053	0.083	0.116	0.153	0.195	0.243	0.301
			0.7	0.036	0.075	0.119	0.170	0.231	0.307	0.407	0.553
		3.6	0.1	0.012	0.025	0.038	0.052	0.067	0.082	0.099	0.116
			0.3	0.036	0.077	0.123	0.177	0.243	0.326	0.439	0.613
			0.5	0.059	0.131	0.220	0.340	0.522	0.854	1.648	3.935
			0.7	0.082	0.186	0.330	0.558	1.008	2.175	5.590	-
	5	3.2	0.1	0.006	0.012	0.018	0.024	0.030	0.036	0.042	0.048
			0.3	0.017	0.035	0.054	0.073	0.092	0.111	0.131	0.151
			0.5	0.030	0.060	0.091	0.124	0.157	0.192	0.228	0.266
			0.7	0.041	0.083	0.128	0.175	0.224	0.277	0.333	0.394
		3.4	0.1	0.012	0.024	0.037	0.049	0.062	0.075	0.089	0.102
			0.3	0.036	0.074	0.113	0.154	0.198	0.244	0.294	0.344
			0.5	0.059	0.123	0.193	0.268	0.352	0.446	0.555	0.684
			0.7	0.083	0.174	0.276	0.393	0.531	0.700	0.923	1.247
		3.6	0.1	0.029	0.058	0.089	0.121	0.154	0.189	0.226	0.264
			0.3	0.084	0.177	0.282	0.404	0.548	0.728	0.970	1.336
			0.5	0.137	0.299	0.500	0.764	1.152	1.844	3.464	8.025
			0.7	0.189	0.426	0.745	1.236	2.712	4.508	-	-
	7	3.2	0.1	0.011	0.022	0.034	0.044	0.056	0.067	0.078	0.090
			0.3	0.033	0.067	0.101	0.135	0.171	0.208	0.244	0.282
			0.5	0.055	0.111	0.169	0.229	0.291	0.356	0.423	0.495
			0.7	0.076	0.155	0.238	0.326	0.418	0.516	0.621	0.734
		3.4	0.1	0.023	0.045	0.068	0.091	0.115	0.139	0.163	0.188
			0.3	0.066	0.135	0.208	0.284	0.364	0.449	0.539	0.636
			0.5	0.106	0.226	0.354	0.493	0.646	0.819	1.019	1.256
			0.7	0.152	0.320	0.507	0.721	0.973	1.283	1.692	2.256
		3.56	0.1	0.042	0.087	0.131	0.178	0.226	0.276	0.328	0.382
			0.3	0.126	0.263	0.414	0.583	0.773	0.996	1.266	1.613
			0.5	0.207	0.444	0.724	1.068	1.519	2.185	3.382	6.219
			0.7	0.285	0.629	1.065	1.665	2.625	4.590	-	-

Table A.14 *Differential swell occurring at the perimeter of a slab for a center lift swelling condition in predominantly Montmorillonite clay soil (50 percent clay).*

Percent Clay (%)	Depth of Constant Suction (FT)	Constant Suction (pF)	Velocity of Moisture Flow (inches/month)	DIFFERENTIAL SWELL (INCHES) EDGE DISTANCE PENETRATION (FT) 1 FT	2 FT	3 FT	4 FT	5 FT	6 FT	7 FT	8 FT
60	3	3.2	0.1	0.003	0.006	0.009	0.013	0.016	0.019	0.022	0.026
			0.3	0.009	0.019	0.028	0.038	0.048	0.059	0.069	0.080
			0.5	0.015	0.031	0.048	0.065	0.082	0.101	0.120	0.140
			0.7	0.022	0.044	0.068	0.092	0.119	0.147	0.176	0.209
		3.4	0.1	0.006	0.013	0.019	0.026	0.033	0.040	0.047	0.054
			0.3	0.019	0.039	0.060	0.082	0.105	0.130	0.157	0.186
			0.5	0.031	0.065	0.102	0.143	0.189	0.240	0.300	0.372
			0.7	0.044	0.093	0.147	0.211	0.286	0.380	0.503	0.683
		3.6	0.1	0.015	0.031	0.048	0.065	0.083	0.102	0.122	0.144
			0.3	0.044	0.095	0.152	0.219	0.300	0.403	0.543	0.758
			0.5	0.073	0.161	0.272	0.420	0.645	1.056	2.037	4.865
			0.7	0.101	0.229	0.407	0.689	1.246	2.689	6.912	-
	5	3.2	0.1	0.008	0.015	0.023	0.030	0.038	0.045	0.053	0.060
			0.3	0.022	0.044	0.067	0.090	0.113	0.138	0.162	0.187
			0.5	0.037	0.074	0.113	0.153	0.194	0.237	0.282	0.329
			0.7	0.050	0.103	0.158	0.217	0.278	0.343	0.412	0.487
		3.4	0.1	0.015	0.030	0.043	0.061	0.077	0.093	0.109	0.126
			0.3	0.045	0.091	0.140	0.191	0.245	0.302	0.363	0.426
			0.5	0.073	0.152	0.237	0.331	0.435	0.551	0.686	0.846
			0.7	0.102	0.214	0.341	0.485	0.655	0.865	1.140	1.541
		3.6	0.1	0.035	0.071	0.109	0.149	0.190	0.233	0.279	0.326
			0.3	0.103	0.219	0.349	0.499	0.678	0.901	1.200	1.652
			0.5	0.169	0.370	0.618	0.945	1.425	2.280	4.284	9.923
			0.7	0.234	0.526	0.922	1.528	2.686	5.574	-	-
	7	3.2	0.1	0.013	0.027	0.041	0.055	0.069	0.083	0.097	0.111
			0.3	0.041	0.082	0.125	0.168	0.212	0.256	0.302	0.349
			0.5	0.068	0.137	0.209	0.283	0.360	0.441	0.524	0.612
			0.7	0.093	0.191	0.294	0.402	0.516	0.637	0.767	0.907
		3.4	0.1	0.027	0.055	0.083	0.112	0.142	0.171	0.201	0.232
			0.3	0.082	0.167	0.256	0.351	0.449	0.555	0.666	0.786
			0.5	0.135	0.280	0.438	0.609	0.799	1.013	1.260	1.553
			0.7	0.188	0.395	0.627	0.892	1.204	1.587	2.092	2.840
		3.56	0.1	0.053	0.107	0.163	0.221	0.281	0.342	0.407	0.474
			0.3	0.156	0.326	0.512	0.720	0.957	1.232	1.566	1.994
			0.5	0.256	0.549	0.895	1.320	1.879	2.702	4.182	8.216
			0.7	0.354	0.779	1.317	2.059	3.247	5.677	-	-

Table A.15 *Differential swell occurring at the perimeter of a slab for a center lift swelling condition in predominantly Montmorillonite clay soil (60 percent clay).*

Percent Clay (%)	Depth to Constant Suction (FT)	Constant Suction (pF)	Velocity of Moisture Flow (inches/month)	DIFFERENTIAL SWELL (INCHES) EDGE DISTANCE PENETRATION (FT) 1 FT	2 FT	3 FT	4 FT	5 FT	6 FT	7 FT	8 FT
70	3	3.2	0.1	0.004	0.007	0.011	0.015	0.019	0.023	0.026	0.030
			0.3	0.011	0.022	0.034	0.046	0.057	0.070	0.082	0.095
			0.5	0.018	0.037	0.057	0.077	0.098	0.120	0.143	0.167
			0.7	0.026	0.052	0.082	0.110	0.141	0.174	0.210	0.248
		3.4	0.1	0.007	0.015	0.023	0.031	0.039	0.048	0.056	0.065
			0.3	0.023	0.047	0.072	0.098	0.126	0.156	0.188	0.222
			0.5	0.038	0.078	0.122	0.171	0.225	0.287	0.358	0.443
			0.7	0.052	0.110	0.176	0.251	0.341	0.452	0.600	0.814
		3.6	0.1	0.018	0.037	0.056	0.077	0.098	0.121	0.145	0.171
			0.3	0.054	0.114	0.182	0.262	0.359	0.481	0.648	0.903
			0.5	0.088	0.192	0.324	0.502	0.769	1.258	2.428	5.796
			0.7	0.120	0.273	0.485	0.821	1.485	3.203	8.234	-
	5	3.2	0.1	0.009	0.017	0.026	0.035	0.044	0.053	0.062	0.071
			0.3	0.026	0.053	0.080	0.107	0.135	0.164	0.193	0.223
			0.5	0.044	0.088	0.134	0.182	0.231	0.283	0.336	0.392
			0.7	0.060	0.123	0.189	0.258	0.331	0.409	0.491	0.580
		3.4	0.1	0.018	0.036	0.055	0.073	0.092	0.111	0.131	0.151
			0.3	0.053	0.109	0.167	0.228	0.292	0.360	0.433	0.511
			0.5	0.088	0.182	0.284	0.395	0.519	0.658	0.818	1.008
			0.7	0.121	0.256	0.406	0.578	0.781	1.031	1.358	1.837
		3.6	0.1	0.042	0.086	0.131	0.178	0.227	0.278	0.332	0.389
			0.3	0.123	0.260	0.415	0.594	0.807	1.073	1.429	1.968
			0.5	0.202	0.441	0.737	1.126	1.698	2.717	5.104	11.822
			0.7	0.278	0.627	1.098	1.820	3.199	6.640	-	-
	7	3.2	0.1	0.016	0.032	0.049	0.065	0.082	0.098	0.115	0.132
			0.3	0.048	0.097	0.148	0.199	0.251	0.305	0.359	0.415
			0.5	0.081	0.163	0.249	0.338	0.429	0.525	0.624	0.729
			0.7	0.112	0.229	0.351	0.480	0.616	0.759	0.915	1.081
		3.4	0.1	0.032	0.066	0.099	0.134	0.168	0.204	0.240	0.276
			0.3	0.098	0.199	0.306	0.418	0.536	0.661	0.794	0.937
			0.5	0.162	0.334	0.522	0.727	0.952	1.207	1.501	1.851
			0.7	0.224	0.470	0.747	1.063	1.435	1.891	2.492	3.383
		3.56	0.1	0.063	0.128	0.194	0.263	0.334	0.407	0.484	0.563
			0.3	0.185	0.387	0.609	0.857	1.139	1.468	1.865	2.376
			0.5	0.305	0.655	1.067	1.573	2.239	3.219	4.983	9.162
			0.7	0.421	0.928	1.569	2.453	3.868	6.763	-	-

Table A.16 *Differential swell occurring at the perimeter of a slab for a center lift swelling condition in predominantly Montmorillonite clay soil (70 percent clay).*

Percent Clay (%)	Depth to Constant Suction (FT)	Constant Suction (pF)	Velocity of Moisture Flow (inches/month)	DIFFERENTIAL SWELL (INCHES) EDGE DISTANCE PENETRATION (FT) 1 FT	2 FT	3 FT	4 FT	5 FT	6 FT	7 FT	8 FT
30	3	3.2	0.1	0.001	0.001	0.002	0.002	0.003	0.003	0.004	0.004
			0.3	0.002	0.003	0.004	0.006	0.007	0.009	0.010	0.012
			0.5	0.003	0.005	0.007	0.010	0.012	0.014	0.017	0.019
			0.7	0.004	0.007	0.010	0.013	0.017	0.020	0.023	0.026
		3.4	0.1	0.001	0.002	0.003	0.004	0.005	0.006	0.007	0.008
			0.3	0.003	0.006	0.009	0.012	0.015	0.018	0.021	0.024
			0.5	0.005	0.010	0.015	0.020	0.025	0.029	0.033	0.037
			0.7	0.007	0.014	0.021	0.027	0.033	0.039	0.045	0.050
		3.6	0.1	0.003	0.005	0.008	0.010	0.013	0.015	0.017	0.020
			0.3	0.008	0.015	0.022	0.029	0.035	0.041	0.047	0.053
			0.5	0.013	0.025	0.035	0.046	0.055	0.064	0.072	0.080
			0.7	0.018	0.034	0.048	0.061	0.073	0.084	0.094	0.104
		3.8	0.1	0.006	0.013	0.018	0.024	0.030	0.035	0.040	0.045
			0.3	0.019	0.035	0.050	0.064	0.076	0.088	0.099	0.109
			0.5	0.031	0.056	0.079	0.097	0.114	0.129	0.143	0.156
			0.7	0.042	0.075	0.102	0.125	0.145	0.164	0.180	0.195
	5	3.2	0.1	0.001	0.002	0.003	0.004	0.005	0.007	0.008	0.009
			0.3	0.003	0.007	0.010	0.013	0.016	0.019	0.022	0.025
			0.5	0.006	0.011	0.016	0.021	0.027	0.032	0.037	0.041
			0.7	0.008	0.015	0.022	0.030	0.037	0.044	0.050	0.057
		3.4	0.1	0.002	0.005	0.007	0.009	0.012	0.014	0.016	0.018
			0.3	0.007	0.014	0.021	0.027	0.034	0.040	0.046	0.053
			0.5	0.012	0.023	0.034	0.045	0.055	0.065	0.075	0.084
			0.7	0.016	0.032	0.047	0.061	0.075	0.089	0.101	0.114
		3.6	0.1	0.006	0.012	0.017	0.023	0.028	0.034	0.039	0.044
			0.3	0.018	0.034	0.050	0.065	0.080	0.094	0.107	0.120
			0.5	0.029	0.056	0.081	0.104	0.126	0.147	0.167	0.186
			0.7	0.041	0.077	0.110	0.141	0.169	0.195	0.220	0.243
		3.8	0.1	0.014	0.028	0.042	0.055	0.067	0.079	0.091	0.102
			0.3	0.043	0.081	0.116	0.148	0.178	0.205	0.231	0.255
			0.5	0.072	0.131	0.183	0.228	0.270	0.307	0.342	0.374
			0.7	0.100	0.178	0.244	0.300	0.350	0.395	0.436	0.474
	7	3.2	0.1	0.002	0.004	0.006	0.008	0.009	0.011	0.013	0.015
			0.3	0.006	0.011	0.017	0.022	0.028	0.033	0.039	0.044
			0.5	0.010	0.019	0.028	0.037	0.046	0.055	0.064	0.072
			0.7	0.013	0.026	0.039	0.052	0.064	0.076	0.088	0.099
		3.4	0.1	0.004	0.008	0.012	0.016	0.020	0.024	0.028	0.032
			0.3	0.012	0.024	0.036	0.048	0.059	0.071	0.082	0.093
			0.5	0.021	0.041	0.060	0.079	0.097	0.115	0.132	0.149
			0.7	0.029	0.056	0.083	0.108	0.133	0.157	0.180	0.202
		3.6	0.1	0.010	0.020	0.030	0.040	0.050	0.059	0.069	0.078
			0.3	0.031	0.061	0.089	0.116	0.142	0.167	0.192	0.215
			0.5	0.052	0.100	0.145	0.187	0.227	0.264	0.300	0.335
			0.7	0.073	0.138	0.198	0.254	0.305	0.353	0.399	0.441
		3.8	0.1	-	-	-	-	-	-	-	-

Table A.17 *Differential swell occurring at the perimeter of a slab for an edge lift swelling condition in predominantly Kaolinite clay soil (30 percent clay).*

Percent Clay	Depth to Constant Suction	Constant Suction	Velocity of Moisture Flow	DIFFERENTIAL SWELL (INCHES)							
				EDGE DISTANCE PENETRATION (FT)							
(%)	(FT)	(pF)	(inches /month)	1 FT	2 FT	3 FT	4 FT	5 FT	6 FT	7 FT	8 FT
40	3	3.2	0.1	0.001	0.001	0.002	0.003	0.003	0.004	0.005	0.005
			0.3	0.002	0.004	0.006	0.008	0.010	0.012	0.014	0.015
			0.5	0.003	0.007	0.010	0.013	0.016	0.019	0.022	0.025
			0.7	0.005	0.009	0.014	0.018	0.022	0.026	0.030	0.034
		3.4	0.1	0.001	0.003	0.004	0.006	0.007	0.008	0.010	0.011
			0.3	0.004	0.008	0.012	0.016	0.020	0.024	0.028	0.032
			0.5	0.007	0.014	0.020	0.027	0.033	0.039	0.045	0.050
			0.7	0.010	0.019	0.028	0.037	0.045	0.052	0.060	0.067
		3.6	0.1	0.004	0.007	0.010	0.014	0.017	0.020	0.023	0.026
			0.3	0.010	0.020	0.030	0.039	0.047	0.055	0.063	0.071
			0.5	0.017	0.033	0.047	0.061	0.074	0.086	0.097	0.108
			0.7	0.024	0.045	0.064	0.081	0.098	0.112	0.126	0.139
		3.8	0.1	0.009	0.017	0.025	0.032	0.040	0.046	0.053	0.060
			0.3	0.025	0.047	0.067	0.085	0.102	0.118	0.132	0.145
			0.5	0.041	0.075	0.104	0.129	0.152	0.173	0.192	0.209
			0.7	0.056	0.100	0.136	0.167	0.195	0.219	0.241	0.261
	5	3.2	0.1	0.002	0.003	0.004	0.006	0.007	0.009	0.010	0.012
			0.3	0.004	0.009	0.013	0.017	0.022	0.026	0.030	0.034
			0.5	0.007	0.015	0.022	0.029	0.035	0.042	0.049	0.056
			0.7	0.010	0.020	0.030	0.040	0.049	0.058	0.067	0.076
		3.4	0.1	0.003	0.006	0.009	0.012	0.016	0.019	0.022	0.025
			0.3	0.009	0.019	0.028	0.037	0.045	0.054	0.062	0.071
			0.5	0.016	0.031	0.046	0.060	0.074	0.087	0.100	0.113
			0.7	0.022	0.043	0.063	0.082	0.101	0.119	0.136	0.152
		3.6	0.1	0.008	0.016	0.023	0.031	0.038	0.045	0.052	0.059
			0.3	0.023	0.046	0.067	0.087	0.107	0.126	0.144	0.161
			0.5	0.039	0.075	0.108	0.140	0.169	0.197	0.223	0.249
			0.7	0.054	0.103	0.147	0.188	0.226	0.261	0.295	0.326
		3.8	0.1	0.019	0.038	0.056	0.073	0.090	0.106	0.121	0.136
			0.3	0.058	0.109	0.156	0.198	0.238	0.275	0.309	0.342
			0.5	0.096	0.176	0.245	0.306	0.361	0.411	0.458	0.501
			0.7	0.134	0.239	0.326	0.402	0.469	0.529	0.584	0.634
	7	3.2	0.1	0.003	0.005	0.008	0.010	0.013	0.015	0.018	0.020
			0.3	0.008	0.015	0.023	0.030	0.037	0.045	0.052	0.059
			0.5	0.013	0.025	0.038	0.050	0.062	0.074	0.085	0.097
			0.7	0.018	0.035	0.052	0.069	0.085	0.102	0.117	0.133
		3.4	0.1	0.006	0.011	0.016	0.022	0.027	0.032	0.038	0.043
			0.3	0.017	0.033	0.049	0.064	0.080	0.095	0.109	0.124
			0.5	0.028	0.054	0.080	0.105	0.130	0.154	0.177	0.200
			0.7	0.039	0.076	0.111	0.145	0.178	0.210	0.241	0.271
		3.6	0.1	0.014	0.027	0.041	0.054	0.067	0.080	0.092	0.105
			0.3	0.042	0.081	0.119	0.155	0.190	0.224	0.257	0.288
			0.5	0.069	0.134	0.194	0.250	0.304	0.354	0.402	0.448
			0.7	0.098	0.185	0.266	0.340	0.409	0.473	0.534	0.591
		3.8	0.1	-	-	-	-	-	-	-	-

Table A.18 *Differential swell occurring at the perimeter of a slab for an edge lift swelling condition in predominantly Kaolinite clay soil (40 percent clay).*

Percent Clay	Depth to Constant Suction	Constant Suction	Velocity of Moisture Flow	DIFFERENTIAL SWELL (INCHES)							
				EDGE DISTANCE PENETRATION (FT)							
(%)	(FT)	(pF)	(inches /month)	1 FT	2 FT	3 FT	4 FT	5 FT	6 FT	7 FT	8 FT
50	3	3.2	0.1	0.001	0.002	0.003	0.003	0.004	0.005	0.006	0.007
			0.3	0.003	0.005	0.007	0.010	0.012	0.015	0.017	0.019
			0.5	0.004	0.008	0.012	0.016	0.020	0.024	0.028	0.032
			0.7	0.006	0.012	0.017	0.023	0.028	0.033	0.038	0.043
		3.4	0.1	0.002	0.004	0.005	0.007	0.009	0.011	0.012	0.014
			0.3	0.005	0.011	0.016	0.021	0.025	0.030	0.035	0.040
			0.5	0.009	0.017	0.026	0.034	0.041	0.049	0.056	0.063
			0.7	0.012	0.024	0.035	0.046	0.056	0.066	0.075	0.084
		3.6	0.1	0.004	0.009	0.013	0.017	0.021	0.025	0.029	0.033
			0.3	0.013	0.025	0.037	0.048	0.059	0.069	0.079	0.089
			0.5	0.022	0.042	0.059	0.076	0.092	0.107	0.121	0.135
			0.7	0.030	0.056	0.080	0.102	0.122	0.141	0.158	0.175
		3.8	0.1	0.011	0.021	0.031	0.040	0.050	0.058	0.067	0.075
			0.3	0.032	0.059	0.084	0.107	0.128	0.147	0.165	0.182
			0.5	0.052	0.094	0.130	0.162	0.191	0.217	0.240	0.262
			0.7	0.071	0.126	0.171	0.210	0.244	0.275	0.302	0.328
	5	3.2	0.1	0.002	0.004	0.006	0.007	0.009	0.011	0.013	0.015
			0.3	0.006	0.011	0.016	0.022	0.027	0.032	0.037	0.043
			0.5	0.009	0.018	0.027	0.036	0.044	0.053	0.061	0.070
			0.7	0.013	0.025	0.038	0.050	0.062	0.073	0.084	0.096
		3.4	0.1	0.004	0.008	0.012	0.016	0.019	0.023	0.027	0.031
			0.3	0.012	0.023	0.035	0.046	0.057	0.068	0.078	0.088
			0.5	0.020	0.039	0.057	0.075	0.092	0.109	0.126	0.142
			0.7	0.028	0.054	0.079	0.103	0.126	0.149	0.170	0.191
		3.6	0.1	0.010	0.019	0.029	0.038	0.047	0.056	0.065	0.074
			0.3	0.029	0.057	0.084	0.109	0.134	0.158	0.180	0.202
			0.5	0.049	0.094	0.136	0.175	0.212	0.247	0.280	0.312
			0.7	0.068	0.129	0.185	0.236	0.283	0.328	0.369	0.408
		3.8	0.1	0.024	0.048	0.070	0.092	0.112	0.132	0.152	0.171
			0.3	0.072	0.137	0.195	0.249	0.298	0.344	0.387	0.428
			0.5	0.120	0.220	0.307	0.384	0.453	0.516	0.574	0.628
			0.7	0.168	0.299	0.409	0.504	0.588	0.663	0.732	0.795
	7	3.2	0.1	0.003	0.006	0.010	0.013	0.016	0.019	0.022	0.025
			0.3	0.010	0.019	0.028	0.038	0.047	0.056	0.065	0.074
			0.5	0.016	0.032	0.047	0.062	0.077	0.092	0.107	0.121
			0.7	0.022	0.044	0.066	0.087	0.107	0.127	0.147	0.167
		3.4	0.1	0.007	0.014	0.021	0.027	0.034	0.041	0.047	0.054
			0.3	0.021	0.041	0.061	0.080	0.100	0.119	0.137	0.156
			0.5	0.035	0.068	0.100	0.132	0.163	0.193	0.222	0.250
			0.7	0.048	0.095	0.139	0.182	0.223	0.263	0.302	0.339
		3.6	0.1	0.017	0.034	0.051	0.068	0.084	0.100	0.116	0.131
			0.3	0.052	0.102	0.149	0.195	0.239	0.281	0.322	0.361
			0.5	0.087	0.168	0.243	0.314	0.381	0.444	0.504	0.562
			0.7	0.122	0.232	0.333	0.426	0.512	0.593	0.669	0.741
		3.8	0.1	-	-	-	-	-	-	-	-

Table A.19 *Differential swell occurring at the perimeter of a slab for an edge lift swelling condition in predominantly Kaolinite clay soil (50 percent clay).*

Percent Clay (%)	Depth to Constant Suction (FT)	Constant Suction (pF)	Velocity of Moisture Flow (inches /month)	DIFFERENTIAL SWELL (INCHES)							
				EDGE DISTANCE PENETRATION (FT)							
				1 FT	2 FT	3 FT	4 FT	5 FT	6 FT	7 FT	8 FT
60	3	3.2	0.1	0.001	0.002	0.003	0.004	0.005	0.006	0.007	0.008
			0.3	0.003	0.006	0.009	0.012	0.015	0.018	0.021	0.023
			0.5	0.005	0.010	0.015	0.020	0.024	0.029	0.033	0.038
			0.7	0.007	0.014	0.021	0.027	0.033	0.040	0.046	0.052
		3.4	0.1	0.002	0.004	0.006	0.009	0.011	0.013	0.015	0.017
			0.3	0.006	0.013	0.019	0.025	0.031	0.036	0.042	0.048
			0.5	0.011	0.021	0.031	0.040	0.049	0.058	0.067	0.076
			0.7	0.015	0.029	0.042	0.055	0.067	0.079	0.090	0.101
		3.6	0.1	0.005	0.011	0.016	0.021	0.025	0.030	0.035	0.040
			0.3	0.016	0.031	0.045	0.058	0.071	0.083	0.095	0.106
			0.5	0.026	0.050	0.071	0.092	0.111	0.129	0.146	0.162
			0.7	0.036	0.068	0.097	0.123	0.147	0.169	0.190	0.210
		3.8	0.1	0.013	0.025	0.037	0.049	0.059	0.070	0.080	0.090
			0.3	0.038	0.071	0.101	0.129	0.154	0.177	0.199	0.219
			0.5	0.062	0.113	0.157	0.195	0.229	0.260	0.289	0.315
			0.7	0.085	0.151	0.205	0.252	0.293	0.330	0.363	0.394
	5	3.2	0.1	0.002	0.004	0.007	0.009	0.011	0.013	0.015	0.017
			0.3	0.007	0.013	0.020	0.026	0.032	0.039	0.045	0.051
			0.5	0.011	0.022	0.033	0.043	0.053	0.064	0.074	0.084
			0.7	0.015	0.031	0.045	0.060	0.074	0.088	0.101	0.115
		3.4	0.1	0.005	0.009	0.014	0.019	0.023	0.028	0.033	0.037
			0.3	0.014	0.028	0.042	0.055	0.068	0.081	0.094	0.106
			0.5	0.024	0.046	0.069	0.090	0.111	0.131	0.151	0.170
			0.7	0.033	0.065	0.095	0.124	0.152	0.179	0.205	0.230
		3.6	0.1	0.012	0.023	0.035	0.046	0.057	0.068	0.079	0.089
			0.3	0.035	0.069	0.101	0.132	0.161	0.189	0.217	0.243
			0.5	0.059	0.113	0.163	0.210	0.255	0.297	0.337	0.375
			0.7	0.082	0.155	0.222	0.284	0.341	0.394	0.444	0.491
		3.8	0.1	0.029	0.057	0.084	0.110	0.135	0.159	0.183	0.206
			0.3	0.087	0.164	0.235	0.299	0.358	0.414	0.466	0.515
			0.5	0.144	0.265	0.369	0.461	0.544	0.620	0.690	0.754
			0.7	0.201	0.360	0.492	0.606	0.707	0.798	0.880	0.956
	7	3.2	0.1	0.004	0.008	0.011	0.015	0.019	0.023	0.027	0.030
			0.3	0.011	0.023	0.034	0.045	0.056	0.067	0.078	0.089
			0.5	0.019	0.038	0.057	0.075	0.093	0.111	0.128	0.146
			0.7	0.027	0.053	0.079	0.104	0.129	0.153	0.177	0.201
		3.4	0.1	0.008	0.017	0.025	0.033	0.041	0.049	0.057	0.065
			0.3	0.025	0.049	0.073	0.097	0.120	0.143	0.165	0.187
			0.5	0.042	0.082	0.121	0.159	0.196	0.232	0.267	0.301
			0.7	0.058	0.114	0.167	0.219	0.268	0.316	0.363	0.408
		3.6	0.1	0.021	0.041	0.061	0.081	0.101	0.129	0.139	0.158
			0.3	0.063	0.122	0.179	0.234	0.287	0.338	0.387	0.434
			0.5	0.105	0.201	0.292	0.377	0.457	0.534	0.606	0.675
			0.7	0.147	0.279	0.400	0.512	0.616	0.713	0.804	0.891
		3.8	0.1	-	-	-	-	-	-	-	-

Table A.20 *Differential swell occurring at the perimeter of a slab for an edge lift swelling condition in predominantly Kaolinite clay soil (60 percent clay).*

Percent Clay (%)	Depth to Constant Suction (FT)	Constant Suction (pF)	Velocity of Moisture Flow (inches /month)	DIFFERENTIAL SWELL (INCHES)							
				EDGE DISTANCE PENETRATION (FT)							
				1 FT	2 FT	3 FT	4 FT	5 FT	6 FT	7 FT	8 FT
70	3	3.2	0.1	0.001	0.002	0.004	0.005	0.006	0.007	0.008	0.009
			0.3	0.004	0.007	0.011	0.014	0.017	0.021	0.024	0.027
			0.5	0.006	0.012	0.017	0.023	0.028	0.034	0.039	0.044
			0.7	0.008	0.016	0.024	0.032	0.039	0.046	0.054	0.061
		3.4	0.1	0.003	0.005	0.007	0.010	0.012	0.015	0.017	0.020
			0.3	0.008	0.015	0.022	0.029	0.036	0.042	0.049	0.056
			0.5	0.012	0.024	0.036	0.047	0.058	0.068	0.078	0.088
			0.7	0.017	0.034	0.049	0.064	0.079	0.092	0.106	0.118
		3.6	0.1	0.006	0.012	0.018	0.024	0.030	0.035	0.041	0.046
			0.3	0.018	0.036	0.052	0.068	0.083	0.097	0.111	0.124
			0.5	0.030	0.058	0.084	0.107	0.130	0.151	0.170	0.189
			0.7	0.042	0.079	0.113	0.143	0.172	0.198	0.222	0.245
		3.8	0.1	0.015	0.030	0.043	0.057	0.069	0.082	0.094	0.105
			0.3	0.044	0.083	0.119	0.150	0.180	0.207	0.232	0.256
			0.5	0.072	0.132	0.183	0.228	0.268	0.304	0.338	0.369
			0.7	0.099	0.176	0.240	0.295	0.343	0.386	0.425	0.460
	5	3.2	0.1	0.003	0.005	0.008	0.010	0.013	0.015	0.018	0.020
			0.3	0.008	0.015	0.023	0.031	0.038	0.045	0.053	0.060
			0.5	0.013	0.026	0.038	0.050	0.062	0.074	0.086	0.098
			0.7	0.018	0.036	0.053	0.070	0.086	0.103	0.119	0.134
		3.4	0.1	0.006	0.011	0.017	0.022	0.027	0.033	0.038	0.043
			0.3	0.017	0.033	0.049	0.064	0.080	0.095	0.110	0.124
			0.5	0.028	0.054	0.080	0.105	0.130	0.153	0.176	0.199
			0.7	0.039	0.075	0.111	0.145	0.177	0.209	0.239	0.268
		3.6	0.1	0.014	0.027	0.041	0.054	0.067	0.079	0.092	0.104
			0.3	0.041	0.080	0.118	0.154	0.188	0.221	0.253	0.284
			0.5	0.068	0.132	0.191	0.246	0.298	0.347	0.393	0.438
			0.7	0.096	0.181	0.259	0.331	0.398	0.460	0.518	0.574
		3.8	0.1	0.034	0.067	0.098	0.128	0.158	0.186	0.213	0.240
			0.3	0.102	0.192	0.274	0.349	0.419	0.483	0.544	0.601
			0.5	0.169	0.309	0.431	0.539	0.636	0.724	0.806	0.881
			0.7	0.235	0.420	0.575	0.708	0.826	0.932	1.028	1.117
	7	3.2	0.1	0.005	0.009	0.013	0.018	0.022	0.027	0.031	0.035
			0.3	0.013	0.027	0.040	0.053	0.066	0.079	0.091	0.104
			0.5	0.022	0.044	0.066	0.087	0.109	0.129	0.150	0.170
			0.7	0.031	0.062	0.092	0.121	0.150	0.179	0.207	0.234
		3.4	0.1	0.010	0.019	0.029	0.038	0.048	0.057	0.066	0.076
			0.3	0.029	0.058	0.086	0.113	0.140	0.167	0.193	0.218
			0.5	0.048	0.095	0.141	0.185	0.228	0.270	0.311	0.351
			0.7	0.068	0.133	0.195	0.256	0.314	0.370	0.424	0.476
		3.6	0.1	0.024	0.048	0.072	0.095	0.118	0.140	0.162	0.184
			0.3	0.073	0.143	0.209	0.274	0.335	0.395	0.452	0.508
			0.5	0.122	0.235	0.341	0.441	0.534	0.623	0.708	0.789
			0.7	0.172	0.326	0.468	0.598	0.719	0.833	0.940	1.041
		3.8	0.1	-	-	-	-	-	-	-	-

Table A.21 *Differential swell occurring at the perimeter of a slab for an edge lift swelling condition in predominantly Kaolinite clay soil (70 percent clay).*

Percent Clay (%)	Depth to Constant Suction (FT)	Constant Suction (pF)	Velocity of Moisture Flow (inches /month)	DIFFERENTIAL SWELL (INCHES) EDGE DISTANCE PENETRATION (FT) 1 FT	2 FT	3 FT	4 FT	5 FT	6 FT	7 FT	8 FT
30	3	3.2	0.1	0.001	0.002	0.003	0.004	0.005	0.006	0.007	0.008
			0.3	0.003	0.006	0.009	0.012	0.015	0.017	0.020	0.023
			0.5	0.005	0.010	0.015	0.019	0.024	0.029	0.033	0.037
			0.7	0.007	0.014	0.020	0.027	0.033	0.039	0.045	0.051
		3.4	0.1	0.002	0.004	0.006	0.008	0.010	0.013	0.015	0.017
			0.3	0.006	0.013	0.019	0.025	0.030	0.036	0.042	0.047
			0.5	0.011	0.021	0.030	0.040	0.049	0.058	0.066	0.075
			0.7	0.015	0.029	0.042	0.054	0.067	0.078	0.089	0.100
		3.6	0.1	0.005	0.010	0.015	0.020	0.025	0.030	0.035	0.039
			0.3	0.016	0.030	0.044	0.057	0.070	0.082	0.094	0.105
			0.5	0.026	0.049	0.071	0.091	0.110	0.128	0.144	0.160
			0.7	0.036	0.067	0.096	0.121	0.145	0.168	0.188	0.208
		3.8	0.1	0.013	0.025	0.037	0.048	0.059	0.069	0.079	0.089
			0.3	0.038	0.071	0.100	0.127	0.152	0.175	0.197	0.217
			0.5	0.061	0.112	0.155	0.193	0.227	0.258	0.286	0.312
			0.7	0.084	0.149	0.203	0.250	0.290	0.327	0.360	0.390
	5	3.2	0.1	0.002	0.004	0.007	0.009	0.011	0.013	0.015	0.017
			0.3	0.007	0.013	0.019	0.026	0.032	0.038	0.045	0.051
			0.5	0.011	0.022	0.032	0.043	0.053	0.063	0.073	0.083
			0.7	0.015	0.030	0.045	0.059	0.073	0.087	0.100	0.114
		3.4	0.1	0.005	0.009	0.014	0.019	0.023	0.028	0.032	0.037
			0.3	0.014	0.028	0.041	0.055	0.068	0.080	0.093	0.105
			0.5	0.023	0.046	0.068	0.089	0.110	0.130	0.149	0.168
			0.7	0.033	0.064	0.094	0.123	0.150	0.177	0.202	0.227
		3.6	0.1	0.012	0.023	0.034	0.046	0.056	0.067	0.078	0.088
			0.3	0.035	0.068	0.100	0.130	0.159	0.187	0.214	0.241
			0.5	0.058	0.112	0.161	0.208	0.252	0.294	0.333	0.371
			0.7	0.081	0.154	0.220	0.281	0.337	0.390	0.439	0.486
		3.8	0.1	0.029	0.057	0.083	0.109	0.134	0.158	0.181	0.203
			0.3	0.086	0.163	0.232	0.296	0.355	0.409	0.461	0.509
			0.5	0.143	0.262	0.365	0.456	0.539	0.613	0.682	0.746
			0.7	0.199	0.356	0.487	0.600	0.699	0.789	0.871	0.946
	7	3.2	0.1	0.004	0.008	0.011	0.015	0.019	0.023	0.026	0.030
			0.3	0.011	0.023	0.034	0.045	0.056	0.067	0.077	0.088
			0.5	0.019	0.038	0.056	0.074	0.092	0.110	0.127	0.144
			0.7	0.027	0.052	0.078	0.103	0.127	0.152	0.175	0.198
		3.4	0.1	0.008	0.016	0.024	0.032	0.040	0.048	0.056	0.064
			0.3	0.025	0.049	0.072	0.096	0.119	0.141	0.163	0.185
			0.5	0.041	0.081	0.119	0.157	0.194	0.229	0.264	0.298
			0.7	0.058	0.113	0.166	0.217	0.266	0.313	0.359	0.404
		3.6	0.1	0.021	0.041	0.061	0.080	0.100	0.119	0.137	0.156
			0.3	0.062	0.121	0.177	0.232	0.284	0.334	0.383	0.430
			0.5	0.103	0.199	0.289	0.373	0.453	0.528	0.600	0.668
			0.7	0.145	0.276	0.396	0.507	0.609	0.706	0.796	0.882
		3.8	0.1	-	-	-	-	-	-	-	-

Table A.22 *Differential swell occurring at the perimeter of a slab for an edge lift swelling condition in predominantly Illite clay soil (30 percent clay).*

Percent Clay (%)	Depth to Constant Suction (FT)	Constant Suction (pF)	Velocity of Moisture Flow (inches /month)	DIFFERENTIAL SWELL (INCHES) EDGE DISTANCE PENETRATION (FT) 1 FT	2 FT	3 FT	4 FT	5 FT	6 FT	7 FT	8 FT
40	3	3.2	0.1	0.001	0.003	0.004	0.006	0.007	0.009	0.010	0.011
			0.3	0.004	0.009	0.013	0.017	0.021	0.025	0.029	0.033
			0.5	0.007	0.014	0.021	0.028	0.035	0.041	0.048	0.054
			0.7	0.010	0.020	0.029	0.039	0.048	0.057	0.065	0.074
		3.4	0.1	0.003	0.006	0.009	0.012	0.015	0.018	0.021	0.024
			0.3	0.009	0.018	0.027	0.035	0.044	0.052	0.060	0.068
			0.5	0.015	0.030	0.044	0.058	0.071	0.083	0.096	0.108
			0.7	0.021	0.041	0.060	0.079	0.096	0.113	0.129	0.145
		3.6	0.1	0.008	0.015	0.022	0.029	0.036	0.043	0.050	0.057
			0.3	0.022	0.044	0.064	0.083	0.101	0.119	0.136	0.152
			0.5	0.037	0.071	0.102	0.131	0.158	0.184	0.208	0.231
			0.7	0.051	0.097	0.138	0.175	0.210	0.242	0.272	0.300
		3.8	0.1	0.019	0.036	0.053	0.069	0.085	0.100	0.115	0.129
			0.3	0.054	0.102	0.145	0.184	0.220	0.253	0.284	0.313
			0.5	0.088	0.161	0.224	0.278	0.328	0.372	0.413	0.451
			0.7	0.122	0.216	0.294	0.360	0.419	0.472	0.519	0.563
	5	3.2	0.1	0.003	0.006	0.009	0.013	0.016	0.019	0.022	0.025
			0.3	0.009	0.019	0.028	0.037	0.046	0.055	0.064	0.073
			0.5	0.016	0.031	0.047	0.062	0.076	0.091	0.105	0.120
			0.7	0.022	0.044	0.065	0.085	0.106	0.125	0.145	0.164
		3.4	0.1	0.007	0.014	0.020	0.027	0.033	0.040	0.046	0.053
			0.3	0.020	0.040	0.060	0.079	0.097	0.116	0.134	0.152
			0.5	0.034	0.066	0.098	0.129	0.158	0.187	0.216	0.243
			0.7	0.047	0.092	0.135	0.177	0.217	0.255	0.292	0.328
		3.6	0.1	0.017	0.033	0.050	0.066	0.081	0.097	0.112	0.127
			0.3	0.050	0.098	0.144	0.188	0.230	0.271	0.310	0.347
			0.5	0.084	0.161	0.233	0.300	0.364	0.424	0.481	0.535
			0.7	0.117	0.222	0.317	0.405	0.487	0.563	0.634	0.701
		3.8	0.1	0.042	0.082	0.120	0.157	0.193	0.228	0.261	0.294
			0.3	0.124	0.235	0.335	0.427	0.512	0.591	0.665	0.735
			0.5	0.206	0.378	0.527	0.659	0.777	0.886	0.985	1.078
			0.7	0.288	0.514	0.703	0.866	1.010	1.139	1.257	1.366
	7	3.2	0.1	0.005	0.011	0.016	0.022	0.027	0.033	0.038	0.043
			0.3	0.016	0.033	0.049	0.065	0.080	0.096	0.112	0.127
			0.5	0.027	0.054	0.081	0.107	0.133	0.158	0.183	0.208
			0.7	0.038	0.076	0.112	0.149	0.184	0.219	0.253	0.286
		3.4	0.1	0.012	0.024	0.035	0.047	0.058	0.070	0.081	0.093
			0.3	0.036	0.070	0.105	0.138	0.171	0.204	0.236	0.267
			0.5	0.059	0.117	0.172	0.227	0.279	0.331	0.381	0.430
			0.7	0.083	0.163	0.239	0.313	0.384	0.452	0.518	0.583
		3.6	0.1	0.030	0.059	0.088	0.116	0.144	0.171	0.198	0.225
			0.3	0.089	0.175	0.256	0.335	0.410	0.483	0.553	0.621
			0.5	0.149	0.288	0.417	0.539	0.654	0.762	0.866	0.965
			0.7	0.210	0.399	0.572	0.731	0.880	1.019	1.149	1.273
		3.8	0.1	-	-	-	-	-	-	-	-

Table A.23 *Differential swell occurring at the perimeter of a slab for an edge lift swelling condition in predominantly Illite clay soil (40 percent clay).*

Percent Clay (%)	Depth to Constant Suction (FT)	Constant Suction (pF)	Velocity of Moisture Flow (inches/month)	DIFFERENTIAL SWELL (INCHES)							
				EDGE DISTANCE PENETRATION (FT)							
				1 FT	2 FT	3 FT	4 FT	5 FT	6 FT	7 FT	8 FT
50	3	3.2	0.1	0.002	0.004	0.006	0.008	0.009	0.011	0.013	0.015
			0.3	0.006	0.011	0.017	0.022	0.028	0.033	0.038	0.044
			0.5	0.009	0.019	0.028	0.037	0.045	0.054	0.062	0.071
			0.7	0.013	0.026	0.038	0.051	0.062	0.074	0.086	0.097
		3.4	0.1	0.004	0.008	0.012	0.016	0.020	0.024	0.027	0.031
			0.3	0.012	0.024	0.035	0.046	0.057	0.068	0.078	0.089
			0.5	0.020	0.039	0.057	0.075	0.092	0.109	0.125	0.141
			0.7	0.028	0.054	0.079	0.103	0.126	0.148	0.169	0.189
		3.6	0.1	0.010	0.020	0.029	0.038	0.048	0.057	0.065	0.074
			0.3	0.029	0.057	0.083	0.108	0.132	0.155	0.178	0.199
			0.5	0.048	0.093	0.134	0.172	0.207	0.241	0.272	0.303
			0.7	0.067	0.127	0.180	0.229	0.274	0.316	0.356	0.392
		3.8	0.1	0.024	0.047	0.069	0.091	0.111	0.131	0.150	0.168
			0.3	0.071	0.133	0.189	0.240	0.287	0.331	0.371	0.409
			0.5	0.116	0.211	0.292	0.364	0.428	0.486	0.540	0.589
			0.7	0.159	0.282	0.384	0.471	0.548	0.616	0.679	0.736
	5	3.2	0.1	0.004	0.008	0.012	0.016	0.021	0.025	0.029	0.033
			0.3	0.012	0.025	0.037	0.049	0.061	0.072	0.084	0.096
			0.5	0.021	0.041	0.061	0.080	0.100	0.119	0.138	0.156
			0.7	0.029	0.057	0.085	0.112	0.138	0.164	0.190	0.215
		3.4	0.1	0.009	0.018	0.026	0.035	0.044	0.052	0.061	0.069
			0.3	0.027	0.052	0.078	0.103	0.127	0.152	0.175	0.198
			0.5	0.044	0.087	0.128	0.168	0.207	0.245	0.282	0.318
			0.7	0.062	0.121	0.177	0.231	0.283	0.334	0.382	0.429
		3.6	0.1	0.022	0.044	0.065	0.086	0.106	0.127	0.147	0.166
			0.3	0.066	0.129	0.188	0.246	0.301	0.354	0.405	0.454
			0.5	0.109	0.211	0.305	0.393	0.476	0.554	0.629	0.700
			0.7	0.153	0.290	0.415	0.530	0.636	0.736	0.829	0.917
		3.8	0.1	0.054	0.107	0.157	0.205	0.252	0.297	0.341	0.384
			0.3	0.162	0.307	0.438	0.558	0.669	0.773	0.870	0.961
			0.5	0.269	0.494	0.689	0.861	1.016	1.158	1.288	1.409
			0.7	0.376	0.672	0.919	1.132	1.320	1.490	1.644	1.785
	7	3.2	0.1	0.007	0.014	0.021	0.028	0.035	0.042	0.049	0.056
			0.3	0.021	0.043	0.064	0.085	0.105	0.126	0.146	0.166
			0.5	0.036	0.071	0.106	0.140	0.173	0.207	0.240	0.272
			0.7	0.050	0.099	0.147	0.194	0.240	0.286	0.331	0.375
		3.4	0.1	0.015	0.031	0.046	0.061	0.076	0.091	0.106	0.121
			0.3	0.046	0.092	0.137	0.181	0.224	0.266	0.308	0.349
			0.5	0.077	0.153	0.225	0.296	0.365	0.432	0.498	0.562
			0.7	0.109	0.213	0.313	0.409	0.501	0.591	0.678	0.762
		3.6	0.1	0.039	0.077	0.115	0.152	0.188	0.224	0.259	0.294
			0.3	0.117	0.228	0.335	0.437	0.536	0.631	0.723	0.811
			0.5	0.195	0.376	0.545	0.704	0.854	0.997	1.132	1.261
			0.7	0.274	0.522	0.748	0.956	1.150	1.332	1.503	1.664
		3.8	0.1	-	-	-	-	-	-	-	-

Table A.24 *Differential swell occurring at the perimeter of a slab for an edge lift swelling condition in predominantly Illite clay soil (50 percent clay).*

Percent Clay (%)	Depth to Constant Suction (FT)	Constant Suction (pF)	Velocity of Moisture Flow (inches/month)	DIFFERENTIAL SWELL (INCHES)							
				EDGE DISTANCE PENETRATION (FT)							
				1 FT	2 FT	3 FT	4 FT	5 FT	6 FT	7 FT	8 FT
60	3	3.2	0.1	0.002	0.005	0.007	0.009	0.012	0.014	0.016	0.018
			0.3	0.007	0.014	0.021	0.027	0.034	0.041	0.047	0.054
			0.5	0.012	0.023	0.034	0.045	0.056	0.067	0.077	0.087
			0.7	0.016	0.032	0.047	0.062	0.077	0.092	0.106	0.119
		3.4	0.1	0.005	0.010	0.015	0.020	0.024	0.029	0.034	0.039
			0.3	0.015	0.029	0.043	0.057	0.071	0.084	0.097	0.110
			0.5	0.025	0.048	0.071	0.093	0.114	0.135	0.155	0.174
			0.7	0.034	0.067	0.097	0.127	0.155	0.182	0.208	0.234
		3.6	0.1	0.012	0.024	0.036	0.047	0.059	0.070	0.081	0.092
			0.3	0.036	0.070	0.103	0.134	0.164	0.192	0.219	0.245
			0.5	0.060	0.114	0.165	0.212	0.256	0.297	0.337	0.374
			0.7	0.083	0.156	0.223	0.283	0.339	0.391	0.439	0.485
		3.8	0.1	0.030	0.058	0.086	0.112	0.137	0.162	0.185	0.208
			0.3	0.088	0.165	0.234	0.297	0.355	0.409	0.459	0.506
			0.5	0.143	0.260	0.361	0.450	0.529	0.601	0.667	0.728
			0.7	0.196	0.348	0.474	0.582	0.677	0.761	0.838	0.909
	5	3.2	0.1	0.005	0.010	0.015	0.020	0.025	0.030	0.035	0.040
			0.3	0.015	0.030	0.045	0.060	0.075	0.089	0.104	0.118
			0.5	0.025	0.050	0.075	0.099	0.123	0.147	0.170	0.193
			0.7	0.036	0.070	0.104	0.138	0.171	0.203	0.234	0.265
		3.4	0.1	0.011	0.022	0.033	0.043	0.054	0.064	0.075	0.085
			0.3	0.033	0.065	0.096	0.127	0.157	0.187	0.216	0.245
			0.5	0.055	0.107	0.158	0.208	0.256	0.303	0.348	0.393
			0.7	0.076	0.149	0.219	0.286	0.350	0.412	0.472	0.530
		3.6	0.1	0.027	0.054	0.080	0.106	0.132	0.157	0.181	0.206
			0.3	0.081	0.159	0.233	0.303	0.371	0.437	0.500	0.561
			0.5	0.135	0.260	0.376	0.485	0.588	0.685	0.777	0.864
			0.7	0.189	0.358	0.512	0.654	0.786	0.908	1.024	1.133
		3.8	0.1	0.067	0.132	0.194	0.254	0.311	0.367	0.422	0.474
			0.3	0.201	0.379	0.541	0.690	0.827	0.955	1.074	1.187
			0.5	0.333	0.611	0.851	1.064	1.255	1.430	1.591	1.740
			0.7	0.465	0.830	1.135	1.398	1.631	1.840	2.030	2.205
	7	3.2	0.1	0.009	0.018	0.026	0.035	0.044	0.052	0.061	0.070
			0.3	0.026	0.053	0.079	0.104	0.130	0.155	0.180	0.205
			0.5	0.044	0.088	0.130	0.173	0.214	0.255	0.296	0.336
			0.7	0.062	0.122	0.182	0.240	0.297	0.353	0.408	0.463
		3.4	0.1	0.019	0.038	0.057	0.076	0.094	0.113	0.131	0.149
			0.3	0.057	0.114	0.169	0.223	0.276	0.329	0.380	0.431
			0.5	0.096	0.188	0.278	0.366	0.451	0.534	0.615	0.694
			0.7	0.134	0.263	0.386	0.505	0.619	0.730	0.837	0.941
		3.6	0.1	0.048	0.095	0.141	0.187	0.232	0.277	0.320	0.364
			0.3	0.144	0.282	0.414	0.540	0.662	0.779	0.893	1.002
			0.5	0.241	0.465	0.674	0.870	1.055	1.231	1.398	0.558
			0.7	0.339	0.644	0.924	1.181	1.421	1.645	1.856	2.055
		3.8	0.1								

Table A.25 *Differential swell occurring at the perimeter of a slab for an edge lift swelling condition in predominantly Illite clay soil (60 percent clay).*

Percent Clay (%)	Depth to Constant Suction (FT)	Constant Suction (pF)	Velocity of Moisture Flow (inches /month)	DIFFERENTIAL SWELL (INCHES)							
				EDGE DISTANCE PENETRATION (FT)							
				1 FT	2 FT	3 FT	4 FT	5 FT	6 FT	7 FT	8 FT
70	3	3.2	0.1	0.003	0.006	0.008	0.011	0.014	0.017	0.019	0.022
			0.3	0.008	0.017	0.025	0.033	0.041	0.048	0.056	0.064
			0.5	0.014	0.027	0.041	0.054	0.067	0.079	0.092	0.104
			0.7	0.019	0.038	0.056	0.074	0.092	0.109	0.126	0.142
		3.4	0.1	0.006	0.012	0.018	0.023	0.029	0.035	0.040	0.046
			0.3	0.018	0.035	0.052	0.068	0.084	0.100	0.115	0.130
			0.5	0.029	0.057	0.084	0.111	0.136	0.160	0.184	0.207
			0.7	0.041	0.079	0.116	0.151	0.185	0.217	0.248	0.278
		3.6	0.1	0.015	0.029	0.043	0.056	0.070	0.083	0.096	0.109
			0.3	0.043	0.084	0.123	0.159	0.195	0.229	0.261	0.292
			0.5	0.071	0.136	0.196	0.252	0.305	0.354	0.401	0.445
			0.7	0.099	0.186	0.265	0.337	0.403	0.465	0.523	0.577
		3.8	0.1	0.036	0.070	0.102	0.133	0.163	0.192	0.220	0.247
			0.3	0.104	0.196	0.279	0.354	0.422	0.486	0.546	0.602
			0.5	0.170	0.310	0.430	0.535	0.630	0.715	0.794	0.866
			0.7	0.234	0.415	0.564	0.692	0.805	0.906	0.998	1.082
	5	3.2	0.1	0.006	0.012	0.018	0.024	0.030	0.036	0.042	0.048
			0.3	0.018	0.036	0.054	0.072	0.089	0.106	0.124	0.141
			0.5	0.030	0.060	0.089	0.118	0.147	0.175	0.202	0.230
			0.7	0.042	0.084	0.124	0.164	0.203	0.241	0.279	0.315
		3.4	0.1	0.013	0.026	0.039	0.052	0.064	0.077	0.089	0.102
			0.3	0.039	0.077	0.115	0.151	0.187	0.223	0.258	0.292
			0.5	0.065	0.128	0.188	0.247	0.305	0.360	0.414	0.467
			0.7	0.091	0.177	0.260	0.340	0.417	0.490	0.562	0.631
		3.6	0.1	0.032	0.064	0.095	0.126	0.157	0.186	0.216	0.245
			0.3	0.097	0.189	0.277	0.361	0.442	0.520	0.595	0.667
			0.5	0.161	0.309	0.448	0.577	0.699	0.815	0.925	1.029
			0.7	0.225	0.426	0.610	0.779	0.935	1.081	1.219	1.348
		3.8	0.1	0.080	0.157	0.231	0.302	0.371	0.437	0.502	0.564
			0.3	0.239	0.452	0.644	0.821	0.984	1.136	1.279	1.413
			0.5	0.396	0.727	1.013	1.266	1.494	1.702	1.894	2.072
			0.7	0.553	0.988	1.351	1.664	1.941	2.190	2.417	2.625
	7	3.2	0.1	0.011	0.021	0.031	0.042	0.052	0.062	0.073	0.083
			0.3	0.031	0.063	0.094	0.124	0.155	0.185	0.215	0.244
			0.5	0.052	0.104	0.155	0.205	0.255	0.304	0.352	0.400
			0.7	0.074	0.146	0.216	0.285	0.354	0.420	0.486	0.551
		3.4	0.1	0.023	0.045	0.068	0.090	0.112	0.134	0.156	0.178
			0.3	0.068	0.135	0.201	0.266	0.329	0.391	0.453	0.513
			0.5	0.114	0.224	0.331	0.436	0.537	0.636	0.732	0.826
			0.7	0.160	0.313	0.459	0.601	0.737	0.869	0.996	1.120
		3.6	0.1	0.057	0.113	0.168	0.223	0.276	0.329	0.381	0.433
			0.3	0.172	0.335	0.492	0.643	0.788	0.928	1.063	1.193
			0.5	0.287	0.553	0.802	1.036	1.256	1.465	1.664	1.854
			0.7	0.403	0.767	1.099	1.406	1.691	1.958	2.209	2.447
		3.8	0.1	-	-	-	-	-	-	-	-

Table A.26 *Differential swell occurring at the perimeter of a slab for an edge lift swelling condition in predominantly Illite clay soil (70 percent clay).*

Percent Clay (%)	Depth to Constant Suction (FT)	Constant Suction (pF)	Velocity of Moisture Flow (inches /month)	DIFFERENTIAL SWELL (INCHES)							
				EDGE DISTANCE PENETRATION (FT)							
				1 FT	2 FT	3 FT	4 FT	5 FT	6 FT	7 FT	8 FT
30	3	3.2	0.1	0.001	0.002	0.004	0.005	0.006	0.007	0.008	0.009
			0.3	0.004	0.007	0.010	0.014	0.017	0.021	0.024	0.027
			0.5	0.006	0.012	0.017	0.023	0.028	0.034	0.039	0.044
			0.7	0.008	0.016	0.024	0.032	0.039	0.046	0.053	0.060
		3.4	0.1	0.003	0.005	0.007	0.010	0.012	0.015	0.017	0.019
			0.3	0.007	0.015	0.022	0.029	0.036	0.042	0.049	0.055
			0.5	0.012	0.024	0.036	0.047	0.058	0.068	0.078	0.088
			0.7	0.017	0.034	0.049	0.064	0.078	0.092	0.105	0.118
		3.6	0.1	0.006	0.012	0.018	0.024	0.030	0.035	0.041	0.046
			0.3	0.018	0.036	0.052	0.068	0.083	0.097	0.111	0.124
			0.5	0.030	0.058	0.083	0.107	0.129	0.150	0.170	0.189
			0.7	0.042	0.079	0.112	0.143	0.171	0.197	0.222	0.245
		3.8	0.1	0.015	0.030	0.043	0.057	0.069	0.082	0.093	0.105
			0.3	0.044	0.083	0.118	0.150	0.179	0.206	0.232	0.255
			0.5	0.072	0.132	0.192	0.227	0.267	0.303	0.337	0.368
			0.7	0.099	0.176	0.239	0.294	0.342	0.385	0.423	0.459
	5	3.2	0.1	0.003	0.005	0.008	0.010	0.013	0.015	0.018	0.020
			0.3	0.008	0.015	0.023	0.030	0.038	0.045	0.053	0.060
			0.5	0.013	0.026	0.038	0.050	0.062	0.074	0.086	0.098
			0.7	0.018	0.036	0.053	0.070	0.086	0.102	0.118	0.134
		3.4	0.1	0.006	0.011	0.016	0.022	0.027	0.033	0.038	0.043
			0.3	0.017	0.033	0.049	0.064	0.080	0.095	0.109	0.124
			0.5	0.028	0.054	0.080	0.105	0.129	0.153	0.176	0.198
			0.7	0.039	0.075	0.110	0.144	0.177	0.208	0.238	0.268
		3.6	0.1	0.014	0.027	0.041	0.054	0.066	0.079	0.092	0.104
			0.3	0.041	0.080	0.118	0.153	0.188	0.221	0.252	0.283
			0.5	0.068	0.131	0.190	0.245	0.297	0.346	0.392	0.437
			0.7	0.095	0.181	0.259	0.330	0.397	0.459	0.517	0.572
		3.8	0.1	0.034	0.067	0.098	0.128	0.157	0.186	0.213	0.239
			0.3	0.101	0.192	0.273	0.348	0.418	0.482	0.543	0.600
			0.5	0.168	0.308	0.430	0.537	0.634	0.722	0.804	0.879
			0.7	0.235	0.419	0.573	0.706	0.824	0.929	1.025	1.114
	7	3.2	0.1	0.004	0.009	0.013	0.018	0.022	0.027	0.031	0.035
			0.3	0.013	0.027	0.040	0.053	0.066	0.078	0.091	0.104
			0.5	0.022	0.044	0.066	0.087	0.108	0.129	0.150	0.170
			0.7	0.031	0.062	0.092	0.121	0.150	0.178	0.206	0.234
		3.4	0.1	0.010	0.019	0.029	0.038	0.048	0.057	0.066	0.075
			0.3	0.029	0.057	0.085	0.113	0.140	0.166	0.192	0.218
			0.5	0.048	0.095	0.141	0.185	0.228	0.270	0.311	0.351
			0.7	0.068	0.133	0.195	0.255	0.313	0.369	0.423	0.475
		3.6	0.1	0.024	0.043	0.071	0.095	0.117	0.140	0.162	0.184
			0.3	0.073	0.142	0.209	0.273	0.334	0.394	0.451	0.506
			0.5	0.122	0.235	0.340	0.439	0.533	0.622	0.706	0.787
			0.7	0.171	0.326	0.466	0.597	0.718	0.831	0.937	1.038
		3.8	0.1	-	-	-	-	-	-	-	-

Table A.27 *Differential swell occurring at the perimeter of a slab for an edge lift swelling condition in predominantly Montmorillonite clay soil (30 percent clay).*

Percent Clay (%)	Depth to Constant Suction (FT)	Constant Suction (pF)	Velocity of Moisture Flow (inches /month)	DIFFERENTIAL SWELL (INCHES) EDGE DISTANCE PENETRATION (FT) 1 FT	2 FT	3 FT	4 FT	5 FT	6 FT	7 FT	8 FT
40	3	3.2	0.1	0.002	0.003	0.005	0.007	0.008	0.010	0.012	0.013
			0.3	0.005	0.010	0.015	0.020	0.025	0.030	0.035	0.039
			0.5	0.009	0.017	0.025	0.033	0.041	0.049	0.056	0.064
			0.7	0.012	0.023	0.035	0.046	0.056	0.067	0.077	0.087
		3.4	0.1	0.004	0.007	0.011	0.014	0.018	0.021	0.025	0.028
			0.3	0.011	0.021	0.032	0.042	0.052	0.061	0.071	0.080
			0.5	0.018	0.035	0.052	0.068	0.084	0.099	0.113	0.128
			0.7	0.025	0.049	0.071	0.093	0.114	0.133	0.153	0.171
		3.6	0.1	0.009	0.018	0.026	0.035	0.043	0.051	0.059	0.067
			0.3	0.027	0.052	0.075	0.098	0.120	0.141	0.160	0.180
			0.5	0.044	0.084	0.121	0.155	0.187	0.218	0.246	0.274
			0.7	0.061	0.114	0.163	0.207	0.248	0.286	0.321	0.355
		3.8	0.1	0.022	0.043	0.063	0.082	0.100	0.118	0.135	0.152
			0.3	0.064	0.121	0.171	0.217	0.260	0.299	0.336	0.370
			0.5	0.105	0.191	0.264	0.329	0.387	0.440	0.488	0.533
			0.7	0.144	0.255	0.347	0.426	0.495	0.557	0.614	0.665
	5	3.2	0.1	0.004	0.007	0.011	0.015	0.019	0.022	0.026	0.030
			0.3	0.011	0.022	0.033	0.044	0.055	0.065	0.076	0.087
			0.5	0.019	0.037	0.055	0.073	0.090	0.107	0.125	0.141
			0.7	0.026	0.052	0.076	0.101	0.125	0.148	0.171	0.194
		3.4	0.1	0.008	0.016	0.024	0.032	0.039	0.047	0.055	0.062
			0.3	0.024	0.047	0.070	0.093	0.115	0.137	0.158	0.179
			0.5	0.040	0.078	0.116	0.152	0.187	0.221	0.255	0.287
			0.7	0.056	0.109	0.160	0.209	0.256	0.302	0.345	0.388
		3.6	0.1	0.020	0.039	0.059	0.078	0.096	0.115	0.133	0.150
			0.3	0.060	0.116	0.170	0.222	0.272	0.320	0.366	0.410
			0.5	0.099	0.190	0.275	0.355	0.430	0.501	0.568	0.633
			0.7	0.138	0.262	0.375	0.479	0.575	0.665	0.749	0.829
		3.8	0.1	0.049	0.096	0.142	0.186	0.228	0.269	0.309	0.347
			0.3	0.147	0.278	0.396	0.505	0.605	0.699	0.786	0.869
			0.5	0.244	0.447	0.623	0.779	0.919	1.047	1.164	1.274
			0.7	0.340	0.607	0.831	1.023	1.193	1.347	1.486	1.614
	7	3.2	0.1	0.006	0.013	0.019	0.026	0.032	0.038	0.045	0.051
			0.3	0.019	0.039	0.058	0.076	0.095	0.114	0.132	0.150
			0.5	0.032	0.064	0.095	0.126	0.157	0.187	0.217	0.246
			0.7	0.045	0.089	0.133	0.176	0.217	0.258	0.299	0.339
		3.4	0.1	0.014	0.028	0.042	0.055	0.069	0.083	0.096	0.109
			0.3	0.042	0.083	0.124	0.163	0.202	0.241	0.278	0.316
			0.5	0.070	0.138	0.204	0.268	0.330	0.391	0.450	0.508
			0.7	0.098	0.192	0.283	0.369	0.453	0.534	0.613	0.689
		3.6	0.1	0.035	0.070	0.104	0.137	0.170	0.202	0.235	0.266
			0.3	0.106	0.206	0.303	0.395	0.484	0.570	0.653	0.734
			0.5	0.176	0.340	0.493	0.637	0.772	0.901	1.023	1.140
			0.7	0.248	0.472	0.676	0.864	1.040	1.204	1.358	1.504
		3.8	0.1	-	-	-	-	-	-	-	-

Table A.28 *Differential swell occurring at the perimeter of a slab for an edge lift swelling condition in predominantly Montmorillonite clay soil (40 percent clay).*

Percent Clay (%)	Depth to Constant Suction (FT)	Constant Suction (pF)	Velocity of Moisture Flow (inches /month)	DIFFERENTIAL SWELL (INCHES) EDGE DISTANCE PENETRATION (FT) 1 FT	2 FT	3 FT	4 FT	5 FT	6 FT	7 FT	8 FT
50	3	3.2	0.1	0.002	0.004	0.007	0.009	0.011	0.013	0.015	0.018
			0.3	0.007	0.013	0.020	0.026	0.033	0.039	0.045	0.052
			0.5	0.011	0.022	0.033	0.043	0.054	0.064	0.074	0.084
			0.7	0.016	0.031	0.045	0.060	0.074	0.088	0.101	0.115
		3.4	0.1	0.005	0.009	0.014	0.019	0.023	0.028	0.032	0.037
			0.3	0.014	0.028	0.042	0.055	0.068	0.080	0.093	0.105
			0.5	0.024	0.046	0.068	0.089	0.109	0.129	0.148	0.167
			0.7	0.033	0.064	0.093	0.122	0.149	0.175	0.200	0.224
		3.6	0.1	0.012	0.023	0.034	0.045	0.056	0.067	0.077	0.088
			0.3	0.035	0.068	0.099	0.128	0.157	0.184	0.210	0.235
			0.5	0.057	0.110	0.158	0.203	0.245	0.285	0.323	0.358
			0.7	0.079	0.150	0.213	0.271	0.325	0.375	0.421	0.465
		3.8	0.1	0.029	0.056	0.082	0.107	0.132	0.155	0.177	0.199
			0.3	0.084	0.158	0.224	0.285	0.340	0.392	0.440	0.485
			0.5	0.137	0.250	0.346	0.431	0.507	0.576	0.639	0.698
			0.7	0.188	0.334	0.454	0.558	0.649	0.730	0.804	0.871
	5	3.2	0.1	0.005	0.010	0.015	0.019	0.024	0.029	0.034	0.039
			0.3	0.015	0.029	0.044	0.058	0.072	0.086	0.100	0.133
			0.5	0.024	0.048	0.072	0.095	0.118	0.141	0.163	0.185
			0.7	0.034	0.067	0.100	0.132	0.163	0.194	0.224	0.254
		3.4	0.1	0.011	0.021	0.031	0.041	0.052	0.062	0.072	0.082
			0.3	0.031	0.062	0.092	0.122	0.151	0.179	0.207	0.235
			0.5	0.052	0.103	0.152	0.199	0.245	0.290	0.334	0.376
			0.7	0.073	0.143	0.210	0.274	0.335	0.395	0.452	0.508
		3.6	0.1	0.026	0.052	0.077	0.102	0.126	0.150	0.174	0.197
			0.3	0.078	0.152	0.223	0.291	0.356	0.419	0.479	0.538
			0.5	0.130	0.249	0.361	0.465	0.563	0.656	0.745	0.829
			0.7	0.181	0.343	0.491	0.627	0.753	0.871	0.981	1.086
		3.8	0.1	0.064	0.126	0.186	0.243	0.299	0.352	0.404	0.454
			0.3	0.192	0.364	0.519	0.661	0.793	0.915	1.030	1.138
			0.5	0.319	0.585	0.816	1.020	1.204	1.371	1.525	1.668
			0.7	0.445	0.796	1.088	1.340	1.563	1.764	1.946	2.114
	7	3.2	0.1	0.008	0.017	0.025	0.034	0.042	0.050	0.059	0.067
			0.3	0.025	0.050	0.075	0.100	0.125	0.149	0.173	0.197
			0.5	0.042	0.084	0.125	0.165	0.205	0.245	0.284	0.322
			0.7	0.059	0.117	0.174	0.230	0.285	0.339	0.391	0.443
		3.4	0.1	0.018	0.037	0.055	0.073	0.090	0.108	0.126	0.143
			0.3	0.055	0.109	0.162	0.214	0.265	0.315	0.365	0.413
			0.5	0.092	0.181	0.267	0.351	0.433	0.512	0.590	0.665
			0.7	0.129	0.252	0.370	0.484	0.594	0.700	0.802	0.902
		3.6	0.1	0.046	0.091	0.136	0.179	0.223	0.265	0.307	0.349
			0.3	0.138	0.270	0.397	0.518	0.635	0.747	0.856	0.961
			0.5	0.231	0.446	0.646	0.834	1.012	1.180	1.341	1.494
			0.7	0.325	0.618	0.885	1.132	1.362	1.577	1.779	1.970
		3.8	0.1	-	-	-	-	-	-	-	-

Table A.29 *Differential swell occurring at the perimeter of a slab for an edge lift swelling condition in predominantly Montmorillonite clay soil (50 percent clay).*

Percent Clay (%)	Depth to Constant Suction (FT)	Constant Suction (pF)	Velocity of Moisture Flow (inches/month)	DIFFERENTIAL SWELL (INCHES) EDGE DISTANCE PENETRATION (FT) 1 FT	2 FT	3 FT	4 FT	5 FT	6 FT	7 FT	8 FT
60	3	3.2	0.1	0.003	0.006	0.008	0.011	0.014	0.016	0.019	0.022
			0.3	0.008	0.016	0.025	0.033	0.040	0.048	0.056	0.064
			0.5	0.014	0.027	0.041	0.054	0.066	0.079	0.091	0.104
			0.7	0.019	0.038	0.056	0.074	0.091	0.109	0.125	0.142
		3.4	0.1	0.006	0.012	0.017	0.023	0.029	0.035	0.040	0.046
			0.3	0.018	0.035	0.051	0.068	0.084	0.099	0.115	0.130
			0.5	0.029	0.057	0.084	0.110	0.135	0.160	0.183	0.206
			0.7	0.041	0.079	0.116	0.151	0.184	0.216	0.247	0.277
		3.6	0.1	0.014	0.029	0.043	0.056	0.070	0.083	0.096	0.109
			0.3	0.043	0.083	0.122	0.159	0.194	0.228	0.260	0.291
			0.5	0.071	0.136	0.195	0.251	0.303	0.352	0.399	0.433
			0.7	0.098	0.185	0.264	0.336	0.402	0.463	0.521	0.575
		3.8	0.1	0.035	0.069	0.102	0.133	0.163	0.191	0.219	0.246
			0.3	0.104	0.195	0.277	0.352	0.421	0.484	0.544	0.599
			0.5	0.169	0.309	0.428	0.533	0.627	0.712	0.790	0.863
			0.7	0.233	0.413	0.562	0.690	0.802	0.903	0.994	1.077
	5	3.2	0.1	0.006	0.012	0.018	0.024	0.030	0.036	0.042	0.048
			0.3	0.018	0.036	0.054	0.071	0.089	0.106	0.123	0.140
			0.5	0.030	0.060	0.090	0.118	0.146	0.174	0.202	0.229
			0.7	0.042	0.083	0.124	0.163	0.202	0.240	0.278	0.314
		3.4	0.1	0.013	0.026	0.039	0.051	0.064	0.076	0.089	0.101
			0.3	0.039	0.077	0.114	0.151	0.187	0.222	0.256	0.291
			0.5	0.065	0.127	0.188	0.246	0.303	0.359	0.413	0.465
			0.7	0.090	0.177	0.259	0.339	0.415	0.488	0.559	0.628
		3.6	0.1	0.032	0.064	0.095	0.126	0.156	0.186	0.215	0.244
			0.3	0.096	0.188	0.276	0.360	0.440	0.518	0.593	0.665
			0.5	0.160	0.308	0.446	0.575	0.697	0.812	0.921	1.025
			0.7	0.224	0.425	0.607	0.775	0.931	1.077	1.214	1.343
		3.8	0.1	0.080	0.156	0.230	0.301	0.369	0.436	0.500	0.562
			0.3	0.238	0.450	0.642	0.817	0.980	1.132	1.274	1.407
			0.5	0.395	0.724	1.009	1.261	1.488	1.695	1.886	2.063
			0.7	0.551	0.984	1.345	1.657	1.933	2.181	2.407	2.614
	7	3.2	0.1	0.010	0.021	0.031	0.042	0.052	0.062	0.072	0.083
			0.3	0.031	0.062	0.093	0.124	0.154	0.184	0.214	0.243
			0.5	0.052	0.104	0.155	0.205	0.254	0.303	0.351	0.398
			0.7	0.073	0.145	0.215	0.284	0.352	0.419	0.484	0.548
		3.4	0.1	0.023	0.045	0.067	0.090	0.112	0.134	0.155	0.177
			0.3	0.068	0.135	0.200	0.264	0.328	0.390	0.451	0.511
			0.5	0.113	0.223	0.330	0.434	0.535	0.633	0.729	0.823
			0.7	0.159	0.311	0.458	0.598	0.734	0.865	0.992	1.115
		3.6	0.1	0.057	0.113	0.168	0.222	0.275	0.328	0.380	0.431
			0.3	0.171	0.334	0.490	0.640	0.785	0.924	1.058	1.188
			0.5	0.286	0.551	0.799	1.031	1.251	1.459	1.658	1.847
			0.7	0.402	0.764	1.095	1.400	1.684	1.950	2.200	2.437
		3.8	0.1	-	-	-	-	-	-	-	-

Table A.30 *Differential swell occurring at the perimeter of a slab for an edge lift swelling condition in predominantly Montmorillonite clay soil (60 percent clay).*

Percent Clay (%)	Depth to Constant Suction (FT)	Constant Suction (pF)	Velocity of Moisture Flow (inches/month)	DIFFERENTIAL SWELL (INCHES) EDGE DISTANCE PENETRATION (FT) 1 FT	2 FT	3 FT	4 FT	5 FT	6 FT	7 FT	8 FT
70	3	3.2	0.1	0.003	0.007	0.010	0.013	0.016	0.020	0.023	0.026
			0.3	0.010	0.020	0.029	0.039	0.048	0.058	0.067	0.076
			0.5	0.016	0.032	0.048	0.064	0.079	0.094	0.109	0.123
			0.7	0.023	0.045	0.067	0.088	0.109	0.129	0.149	0.169
		3.4	0.1	0.007	0.014	0.021	0.028	0.034	0.041	0.048	0.054
			0.3	0.021	0.041	0.061	0.081	0.100	0.118	0.137	0.155
			0.5	0.035	0.068	0.100	0.131	0.161	0.190	0.219	0.246
			0.7	0.048	0.094	0.138	0.179	0.219	0.258	0.294	0.330
		3.6	0.1	0.017	0.034	0.051	0.067	0.083	0.099	0.114	0.129
			0.3	0.051	0.099	0.145	0.189	0.231	0.271	0.310	0.347
			0.5	0.084	0.162	0.233	0.299	0.361	0.420	0.475	0.528
			0.7	0.117	0.221	0.314	0.400	0.479	0.552	0.620	0.684
		3.8	0.1	0.042	0.082	0.121	0.158	0.194	0.228	0.261	0.293
			0.3	0.124	0.233	0.330	0.419	0.501	0.577	0.648	0.714
			0.5	0.202	0.368	0.510	0.635	0.747	0.849	0.942	1.028
			0.7	0.277	0.492	0.669	0.822	0.955	1.075	1.184	1.284
	5	3.2	0.1	0.007	0.014	0.022	0.029	0.036	0.043	0.050	0.057
			0.3	0.022	0.043	0.064	0.085	0.106	0.126	0.147	0.167
			0.5	0.036	0.071	0.106	0.140	0.174	0.207	0.240	0.273
			0.7	0.050	0.099	0.147	0.195	0.241	0.286	0.331	0.374
		3.4	0.1	0.015	0.031	0.046	0.061	0.076	0.091	0.106	0.121
			0.3	0.046	0.092	0.136	0.179	0.222	0.264	0.306	0.346
			0.5	0.077	0.151	0.223	0.293	0.361	0.427	0.492	0.554
			0.7	0.108	0.210	0.309	0.403	0.494	0.582	0.666	0.748
		3.6	0.1	0.038	0.076	0.113	0.150	0.186	0.221	0.256	0.290
			0.3	0.115	0.224	0.329	0.429	0.525	0.617	0.706	0.792
			0.5	0.191	0.367	0.531	0.685	0.830	0.967	1.097	1.221
			0.7	0.267	0.506	0.724	0.924	1.110	1.283	1.446	1.600
		3.8	0.1	0.095	0.186	0.274	0.358	0.440	0.519	0.595	0.669
			0.3	0.283	0.536	0.764	0.974	1.168	1.348	1.517	1.677
			0.5	0.470	0.862	1.262	1.502	1.773	2.020	2.247	2.458
			0.7	0.656	1.172	1.603	1.974	2.303	2.598	2.867	3.114
	7	3.2	0.1	0.012	0.025	0.037	0.050	0.062	0.074	0.086	0.098
			0.3	0.037	0.074	0.111	0.147	0.183	0.219	0.255	0.290
			0.5	0.062	0.124	0.184	0.244	0.303	0.361	0.418	0.475
			0.7	0.087	0.173	0.256	0.339	0.419	0.499	0.577	0.653
		3.4	0.1	0.027	0.054	0.080	0.107	0.133	0.159	0.185	0.211
			0.3	0.081	0.160	0.238	0.315	0.390	0.464	0.537	0.609
			0.5	0.135	0.266	0.393	0.517	0.637	0.754	0.869	0.980
			0.7	0.189	0.371	0.545	0.713	0.875	1.031	1.182	1.329
		3.6	0.1	0.068	0.134	0.200	0.264	0.328	0.391	0.453	0.514
			0.3	0.204	0.398	0.584	0.763	0.935	1.101	1.261	1.415
			0.5	0.341	0.656	0.951	1.229	1.490	1.739	1.975	2.200
			0.7	0.479	0.910	1.304	1.668	2.006	2.323	2.621	2.903
		3.8	0.1	-	-	-	-	-	-	-	-
			0.3								
			0.5								

Table A.31 *Differential swell occurring at the perimeter of a slab for an edge lift swelling condition in predominantly Montmorillonite clay soil (70 percent clay).*

A.5 — Portland Cement Association and Concrete Reinforcing Steel Institute Tables

Slab thickness in.	Subgrade k* pci	Allowable load, psf** Concrete flexural strength, psi			
		550	600	650	700
5	50	535	585	635	685
	100	760	830	900	965
	200	1,075	1,175	1,270	1,370
6	50	585	640	695	750
	100	830	905	980	1,055
	200	1,175	1,280	1,390	1,495
8	50	680	740	800	865
	100	960	1,045	1,135	1,220
	200	1,355	1,480	1,603	1,725
10	50	760	830	895	965
	100	1,070	1,170	1,265	1,365
	200	1,515	1,655	1,790	1,930
12	50	830	905	980	1,055
	100	1,175	1,280	1,390	1,495
	200	1,660	1,810	1,965	2,115
14	50	895	980	1,060	1,140
	100	1,270	1,385	1,500	1,615
	200	1,795	1,960	2,120	2,285

* *k* of subgrade; disregard increase in k due to subbase.

** For allowable stress equal to 1/2 flexural strength.
Based on aisle and load widths giving maximum stress.

Table A.32 *Allowable distributed load on slabs with unjointed aisles and variable layout, from* Reference 6.

Slab thickness, in.	Working stress, psi	Critical aisle width, ft.[2]	Allowable load, psf					
			At critical aisle width	At other aisle widths				
				6-ft. aisle	8-ft. aisle	10-ft. aisle	12-ft. aisle	14-ft. aisle
Subgrade k = 50 pci[1]								
5	300	5.6	610	615	670	815	1,050	1,215
	350	5.6	710	715	785	950	1,225	1,420
	400	5.6	815	820	895	1,085	1,400	1,620
6	300	6.4	670	675	695	780	945	1,175
	350	6.4	785	785	810	910	1,100	1,370
	400	6.4	895	895	925	1,040	1,260	1,570
8	300	8.0	770	800	770	800	880	1,010
	350	8.0	900	935	900	935	1,025	1,180
	400	8.0	1,025	1,070	1,025	1,065	1,175	1,350
10	300	9.4	845	930	855	850	885	960
	350	9.4	985	1,085	1,000	990	1,035	1,120
	400	9.4	1,130	1,240	1,145	1,135	1,185	1,285
12	300	10.8	915	1,065	955	915	925	965
	350	10.8	1,065	1,240	1,115	1,070	1,080	1,125
	400	10.8	1,220	1,420	1,270	1,220	1,230	1,290
14	300	12.1	980	1,225	1,070	1,000	980	995
	350	12.1	1,145	1,430	1,245	1,170	1,145	1,160
	400	12.1	1,310	1,630	1,425	1,335	1,310	1,330
Subgrade k = 100 pci[1]								
5	300	4.7	865	900	1,090	1,470	1,745	1,810
	350	4.7	1,010	1,050	1,270	1,715	2,035	2,115
	400	4.7	1,155	1,200	1,455	1,955	2,325	2,415
6	300	5.4	950	955	1,065	1,320	1,700	1,925
	350	5.4	1,105	1,115	1,245	1,540	1,985	2,245
	400	5.4	1,265	1,275	1,420	1,760	2,270	2,565
8	300	6.7	1,095	1,105	1,120	1,240	1,465	1,815
	350	6.7	1,280	1,285	1,305	1,445	1,705	2,120
	400	6.7	1,460	1,470	1,495	1,650	1,950	2,420
10	300	7.9	1,215	1,265	1,215	1,270	1,395	1,610
	350	7.9	1,420	1,475	1,420	1,480	1,630	1,880
	400	7.9	1,625	1,645	1,625	1,690	1,860	2,150
12	300	9.1	1,320	1,425	1,325	1,330	1,400	1,535
	350	9.1	1,540	1,665	1,545	1,550	1,635	1,795
	400	9.1	1,755	1,900	1,770	1,770	1,865	2,050
14	300	10.2	1,405	1,590	1,445	1,405	1,435	1,525
	350	10.2	1,640	1,855	1,685	1,640	1,675	1,775
	400	10.2	1,875	2,120	1,925	1,875	1,915	2,030
Subgrade k = 200 pci[1]								
5	300	4.0	1,225	1,400	1,930	2,450	2,565	2,520
	350	4.0	1,425	1,630	2,255	2,860	2,990	2,940
	400	4.0	1,630	1,865	2,575	3,270	3,420	3,360
6	300	4.5	1,340	1,415	1,755	2,395	2,740	2,810
	350	4.5	1,565	1,650	2,050	2,800	3,200	3,275
	400	4.5	1,785	1,890	2,345	3,190	3,655	3,745
8	300	5.6	1,550	1,550	1,695	2,045	2,635	3,070
	350	5.6	1,810	1,810	1,980	2,385	3,075	3,580
	400	5.6	2,065	2,070	2,615	2,730	3,515	4,095
10	300	6.6	1,730	1,745	1,775	1,965	2,330	2,895
	350	6.6	2,020	2,035	2,070	2,290	2,715	3,300
	400	6.6	2,310	2,325	2,365	2,620	3,105	3,860
12	300	7.6	1,890	1,945	1,895	1,995	2,230	2,610
	350	7.6	2,205	2,270	2,210	2,330	2,600	3,045
	400	7.6	2,520	2,595	2,525	2,660	2,972	3,480
14	300	8.6	2,025	2,150	2,030	2,065	2,210	2,480
	350	8.6	2,360	2,510	2,365	2,405	2,580	2,890
	400	8.6	2,700	2,870	2,705	2,750	2,950	3,305

[1] k of subgrade; disregard increase in k due to subbase.

[2] Critical aisle width equals 2.209 times radius of relative stiffness.

Assumed load width = 300 in.; allowable load varies only slightly for other load widths. Allowable stress = one-half flexural strength.

Table A.33 *Allowable distributed loads, unjointed aisle (uniform load, fixed layout), from* Reference 6.

Slab Moment Capacities For One or Two Layers of Rebar at 12-Inch Spacing

Slab Thickness		6-in.						8-in.					
		#3	#4	#5	#6	#7	#8	#3	#4	#5	#6	#7	#8
Number of Layers	A_s (in.²)	0.11	0.20	0.31	0.44	0.60	0.79	0.11	0.20	0.31	0.44	0.60	0.79
	d_b (in.)	0.38	0.50	0.63	0.75	0.88	1.00	0.38	0.50	0.63	0.75	0.88	1.00
ONE	d (in.)	3.00	3.00	3.00	3.00	3.00	3.00	4.00	4.00	4.00	4.00	4.00	4.00
	resulting cover (in.)	2.63	2.50	2.38	2.25	2.13	2.00	3.63	3.50	3.38	3.25	3.13	3.00
	M_u (ft-k/ft)	1.34	2.43	3.77	5.35	7.29	9.60	1.78	3.24	5.02	7.13	9.72	12.80
	p (%)	0.15	0.28	0.43	0.61	0.83	1.10	0.11	0.21	0.32	0.46	0.62	0.82
	wgt (psf)	0.75	1.37	2.12	3.01	4.10	5.40	0.75	1.37	2.12	3.01	4.10	5.40
TWO*	d (in.)	4.38	4.25	4.13	4.00	3.88	3.75	6.38	6.25	6.13	6.00	5.88	5.75
	M_u (ft-k/ft)	1.95	3.44	5.18	7.13.	9.42	12.00	2.84	5.06	7.69	10.69	14.28	18.40
	p (%)	0.31	0.56	0.86	1.22	1.67	2.19	0.23	0.42	0.65	0.92	1.25	1.65
	wgt (psf)	1.50	2.73	4.24	6.01	8.20	10.79	1.50	2.73	4.24	6.01	8.20	10.79

* *Not recommended for 6-in. slab. Shown only for information purposes.*

Slab Thickness		10-in.						12-in.					
		#3	#4	#5	#6	#7	#8	#3	#4	#5	#6	#7	#8
Number of Layers	A_s (in.²)	0.11	0.20	0.31	0.44	0.60	0.79	0.11	0.20	0.31	0.44	0.60	0.79
	d_b (in.)	0.38	0.50	0.63	0.75	0.88	1.00	0.38	0.50	0.63	0.75	0.88	1.00
ONE	d (in.)	5.00	5.00	5.00	5.00	5.00	5.00	6.00	6.00	6.00	6.00	6.00	6.00
	resulting cover (in.)	4.63	4.50	4.38	4.25	4.13	4.00	5.63	5.50	5.38	5.25	5.13	5.00
	M_u (ft-k/ft)	2.23	4.05	6.28	8.91	12.15	16.00	2.67	4.86	7.53	10.69	14.58	19.20
	p (%)	0.09	0.17	0.26	0.37	0.50	0.66	0.08	0.14	0.22	0.31	0.42	0.55
	wgt (psf)	0.75	1.37	2.12	3.01	4.10	5.40	0.75	1.37	2.12	3.01	4.10	5.40
TWO	d (in.)	8.38	8.25	8.13	8.00	7.88	7.75	10.38	10.25	10.13	10.00	9.88	9.75
	M_u (ft-k/ft)	3.73	6.68	10.20	14.26	19.14	24.80	4.62	8.30	12.71	17.82	24.00	31.20
	p (%)	0.18	0.33	0.52	0.73	1.00	1.32	0.15	0.28	0.43	0.61	0.83	1.10
	wgt (psf)	1.50	2.73	4.24	6.01	8.20	10.79	1.50	2.73	4.24	6.01	8.20	10.79

NOTES:

1. $f'_c = 4000$ *psi;* $f_y = 60{,}000$ *psi; cover = 1.25 in; b = 12 in;* $\phi = 0.90$

2. Design assumptions made in table:

	One Layer	Two Layers
d	$t/2$	t-(cover + d_b)
$j_u d$	$0.9d$	
M_u	$\phi A_s f_y (j_u d)$	

3. Percentage of reinforcement based on gross section, $b \times t$

Table A.34 *Slab moment capacities (resistance) in foot-kips per foot of slab width, using either one or two layers of reinforcing bars, from* Reference 11.

f'_c (psi)	Modulus of rupture (psi)	Slab Thicknesses t (inches) 6	8	10	12
2500	450	2.70	4.80	7.50	10.80
3000	493	2.96	5.26	8.22	11.83
3500	532	3.19	5.67	8.87	12.77
4000	570	3.42	6.08	9.50	13.68
4500	603	3.62	6.43	10.05	14.47
5000	636	3.82	6.78	10.60	15.26

NOTES:

1. f'_c *is the specified design value.*

2. Modulus of rupture MOR is $9 \times \sqrt{f'_c}$

3. M_{cr} *is* $(bt^2/6) \times$ *MOR, divided by 12,000 to give units of foot-kips per foot.*

4. Even though values of M_{cr} *are given to a hundredth of a foot-kip, the authors feel that the practical accuracy is no more than to the nearest 0.1 foot-kip/foot.*

Table A.35 *Value of cracking moment in foot-kips per foot of slab width, for slabs of the indicated strength and thickness.*

CITED REFERENCES

1. Yoder, E. J. and Witczak, M.W., *Principles of Pavement Design*, John Wiley & Sons, New York, 2nd edition, 1975.

2. Corps of Engineers, U. S. Army, "The Unified Soil Classification System," Waterways Experiment Station, *Technical Memorandum* 3-357, Vicksburg, MS.

3. *ASTM Annual Book of Standards*, American Society for Testing and Materials, Philadelphia, PA.

4. Goodyear Tire & Rubber Company, Engineering Reports: "Over-the-Road Tires Engineering Data" (1985) and "Off-the-Road Tires Engineering Data" (1983), The Goodyear Tire & Rubber Company, Akron, OH.

5. Tatnall, P.C., personal communication, Bekaert Corporation, Marietta, GA, 1988.

6. Spears, Ralph, and Panarese, William, *Concrete Floors on Ground*, EB075.02D, Portland Cement Association, Skokie, IL, second edition 1983; revised 1990.

7. ACI Committee 318, "Building Code Requirements for Reinforced Concrete (ACI 318-89),' and "Commentary (ACI 318R-89);" "Building Code Requirements for Structural Plain Concrete (ACI 318.1-89)," and "Commentary (ACI 318.1R-89);", American Concrete Institute, Detroit, MI, 1989.

8. ACI Committee 360, "Design of Slabs on Grade (ACI 360.1R-92)," American Concrete Institute, Detroit, MI, 1992.

9. PCA, "AIRPORT", Concrete Thickness Design for Airport and Industrial Pavements (MC006X), The Portland Cement Association, Skokie, IL, 1987 (computer program).

10. Post-Tensioning Institute, *Design and Construction of Post-Tensioned Slabs on Ground*, The Post-Tensioning Institute, Phoenix, AZ, first edition 1980; second edition scheduled for 1995 release. (Authors used information from the second edition at points where PTI "revisions" are cited.)

11. Ringo, B.C., "The Structurally Reinforced Slab-On-Grade," *Engineering Data Report*, Number 33, The Concrete Reinforcing Steel Institute, Schaumburg, IL, 1989.

12. Anderson, Robert B. "Using Fibers in Slabs-on-Ground: Do it Right," *Civil Engineering News*, April 1994.

13. ACI Committee 223, "Shrinkage Compensating Concrete Design (ACI 223-90)," American Concrete Institute, Detroit, MI, 1990.

14. Packard, R.G., *Slab Thickness Design for Industrial Concrete Floors on Grade*, IS195.01D, The Portland Cement Association, Skokie, IL, 1976.

15. Wire Reinforcement Institute, "Design Procedures for Industrial Slabs," Interim Report (1973),and "Structural Welded Wire Fabric Detailing Manual," Appendix A, The Wire Reinforcement Institute, Reston, VA., 1989.

16. Corps of Engineers, U. S. Army, "Engineering Design: Rigid Pavements for Roads, Streets, Walks and Open Areas," Engineering Manual EM 1110-3-132, Department of The Army, Washington, D.C., 1984.

17. Departments of The Army and The Air Force, "Concrete Floor Slabs on Grade Subjected to Heavy Loads," Technical Manual TM-5-809-12 and AFM 88-3, Chapter 15, U.S. Government Printing Office, Washington, D.C., August 1987.

18. PCA, "MATS", Analysis And Design of Foundation Mats, Combined Footings and Slabs-on-Grade (MC012), The Portland Cement Association, Skokie, IL, 1994 (computer program).

19. Hetenyi, M., *Beams on Elastic Foundations*, University of Michigan Press, Ann Arbor, MI, 1946.

20. Rice, P.F., "Design of Concrete Floors on Ground for Warehouse Loadings," *ACI Journal*, August 1957, pages 105-113.

21. Panak, John J. and Rauhut, J. B., "Behavior and Design of Industrial Slabs on Grade," *ACI Journal*, May 1975, pages 219-224.

22. Ringo, B.C., "Design, Construction, and Performance of Slabs-On-Grade for an Industry," *ACI Journal*, November 1978.

23. Ytterberg, R.F., "Shrinkage and Curling of Slabs on Grade,"(Parts 1, 2, and 3), *Concrete International*, April, May, and June 1987.

24. ACI Committee 544, "Design Considerations for Steel Fiber Reinforced Concrete (ACI 544.4R-88)," American Concrete Institute, Detroit, MI, 1988.

25. Balaguru, P.N. and Shah, S. P., *Fiber Reinforced Cement Composites*, McGraw-Hill, 1992.

26. ACI Committee 302, "Guide for Concrete Floor and Slab Construction (ACI 302.1R-89)," American Concrete Institute, Detroit, MI, 1989.

27. ACI Committee 330, "Guide for Design and Construction of Concrete Parking Lots (ACI 330R-87)," American Concrete Institute, Detroit, MI, 1987.

28. *Fibresteel Technical Manual*, AWI Fibresteel, Five Dock, 2046, Australia, November 1981.

29. ACI Committee 117, "Standard Specifications for Tolerances for Concrete Construction and Materials (ACI 117-90)," American Concrete Institute, Detroit, MI, 1990.

30. Ytterberg, C., "The Waviness Index Compared with Other Floor Tolerancing Systems, Parts I and II," *Concrete International*, October and November 1994.

31. Tipping, E., "Bidding and Building to F-Number Floor Specs," *Concrete Construction*, January 1992, pages 18-19.

32. *Design and Construction of Concrete Slabs on Grade*, SCM-11(86), The American Concrete Institute, Detroit, MI, 1986.

33. American Association of State Highway Officials, "Standard Specification for Highway Materials and Methods of Sampling and Testing," 1961.

34. Federal Aviation Administration, "Airport Paving," AC150/5320-cH, U. S. Department of Transportation, Washington, D.C., 1967.

35. Anderson, Robert B., "Controlling Concrete Shrinkage in Slabs-on-Grade," *Civil Engineering News*, August 1993.

Additional References

The following are not cited in the text but they will be useful for the designer who wants more background information.

Wirand Concrete Design Manual for Factory and Warehouse Floor Slabs, Battelle Development Corp., Columbus, OH, 1975.

ACI Committee 301, "Specifications for Structural Concrete for Buildings (ACI 301-72 Revised 1975)," American Concrete Institute, Detroit, MI, 1975.

"Concrete Floor Flatness and Levelness," ACI Compilation No. 9, American Concrete Institute, Detroit, MI, 1989.

ACI Committee 316, "Recommendations for Construction of Concrete Pavements and Concrete Bases (ACI 316R-82)," American Concrete Institute, Detroit, MI, 1982.

"Slabs and Floors," *Concrete International*, American Concrete Institute, Detroit, MI, June, 1989.

Ringo, Boyd C., "Specifying a Quality Industrial Floor," *The Construction Specifier*, December, 1988.

INDEX